# Studies in Logic
Volume 67

# Many-Valued Logics
A Mathematical and Computational Introduction

Second Edition

Volume 56
Dualities for Structures of Applied Logics
Ewa Orłowska, Anna Maria Radzikowska and Ingrid Rewitzky

Volume 57
Proof-theoretic Semantics
Nissim Francez

Volume 58
Handbook of Mathematical Fuzzy Logic, Volume 3
Petr Cintula, Petr Hajek and Carles Noguera, eds.

Volume 59
The Psychology of Argument. Cognitive Approaches to Argumentation and Persuasion
Fabio Paglieri, Laura Bonelli and Silvia Felletti, eds

Volume 60
Absract Algebraic Logic. An Introductory Textbook
Josep Maria Font

Volume 61
Philosophical Applications of Modal Logic
Lloyd Humberstone

Volume 62
Argumentation and Reasoned Action. Proceedings of the 1st European Conference on Argumentation, Lisbon 2015. Volume I
Dima Mohammed and Marcin Lewiński, eds

Volume 63
Argumentation and Reasoned Action. Proceedings of the 1st European Conference on Argumentation, Lisbon 2015. Volume II
Dima Mohammed and Marcin Lewiński, eds

Volume 64
Logic of Questions in the Wild. Inferential Erotetic Logic in Information Seeking Dialogue Modelling
Paweł Łupkowski

Volume 65
Elementary Logic with Applications. A Procedural Perspective for Computer Scientists
D. M. Gabbay and O. T. Rodrigues

Volume 66
Logical Consequences. Theory and Applications: An Introduction.
Luis M. Augusto

Volume 67
Many-Valued Logics: A Mathematical and Computational Introduction
Luis M. Augusto

Studies in Logic Series Editor
Dov Gabbay                                                        dov.gabbay@kcl.ac.uk

# Many-Valued Logics
## A Mathematical and Computational Introduction

Second Edition

Luis M. Augusto

© Individual author and College Publications 2017. Second edition 2020
All rights reserved.

ISBN 978-1-84890-250-3

College Publications
Scientific Director: Dov Gabbay
Managing Director: Jane Spurr


http://www.collegepublications.co.uk

---

All rights reserved. No part of this publication may be reproduced, stored in a retrieval system or transmitted in any form, or by any means, electronic, mechanical, photocopying, recording or otherwise without prior permission, in writing, from the publisher.

# Contents

Preface to the 1st edition . . . . . . . . . . . . . . . . . . . . . . . xiii

Preface to the 2nd edition . . . . . . . . . . . . . . . . . . . . . . . xvii

**1 Introduction** . . . . . . . . . . . . . . . . . . . . . . . . . . . . . 1
   1.1 Logics, classical and non-classical, among which the many-valued . . . . . . . . . . . . . . . . . . . . . . . . . . . . . . . 1
   1.2 Mathematical logic . . . . . . . . . . . . . . . . . . . . . . . . 3
   1.3 Logic and computation . . . . . . . . . . . . . . . . . . . . . . 6

## I  THINGS LOGICAL     11

**2 Logical languages**     13
   2.1 Formal languages and logical languages . . . . . . . . . 13
   2.2 Propositional and first-order languages . . . . . . . . . . . 17
   2.3 The language of classical logic . . . . . . . . . . . . . . . . 22
   2.4 Clausal and normal forms . . . . . . . . . . . . . . . . . . . 25
      2.4.1 Literals and clauses . . . . . . . . . . . . . . . . . . 25
      2.4.2 Negation normal form . . . . . . . . . . . . . . . . 26
      2.4.3 Prenex normal form . . . . . . . . . . . . . . . . . . 28
      2.4.4 Skolem normal form . . . . . . . . . . . . . . . . . . 30
      2.4.5 Conjunctive and disjunctive normal forms . . . . . 32
   2.5 Signed formalisms . . . . . . . . . . . . . . . . . . . . . . . . 37
      2.5.1 Signed logic . . . . . . . . . . . . . . . . . . . . . . . 37
      2.5.2 Signed clause logic . . . . . . . . . . . . . . . . . . 39
   2.6 Substitutions and unification for FOL . . . . . . . . . . . 40
   Exercises . . . . . . . . . . . . . . . . . . . . . . . . . . . . . . . . . 46

**3 Logical systems**     51
   3.1 Logical consequence and inference . . . . . . . . . . . . . 51
   3.2 Semantics and model theory . . . . . . . . . . . . . . . . . 55
      3.2.1 Truth-functionality and truth-functional completeness . . . . . . . . . . . . . . . . . . . . . . . . . . . . 55

|  |  | 3.2.2 | Semantics and deduction | 59 |
|---|---|---|---|---|
|  |  | 3.2.3 | Matrix semantics | 63 |
|  | 3.3 | Syntax and proof theory | | 66 |
|  |  | 3.3.1 | Inference rules and proof systems | 66 |
|  |  | 3.3.2 | Syntax and deduction | 69 |
|  | 3.4 | Adequateness of a deductive system | | 70 |
|  | 3.5 | The system of classical logic | | 75 |
|  | Exercises | | | 78 |

## 4 Logical decisions — 81

- 4.1 Meeting the decision problem and the SAT ... 81
  - 4.1.1 The Boolean satisfiability problem, or SAT ... 82
  - 4.1.2 Refutation proof procedures ... 84
- 4.2 Some historical notes on automated theorem proving ... 85
- 4.3 Herbrand semantics ... 87
- 4.4 Proving validity and satisfiability ... 95
  - 4.4.1 Truth tables ... 95
  - 4.4.2 Axiom systems ... 95
  - 4.4.3 Natural deduction ... 97
  - 4.4.4 The sequent calculus $\mathcal{LK}$ ... 100
  - 4.4.5 The DPLL procedure ... 104
- 4.5 Refutation I: Analytic tableaux ... 108
  - 4.5.1 Analytic tableaux as a propositional calculus ... 108
  - 4.5.2 Analytic tableaux as a predicate calculus ... 116
    - 4.5.2.1 FOL tableaux without unification ... 118
    - 4.5.2.2 FOL tableaux with unification ... 119
- 4.6 Refutation II: Resolution ... 123
  - 4.6.1 The resolution principle for propositional logic ... 123
  - 4.6.2 The resolution principle for FOL ... 125
  - 4.6.3 Completeness of the resolution principle ... 127
  - 4.6.4 Resolution refinements ... 129
    - 4.6.4.1 A-ordering ... 130
    - 4.6.4.2 Hyper-resolution and semantic resolution ... 134
  - 4.6.5 Implementation of resolution in Prover9-Mace4 ... 137
- Exercises ... 144

## II  MANY-VALUED LOGICS — 149

## 5 Many-valued logics — 151

- 5.1 Some historical notes ... 151
- 5.2 Many-valuedness and interpretation ... 152

|  |  | 5.2.1 | Suszko's Thesis . . . . . . . . . . . . . . . . . . . 152 |
|---|---|---|---|

- 5.2.1 Suszko's Thesis . . . . . . . . . . . . . . . . . . . 152
- 5.2.2 Non-trivial many-valuedness . . . . . . . . . . . . 154
- 5.2.3 Classical generalizations to the many-valued logics 155
- 5.3 Structural properties of many-valued logics . . . . . . . . 160
- 5.4 The Łukasiewicz propositional logics . . . . . . . . . . . . 161
  - 5.4.1 Łukasiewicz's 3-valued propositional logic Ł$_3$ . . . 161
  - 5.4.2 Tautologousness, contradictoriness, and entailment in Ł$_3$ . . . . . . . . . . . . . . . . . . . . . . . . . . 167
  - 5.4.3 $N$-valued generalizations of Ł$_3$ . . . . . . . . . . . 167
- 5.5 Finitely many-valued propositional logics . . . . . . . . . 170
  - 5.5.1 Bochvar's 3-valued system . . . . . . . . . . . . . . 171
  - 5.5.2 Kleene's 3-valued logics . . . . . . . . . . . . . . 174
  - 5.5.3 Finn's 3-valued logic . . . . . . . . . . . . . . . . 176
  - 5.5.4 Logics of nonsense: the 3-valued logics of Halldén, Åqvist, Segerberg, and Piróg-Rzepecka . . . . . . . 177
  - 5.5.5 Heyting's 3-valued logic . . . . . . . . . . . . . . 181
  - 5.5.6 Reichenbach's 3-valued logic . . . . . . . . . . . . 182
  - 5.5.7 Belnap's 4-valued logic . . . . . . . . . . . . . . . 183
  - 5.5.8 The finitely $n$-valued logics of Post and Gödel . . . 186
    - 5.5.8.1 Post logics . . . . . . . . . . . . . . . . . 186
    - 5.5.8.2 Gödel logics . . . . . . . . . . . . . . . . 188
- 5.6 Fuzzy logics . . . . . . . . . . . . . . . . . . . . . . . . . 189
- 5.7 Quantification in many-valued logics . . . . . . . . . . . . 193
  - 5.7.1 Quantification in finitely many-valued logics . . . . 194
  - 5.7.2 Quantification in fuzzy logics . . . . . . . . . . . . 202
- Exercises . . . . . . . . . . . . . . . . . . . . . . . . . . . . . . 207

## III  REFUTATION CALCULI FOR MANY-VALUED LOGICS                                                                 215

## 6  The signed SAT for many-valued logics                                                                        217

- 6.1 From the MV-SAT to the signed MV-SAT . . . . . . . . 217
- 6.2 From many-valued formulae to signed formulae . . . . . . 221
  - 6.2.1 General notions and definitions . . . . . . . . . . . 221
  - 6.2.2 Transformation rules for many-valued connectives  226
  - 6.2.3 Transformation rules for many-valued quantifiers . 229
  - 6.2.4 Transformation rules and preservation of structure 233
  - 6.2.5 Translation to clausal form . . . . . . . . . . . . . 234
- Exercises . . . . . . . . . . . . . . . . . . . . . . . . . . . . . . 238

## 7  Signed tableaux for the MV-SAT                                                                               239

## Contents

    7.1   Introductory remarks . . . . . . . . . . . . . . . . . . . . 239
    7.2   Signed analytic tableaux for classical formulae . . . . . . . 241
    7.3   Surma's algorithm . . . . . . . . . . . . . . . . . . . . . . 243
    7.4   Signed tableaux for finitely many-valued logics . . . . . 249
           7.4.1   Propositional signed tableaux . . . . . . . . . . . . 252
           7.4.2   FO signed tableaux . . . . . . . . . . . . . . . . . 263
    7.5   Signed tableaux for infinitely many-valued logics . . . . . 270
    Exercises . . . . . . . . . . . . . . . . . . . . . . . . . . . . . . 277

## 8  Signed resolution for the MV-SAT     281

    8.1   Introductory remarks . . . . . . . . . . . . . . . . . . . . 281
    8.2   Signed resolution for finitely many-valued logics . . . . . 283
           8.2.1   Signed resolution proof procedures . . . . . . . . . 283
                  8.2.1.1   Main rules . . . . . . . . . . . . . . . . 283
                  8.2.1.2   Refinements of signed resolution . . . . . 286
           8.2.2   The main theorem of signed resolution . . . . . . . 287
    8.3   Signed resolution for infinitely many-valued logics . . . . . 294
    Exercises . . . . . . . . . . . . . . . . . . . . . . . . . . . . . . 307

## IV  APPENDIX     311

## 9  Mathematical notions     313

    9.1   Sets . . . . . . . . . . . . . . . . . . . . . . . . . . . . . . 313
    9.2   Functions, operations, and relations . . . . . . . . . . . . 316
           9.2.1   Functions, ordered intervals, and sets: Crisp sets and fuzzy sets . . . . . . . . . . . . . . . . . . . . 319
    9.3   Algebras and algebraic structures . . . . . . . . . . . . . . 321
    9.4   Lattices . . . . . . . . . . . . . . . . . . . . . . . . . . . . 325
    9.5   Graphs and trees . . . . . . . . . . . . . . . . . . . . . . . 330

## Bibliography     333

## Index     349

# List of Figures

| | | |
|---|---|---|
| 1.1.1 | Some main non-classical logics. . . . . . . . . . . . . . | 4 |
| 2.4.1 | Tseitin transformations for the connectives of L. . . . . . . | 36 |
| 2.5.1 | A partially ordered set. . . . . . . . . . . . . . . . . | 38 |
| 2.6.1 | Unifying the pair $\langle P(a,x,h(g(z))), P(z,h(y),h(y))\rangle$ . . | 45 |
| 3.4.1 | Adequateness of a deductive system $\mathsf{L} = (\mathsf{L}, \Vdash)$. . . . . . | 74 |
| 4.3.1 | Closed semantic tree of $C = \{\mathcal{C}_1, \mathcal{C}_2, \mathcal{C}_3, \mathcal{C}_4, \mathcal{C}_5\}$ in Example 4.3.3. . . . . . . . . . . . . . . . . . . . . . . . | 92 |
| 4.3.2 | A closed semantic tree. . . . . . . . . . . . . . . | 93 |
| 4.4.1 | Proof of $\vdash A \to A$ in $\mathscr{L}$ . . . . . . . . . . . . . . . | 96 |
| 4.4.2 | $A, \forall x A \to B \vdash \forall x B$: Proof in the axiom system $\mathcal{L}^*$. . . . . | 97 |
| 4.4.3 | $\mathcal{NK}$ proof of $\vdash (A \to B) \wedge (A \to C) \to (A \to (B \wedge C))$. . . | 99 |
| 4.4.4 | Proof in $\mathcal{LK}$ of a FO theorem. . . . . . . . . . . . . | 104 |
| 4.4.5 | Proof in $\mathcal{LK}$ of axiom $\mathscr{L}2$ of the axiom system $\mathscr{L}$. . . . | 105 |
| 4.4.6 | A DPLL proof procedure. . . . . . . . . . . . . . . | 107 |
| 4.5.1 | Analytic tableaux expansion rules: $\alpha\beta$-classification. . . . | 111 |
| 4.5.2 | $\vdash ((A \to B) \wedge ((A \wedge B) \to C)) \to (A \to C)$: A propositional tableau proof. . . . . . . . . . . . . . . . . | 113 |
| 4.5.3 | Analytic tableaux expansion rules: $\gamma\delta$-classification. . . . . | 117 |
| 4.5.4 | An FO tableau proof. . . . . . . . . . . . . . . . . | 120 |
| 4.5.5 | An FO tableau with unification. . . . . . . . . . . . . | 122 |
| 4.6.1 | A refutation tree. . . . . . . . . . . . . . . . . . | 126 |
| 4.6.2 | A refutation-failure tree. . . . . . . . . . . . . . . . | 127 |
| 4.6.3 | Hyper-resolution of $\Xi = (\mathcal{C}_3; \mathcal{C}_1, \mathcal{C}_2)$. . . . . . . . . . . | 135 |
| 4.6.4 | Input in Prover9-Mace4. . . . . . . . . . . . . . . | 138 |
| 4.6.5 | Output by Prover9-Mace4. . . . . . . . . . . . . . | 139 |
| 4.6.6 | Output by Prover9-Mace4. . . . . . . . . . . . . . | 140 |
| 4.6.7 | Schubert's steamroller in natural language. . . . . . . . | 141 |
| 4.6.8 | Schubert's steamroller in FOL. . . . . . . . . . . . . | 142 |
| 4.6.9 | Proof of Schubert's steamroller by Prover9-Mace4. . . . . | 143 |
| 5.2.1 | The homomorphic interpretation $h \in Hom(\mathfrak{L}, \mathfrak{A})$ and the valuation $val_h : F \longrightarrow W_2$. . . . . . . . . . . . . . | 154 |

# List of Figures

| | | |
|---|---|---|
| 5.4.1 | Proof of $\vdash_{Ł3} \neg P \to (P \to Q)$ in $\mathscr{L}_3$. | 164 |
| 5.4.2 | Proof of $\vdash_{Ł\aleph} P \to ((P \to Q) \to Q)$ in $\mathscr{L}_\aleph$. | 170 |
| 5.5.1 | The lattices $\mathcal{A}4$ (1.) and $\mathcal{L}4$ (2.). | 184 |
| 5.7.1 | Proof of $\vdash_{Ł_3^*} \forall x P(x) \to \exists x P(x)$ in $\mathscr{L}_3^*$. | 197 |
| 5.7.2 | A fuzzy binary relation. | 204 |
| 6.2.1 | *cnfs* for the connective $\to_{Ł3}$. | 230 |
| 6.2.2 | Transformation rules for signed formulae of $Ł_n$. | 235 |
| 6.2.3 | Clause transformation process of a signed FO many-valued formula. | 237 |
| 7.1.1 | General tableau rule schema for finite tableaux. | 239 |
| 7.2.1 | Signed tableaux expansion rules: $\alpha\beta$-classification. | 242 |
| 7.2.2 | Signed tableaux expansion rules: $\gamma\delta$-classification. | 242 |
| 7.2.3 | A signed tableau proof of a CPL theorem. | 243 |
| 7.4.1 | Signed tableau rules for some connectives of L$_3$. | 257 |
| 7.4.2 | A signed tableau proof in L$_3$ of the classical tautology *modus ponens*. | 259 |
| 7.4.3 | Signed tableau rules for the logical system MK. | 262 |
| 7.4.4 | FO tableau rules for filters (1) and ideals (2). | 268 |
| 7.4.5 | Signed tableau rules for quantified formulae in MK. | 270 |
| 7.5.1 | Signed fuzzy tableaux. | 275 |
| 8.2.1 | A signed resolution procedure. | 290 |
| 9.1.1 | Venn diagram of the set $A$. | 313 |
| 9.1.2 | Venn diagrams of the union (1) and the intersection (2) of two sets. | 315 |
| 9.2.1 | A partially ordered set. | 318 |
| 9.2.2 | Membership functions for young, middle, and old age in humans. | 321 |
| 9.4.1 | Join table of $2^A$. | 327 |
| 9.4.2 | Meet table of $2^A$. | 328 |
| 9.4.3 | The lattice $(\mathcal{S}, \cup, \cap)$. | 328 |
| 9.4.4 | The non-distributive lattices $\mathcal{L}_1$ and $\mathcal{L}_2$. | 329 |

# List of Algorithms

2.1 NNF transformation . . . . . . . . . . . . . . . . . . . . 27
2.2 PNF transformation . . . . . . . . . . . . . . . . . . . . 29
2.3 Skolemization . . . . . . . . . . . . . . . . . . . . . . . . 31
2.4 CNF/DNF transformation . . . . . . . . . . . . . . . . . 34
2.5 Tseitin CNF transformation . . . . . . . . . . . . . . . . 35
2.6 Robinson unification procedure . . . . . . . . . . . . . . 42

4.1 Construction of the Herbrand universe . . . . . . . . . . 89
4.2 DPLL proof procedure . . . . . . . . . . . . . . . . . . . 107
4.3 Analytic tableaux proof procedure . . . . . . . . . . . . 109
4.4 Binary resolution proof procedure . . . . . . . . . . . . . 124

7.1 Signed tableaux proof procedure for MV-SAT . . . . . . 250

8.1 Signed resolution procedure for MV-SAT (a.k.a. MVRES) 293

# Preface to the 1st edition

Although the title of this book indicates two main components, to wit, mathematical and computational, we wish to emphasize that this is first and foremost a book on many-valued logics. As such, the reader who might just be interested in the various many-valued logics will find abundant material in that which is the central part of this book, to wit, Part II. In this, the several main many-valued logics are presented from the viewpoints of their semantical properties, and their semantics are characterized mostly by means of truth tables and logical matrices. From the proof-theoretical viewpoint, for many of these logics partial or full axiomatizations are provided.

This Part II is, mainly for the objective of self-containment, preceded by an extensive introduction to "things" logical: logical languages, systems, and decisions. In this Part I, important notions such as logical consequence, adequateness of a logical system, and matrix semantics, among others, are rigorously defined. Further notions regarding decision procedures in classical logic are introduced, namely the satisfiability and validity problems (abbreviated as SAT and VAL, respectively), and they are put into relation with the several proof systems available. In Part III, we chose to approach the many-valued satisfiability problem (MV-SAT) via refutation, i.e. by proving unsatisfiability of a set of formulae (a theory), because of the many advantages, computational and other, that this proof procedure presents for the many-valued logics. This explains our extensive discussion of the analytic tableaux and resolution calculi for classical logic in this Part I. However, we wanted the reader to acquire a good grasp also of validity- and satisfiability-testing methods, and we thus provide sufficient material on this topic.

Part III deals with well-studied computational, or potentially computerizable, approaches to the many-valued logics, namely as far as automated deduction is concerned. Here, signed (clause) logic has a central place (Chapter 6). In effect, both signed tableaux and signed resolution are automatable proof calculi that can very naturally be applied to many-valued logics, reason why we chose to elaborate on them at length. This elaboration constitutes Chapters 7 and 8, for signed tableaux and

## Preface to the 1st edition

signed resolution, respectively.

We did not wish to emphasize the mathematical component, as this could scare away many a potential reader. Nevertheless, we believe that logic is a branch of mathematics, and as such it is both mathematically motivated and justified. (Incidentally, we do not think that logic is a foundation for mathematics–the reverse is true, in our opinion.) Given these two somehow conflicting views, we decided that the best thing to do would be to provide the mathematically literate, or just simply interested, readers with an Appendix in which the mathematical bases of our approach are expounded to the required extent. This is Part IV.

Logic is a subject that requires (many) years before one grasps satisfactorily what it is actually all about. This stage will arguably not be reachable without hands-on practice, much in the same way, perhaps, that many other fields require extensive, often arduous, practice. In this belief, we provide the readers with a large selection of exercises.[1]

In effect, this book is conceived for the reader who wishes to *do* something with many-valued logics, especially–but by no means only–in a computational context. With this we associate fields as diverse as cognitive modeling and switching theory. The literature on many-valued logics is abundant, but the "standard" monographs (Bolc & Borowik, 1992; Malinowski, 1993; Rescher, 1969; Rosser & Turquette, 1952), while not being obsolete, are now of mainly theoretical and historical interest. Some more recent work claiming to have practical applications in mind (e.g., Bolc & Borowik, 2003) follows to a great extent in the footsteps of these earlier monographs. On the other hand, recent work both in mathematical logic (e.g., Gottwald, 2001; Hähnle, 2001) and in philosophical logic / the philosophy of logic (e.g., Bergmann, 2008) provides little or no material for the practical applications mentioned.

Surprisingly–or just plainly intriguingly–, the last comprehensive discussion of modern applications of many-valued logics dates from 1977 (Dunn & Epstein). The fact is that many-valued logics are today in more demand than ever before, due to the realization that inconsistency in information (and knowledge bases are frequently inconsistent) "is respectable and is most welcome" (Gabbay, 2014). We now know and accept that it is often the case that theories have truth-value gaps (some propositions appear to be neither true nor false), truth-value gluts (some propositions appear to be both true and false), vague concepts (e.g., cold, young, tall, sufficient), indefinite functions, lacunae, etc. Moreover, theories are in constant updating processes, and human and arti-

---

[1] For reasons of timing, solutions to the exercises are not provided in this edition. In due time, there will be an Appendix with these solutions, which may actually be (also) posted online.

ficial reasoners require formal means to keep pace with these often fast and/or unpredictable processes, namely in order to review beliefs: what yesterday was a true/false proposition may today be neither true nor false, or both; what yesterday was clear-cut is now vague or indeterminate. The many-valued logics provide us with a powerful formalism to tackle these cases in terms of reasoning. In fact, this scenario calls for not only a practical-based approach to the many-valued logics per se, but also for hybridizations of these with other formal systems, as well as with more quantitative(-like) cognitive approaches (e.g., D'Avila Garcez, Lamb, & Gabbay, 2009). Importantly, the practical applications of many-valued logics have now gone beyond the mere industrial applications (e.g., fuzzy logics in washing machines), and promise to be of import for "higher," cognitive modeling. This *new logic* entails, in particular, a reappraisal of both "psychologism" and "formalism"–if not also of "logicism"–, especially because it proposes a reappraisal of what a cognitive, or reasoning, agent is (see, e.g. Gabbay & Woods, 2001). In order for this emerging work (Gabbay & Woods, 2003; 2005) to be further carried out and to incorporate the many-valued logics, we need a solid mathematical-computational grasp of these logics.

We hope the present book will provide the motivation for a new comprehensive treatment of the modern applications of many-valued logics that will include the *new-logic* factor as a central aspect of modern logic and its applications. Thus, a book like this on many-valued logics is now very timely, for *the times, they are a-changin'*.

I wish to thank Dov M. Gabbay for including this book in the excellent Studies in Logic series of College Publications. My thanks go also to Jane Spurr for a smooth publication process.

Madrid, Summer 2017

Luis M. S. Augusto

# Preface to the 2nd edition

Generally, the present 2nd edition corrects detected addenda and errata, uniformizes the notation, has improved figures, and introduces and/or defines some concepts that were either missing or not adequately defined in the 1st edition (e.g., *quantifier closure, parameter, free for, fuzzy set*).

The most significant change was in the notation. Because in this book there is a large variety of logical systems with corresponding object languages a meta-metalanguage was thought useful for utmost generality. This meta-metalanguage is easily made evident by the use of Greek letters for formulae and sets of formulae; the use of the same Roman letters for all the object languages gives homogeneity to the metalanguage-level presentation of the diverse object languages. Still with respect to this aspect, a new Proposition (3.4.7) was introduced, as it helps to clarify what appears to be, perhaps mostly for historical reasons, an inevitable notational ambiguity in modern first-order logic.

The Index was completely revised with the aim of being as comprehensive as possible. This said, only for names of authors are pages given exhaustively; for the concepts, only the page in which they are defined is given, unless a concept has more than one definition or occurs in distinct contexts.

Although the original structure of the book is unaltered (i.e. no chapters, sections, etc. were deleted or newly added), throughout the main text changes–often substantial–were made with enhanced clarity in view, so that paragraphs or sentences in the first edition might have been significantly changed or even entirely deleted, while new paragraphs were added. In particular, many new propositions and examples were added in this edition. This, together with some typographical improvements, contributed to the increase in the number of pages.

Another innovation is the isolation of algorithms in shaded boxes. In this edition, I dispense entirely with pseudo-code for the formulation of algorithms, preferring a simple listing or enumeration of steps or rules to be applied. The readers can easily turn this formulation into their own (pseudo-)code.

*Preface to the 2nd edition*

Finally, for the same reason invoked in the 1st edition, no solutions are given for any of the exercises.

I acknowledge the readiness of College Publications to publish the present 2nd edition.

Madrid, June 2020

Luis M. S. Augusto

# 1. Introduction

The *many-valued logics* (sometimes also known as *multiple-valued logics*) are today an indispensable component of the so-called non-classical logics, and this to a large extent due to the many computational applications they have in a large plethora of fields. In what follows, we explain why these logics are non-classical, why a mathematical formulation is important, and what computation means in this book. These three topics reflect the interdisciplinary approach that characterizes these logics. In effect, firstly conceived with philosophical concerns in mind, they were soon thereafter approached from a mathematical perspective, and are today studied from the viewpoint of key technologies. This means that a satisfactory understanding of these logics requires a philosophically-inclined mind, some mathematical literacy, and an interest in computer science and artificial intelligence in general.

## 1.1. Logics, classical and non-classical, among which the many-valued

*Logic* is the science of (the modes of) reasoning, and *deductive logic* is the study of valid reasoning. Although deductive logic is not the whole of logic, this being also constituted by *inductive* and *abductive* main branches, by *logic* one usually means deductive logic. This is a branch of *formal logic*, i.e. logic based on a formal language. There is also, perhaps to the chagrin of many a logician, the/a so-called *informal logic*, whose language mostly coincides with a natural language such as English or Sioux. This book is on formal logic.[1] Henceforth, more often than not we omit the adjective "formal."

Strictly conceived, logic is the *interpretation*, from the viewpoint of logical consequence, of formulae that are well formed as exactly defined in a formal language. The immediate *object* of logic is (classes of) *logical systems* and their *logics*. The *focus* of logic is the notion of *logical consequence*. Logic studies both what makes a particular *symbolic*, or

---
[1]This said, we think that an intuitive, informal understanding of formal logic is useful, and we do encourage such an understanding.

## 1. Introduction

*formal*, system a logical system, and the way it behaves from the viewpoint of logical consequence. Thus, logic comprises a *meta-theory* and a *theory of the techniques* accounting for logical consequence. The former approaches fundamental results that have to do with consistency, completeness, etc., of logical systems. The latter can be of two types, known as *proof theory* if the techniques respect *syntactical* consequence, and *model theory* when *semantical*[2] consequence is considered. (Importantly, both notions of consequence coincide in the case of (strong) completeness.)

Logic is all about what sentences[3] follow from, or are consequences of, what other sentences, and how. This can be *classically*, or *non-classically* so. Thus, this is a main division in the field of logic.

Classical logic has a long history, which makes it difficult to characterize it in an unequivocal way, and especially so if we want to do it concisely. In any case, there are a few features that contribute to this characterization. Firstly, classical logic is *bivalent*: a sentence or statement must be either true or false. These are the two–and only–truth values of classical logic. Thus, classical logic is *two-valued*–or *Boolean*–and rejects any truth value other than these two, which means that there is no third, fourth, or $i$-th, value available. Moreover, for any sentence, either its affirmation or its negation is always true. This is known as the *principle of excluded middle* (abbreviated: PEM), or, in Latin, *tertium non datur*.[4] Secondly, classical logic is deductive: it satisfies the deduction-detachment theorem, which in turn requires that both the deduction theorem and the *modus ponens* rule hold. This means that classical logic is *truth-preserving*: if the premises in an argument are true, then the conclusion is necessarily true. Thirdly, classical logic is *reflexive*, *monotonic*, and *transitive* with respect to logical consequence: any sentence is a consequence of itself, i.e. any sentence follows logically from itself; adding more sentences to the premises does not undo a deductive argument; and derived sentences can be reused as premises. Finally, the connectives of classical logic follow strict rules–sometimes rather paradoxical-sounding–of classicality.[5]

Since the end of the 19th century, classical logic has been an indispens-

---

[2] We prefer the form *semantical* to *semantic*; nevertheless, we use the latter form in "crystallized" expressions or labels (e.g., semantic tree).

[3] Also: statements, propositions, or formulae.

[4] These and the following expressions characteristic of logical jargon have very precise formulations, and hence meanings, in the standard object language of classical logic. See Chapter 2 below.

[5] See Augusto (2017) for a comprehensive discussion of the classicality conditions for the connectives of the standard object language of classical logic.

able tool in both pure and applied mathematics. Especially the latter, embodied to a great extent in the digital computer, has nevertheless motivated a plethora of *new* logics. In fact, issues to do with accuracy, security, etc., of new key technologies were soon verified not to be tractable from the perspective of classical, Boolean logic, calling for logical systems that could formalize such disparate aspects as incompleteness or inconsistency of available information, uncertainty in reasoning, temporal and dynamic requirements, etc. These logics, whether merely *extending* classical logic or actually *replacing* it in ad-hoc applications, are known as *non-classical logics* (see Fig. 1.1.1).[6]

Among these, the *many-valued logics*, whose main distinctive characteristic is the fact that they have more than the two classical truth values, are of particular interest in key applications. Among these we count, for instance, switching theory, logic programming, hardware verification, non-monotonic reasoning, analysis of communication with feedback, and natural language processing. Moreover, the many many-valued logics commonly generalize classical logic, which makes them important tools to investigate central aspects of classical logic systems such as the relations between proof theory and semantics; in particular, they have been found to have important applications in the investigation of the relations between classical logic and intuitionistic logic. As a matter of fact, in an abstract sense–by considering the equivalence classes obtained by a specific notion of logical equivalence as truth values –, every logic can be seen as a many-valued logic.

## 1.2. Mathematical logic

*Mathematical logic* just is the circumscription of (the study of) logic to a mathematical context. This can be by researching the mathematical properties of logical systems, and/or just by applying mathematical language to the study of logic. For some, however, mathematical logic just is logic as applicable to mathematical proof.

Indeed, there is some controversy on whether logic is a part of mathematics, a stance defended by *constructivism*, or mathematics a part of logic, a view known as *logicism*. Historically, L. E. J. Brouwer stands for the former, Gottlob Frege and David Hilbert for the latter. Either way, logic and mathematics are intimately tied up, not the least because

---

[6]Besides extensions and deviations, non-classical logics can be mere *variations*: the content of classical logic is preserved, but notation can change significantly with respect to it (e.g., many-sorted logic).

# 1. Introduction

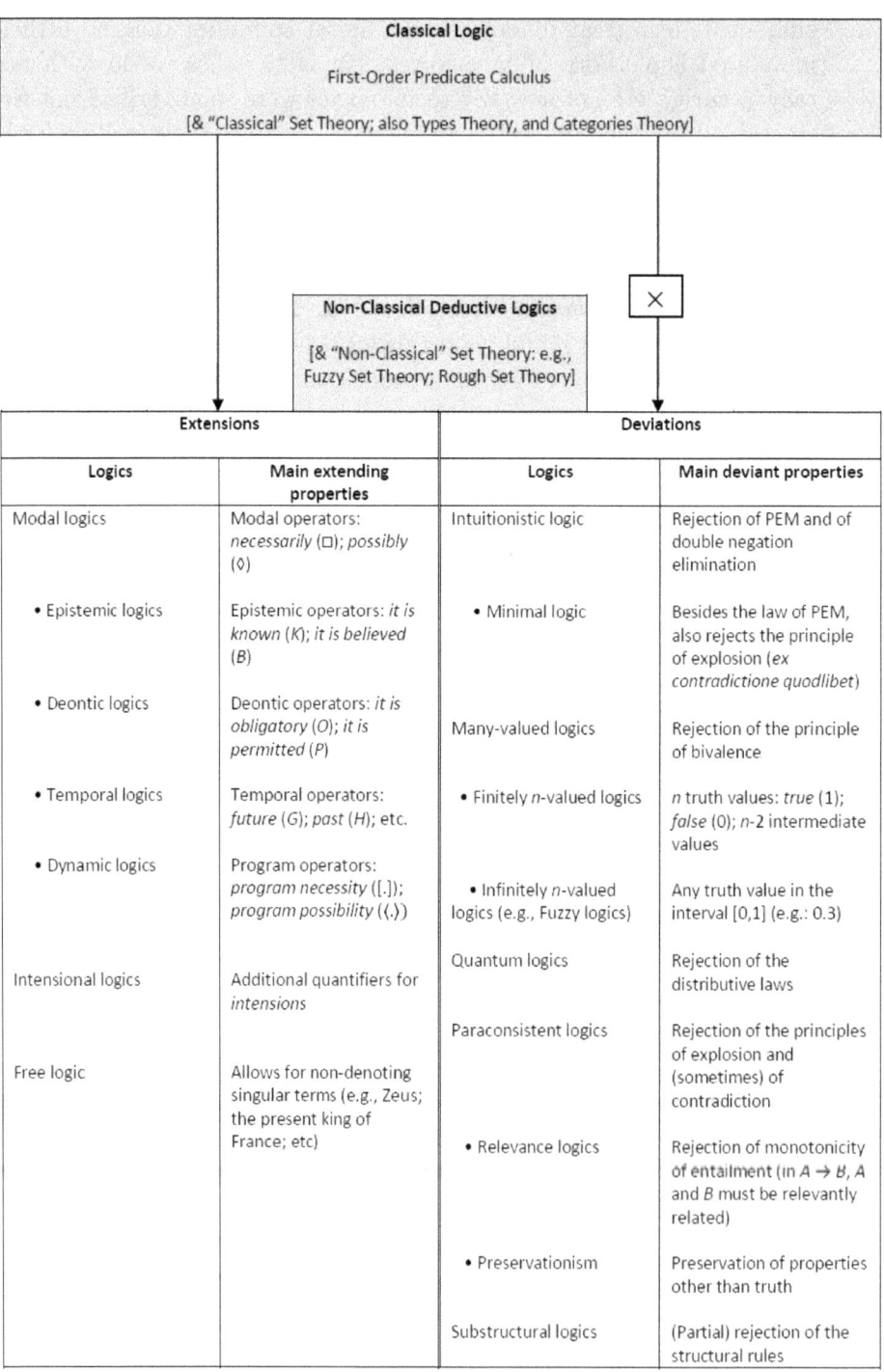

Figure 1.1.1.: Some main non-classical logics.

many proofs in mathematics are logical in essence. But whether logic is a part of mathematics or the reverse is true, logic is essentially amenable to a (meta-)mathematical language.

In very general terms, the meta-mathematical language is typically highly segregated with respect to statements, the most common of which are definitions, theorems, propositions, lemmas, corollaries, and results. In turn, the mathematical language proper has a rigorously defined syntax (grammar) and an unequivocal notation, which lends to (meta-)mathematical statements formal rigor and conciseness. Although the notation and the syntax of the languages of logic typically do not necessarily coincide with those of mathematics, it is often the case that we apply the latter in metalogic, i.e. when we reason *about* (vs. within) the logical languages and systems. This is particularly so with respect to set theory and universal algebra, and abstract mathematics in general. This use of a mathematical (meta)language makes it easier to apply logical results in fields such as computing and artificial intelligence, as mathematics is a common ground for these and related fields.

For these reasons, we use (meta-)mathematical language throughout this book. Although we take the usual segregation of (meta-)mathematical statements in a rather loose way–mainly for the reason that their contents often overlap–, for the sake of both clarity and formal rigor we segregate these statements into definitions, propositions, and theorems, mainly, with the odd corollary or lemma.

Proofs are an important component in mathematical(-based) texts. One provides such proofs when one needs to convince peers with respect to (meta-)mathematical statements such as theorems and propositions. In the case of logic, we find it essential to provide proofs in a manual, a dissertation, or a thoroughly academic book, i.e. with an academic audience in view. This book is a bit of all–or maybe none!–of these. However, we shall be concerned mostly with well-established–indeed often folklore–results. This justifies the relative scarcity of proofs in this book in comparison with other logic books. On the other hand, often proofs help us to understand the statements. *Meden agan*, we provide only such proofs, either complete or sketches thereof; in the latter case, the reader may wish to carry out the proof completely as an exercise.

In any case, we do not expect readers to be versed in mathematical jargon, and we provide this in the Appendix. We do not refer the reader to mathematical statements in the Appendix simply because we do not wish to over-refer in the main text. However, the contents in the Appendix are organized in such a way that the reader searching for a mathematical concept will easily find it.

## 1. Introduction

## 1.3. Logic and computation

*Computational logic* is the research into the computational properties of logical systems and associated logics, i.e. the properties that logical systems possess that allow them to perform computation and/or reason about computation. Important subfields of computational logic are, for instance, automated reasoning and program verification.

We take the term "computation" here in two related senses. Firstly, in the sense applied to functions, i.e. computable functions, and these are, as is well known, those for which there is an *algorithm*, a clearly step-by-step defined procedure that ends in either a "Yes" or "No" output. Secondly, we take this term in the sense that these algorithmic procedures are actually implementable in a machine, namely the digital computer–though, of course, they can be implemented by a human brain, or some other computing device, provided that these can cope with the required levels of computational complexity.

Indeed, for a logical system (or a logic) to be useful in terms of applications, it must meet requirements of *computational tractability*. Two components in particular are required for this end: a logical system should provide a proof system allowing for effective proof strategies, and a corresponding simple and intuitive semantics. The first aspect is central to logic, given that we can view a logical system as a pair $(\vdash, \mathcal{S}_\vdash)$, where $\vdash$ is a consequence relation between (sets of) formulae and $\mathcal{S}_\vdash$ is an algorithmic system or a proof system–ideally both, i.e. an *algorithmic proof system*–for $\vdash$. By this it is meant the algorithmic procedures that generate pairs of (sets of) formulae $(X, Y)$ such that we have $X \vdash Y$ if and only if we have $\mathcal{S}_\vdash(X, Y)$. Given two sets of formulae $X$ and $Y$ and some logical system, we often want, or need, to know whether $X \vdash Y$ holds, i.e. whether $Y$ follows, or is derivable, from $X$. This is a *deduction problem*.

Actually, we can speak of deduction *problems*. Allow for $X$ to be possibly empty and let $Y = \{\phi\}$. Then, we may want to know if $X \vdash \phi$, or if $\phi$ is a theorem (i.e. $\emptyset \vdash \phi$). Appealing to the concept of *truth* (the second aspect above), a concept possibly more intuitive than *proof*, we may want to know whether $X \models \phi$, or $X/\phi$ is a valid argument, or if $\phi$ is a valid formula (i.e. $\emptyset \models \phi$); in other words, we may want to know whether $\phi$ is true in all interpretations or, if all formulae $\chi \in X$ are true, whether $\phi$ is true. Or we may want to know if $X \models \phi$ is satisfiable: if there is at least one interpretation that assigns the value *true* to (all formulae of $X$ and) $\phi$. The former is the *validity problem*, abbreviated VAL, and the latter is the *satisfiability problem*, or SAT.

Once we are faced with a deduction problem, this is accompanied by

## 1.3. Logic and computation

yet another problem, to wit, the *decision problem*. This is the problem of whether there is (or not) an answer to the deduction problem, or, what is the same, whether there is an algorithm–a *decision procedure*–that terminates with a correct "yes/no" answer to the deduction problem at hand. Due to space- and time-complexity issues, a decision procedure for a deduction problem may be useless to humans without the help of computing machinery other than the human brain; thus, we require other ways to implement the decision procedures, namely in a computer program. The objective is thus to *automate*, totally or partially, the algorithms that answer the decision problem. In the first case, we talk of *provers*, and we talk of *assistants* when the automation is partial.

*Automated deduction* is thus a key requirement in any logical system, as deduction strategies can be laborious and prone to errors, in particular when high levels of complexity are unavoidable. Although automated deduction was developed mainly with mathematical proof in mind, it is a subfield of *automated reasoning*, and thus it serves more and more deduction outside the immediate realm of mathematics. In order to be formalized a *theory* has to be constituted by sentences that are either axioms or are proven from these and/or from derived sentences by means of rules, i.e. the theorems. This is a theory in the logical sense of the word, and any theory can potentially be "computerized" by means of a formal language. It is often the case that we want to know if a sentence belongs to the theory; this is basically what is done in symbolic, or logic-based, computational modeling for a large plethora of fields, among which we mention robotics and human cognition. In effect, any computer program requires verification, so as to guarantee its correctness, and this is done by means of automated deduction.

Classical logic–both propositional and first order–has long been successfully automated in terms of deduction and now many provers are available. Examples of such computer-implemented algorithmic systems are truth tables (many online implementations), resolution (e.g. Prover9-Mace4, Vampire), analytic tableaux (e.g., TableauxProver), as well as hybrid systems such as resolution-tableaux (e.g., SETHEO). However, the terrain of non-classical logics has only recently been a focus for automated deduction, and much is still to be investigated and done. Nevertheless, VAL and SAT, first and foremost problems in classical propositional logic, generalize naturally to many-valued logics:[7] we often want to know if it is the case that a many-valued formula is evaluated as true in all interpretations, or in some interpretation. These are known as the *MV-VAL* and *MV-SAT*, respectively, where *MV* stands

---

[7] As well as to first-order logics, classical or many-valued.

7

*1. Introduction*

for *many-valued*.

VAL and SAT are related in the following way: when one needs to find out whether a formula is valid in a logical system, one can check for unsatisfiability instead, given that a formula $\phi$ is valid if and only if its negation ($\neg\phi$) is unsatisfiable (and vice-versa). This is known as a *refutation procedure*, and it explains why in the literature the VAL is often referred to as $\overline{SAT}$, where the upper line indicates duality.

The importance of this duality is that the methods for (in)validity checking are typically less efficiently implementable when compared to those for (un)satisfiability checking: the former relies on truth tables, as well as on proof systems such as Hilbert systems, natural deduction, and sequent calculi, whereas the latter relies mostly on the DPLL method, resolution and analytic tableaux. The latter proof systems are more efficient in terms of computational implementation, because they use conjunctive normal forms (DPLL; resolution) or disjunctive normal forms (analytic tableaux), which can be easily computerized with fewer computational costs compared to, for instance, truth tables. These, as well as Hilbert systems, natural deduction, and sequent calculi, are, to put it informally but clearly, "not easy to use." Moreover, proofs in the former systems are easier to grasp by humans. But, above all else, tableaux and resolution are *analytic* proof procedures: they decompose (analyze) formulae into ever smaller (sub)formulae up to atomic (sub)formulae. This might actually not help (much) in terms of proof complexity, but it assures us that at least for finite sets there is a final "yes/no" answer to the decision problem.

The truth-table method, as well as axiom- and rule-based proof systems, for many-valued logics is discussed already in Part II, along with the very discussion on the interpretation and structural properties of many-valued logics. Resolution and analytic tableaux are the computerizable refutation methods of excellence to check for (un)satisfiability of many-valued formulae, and we thus concentrate our discussion on them: in Part III we approach an automation method for the MV-SAT based on resolution, to wit, signed resolution (Chapter 8); Chapter 7 is dedicated to the application of signed tableaux to check for the MV-SAT. In choosing to concentrate on these two refutation methods for many-valued logics and with self-containment in mind, we decided to elaborate on these methods for classical logic to the extent that is thought adequate in this text (Part I, Sections 4.5-6); the reader is referred to, for example, Fitting (1996) and Chang & Lee (1973) for comprehensive treatments of analytic tableaux and resolution from the viewpoint of classical logic.

A final remark: Although this book is a computational introduction to many-valued logics, it would be impractical to provide an introduction to computation and computability theory here. Of these, we provide here only the concepts and results *sine quibus non*. These do not include, for instance, mathematical definitions of algorithm or Turing machine(s), or a mathematical treatment of the complexity classes. But we do not assume that the reader is knowledgeable of these areas, reason why we refer them to, for instance, Augusto (2020a, b) or, more advanced, Boolos, Burgess, & Jeffrey (2007).

# Part I.
# THINGS LOGICAL

# 2. Logical languages

As stated in the Introduction, at the root of the study of formal logic lies the notion of a *formal language*. In fact, a *logical language* is first and foremost a formal language. We begin our study of formal logic by elaborating on the relations between a formal language and a logical language (Section 2.1). We then elaborate on the *standard propositional* and *first-order languages* (Sections 2.2-3), and finish this Chapter with further topics that are essential for the *automation of logical reasoning*, both in classical logic and in many-valued logics (Sections 2.4-6).

## 2.1. Formal languages and logical languages

**2.1.1. (Def.)** A *formal language* is a pair $\mathsf{F} = (\Sigma, \Omega)$, where $\Sigma$ is a possibly infinite set of symbols, called the *alphabet* (of $\mathsf{F}$), and $\Omega$ is a finite set of *rules* for making well-formed strings of symbols, called the *syntax* (of $\mathsf{F}$).

**2.1.2. (Def.)** A formal language $\mathsf{F}$ whose alphabet $\Sigma$, denoted by $\Sigma_\mathsf{F}$, contains *logical constants* is a *logical language*. For $\mathsf{F}$ a logical language, $\Sigma_\mathsf{F}$ can consist of

1. an infinite supply of symbols for *variables*, as well as possibly of symbols for *predicates* and *functions* of arity $n \geq 0$ (i.e. $n$-place arguments), and

2. a finite number of (symbols for) logical constants, namely *operators* $\heartsuit_1, ..., \heartsuit_r$ and *quantifiers* $\blacklozenge_1, ..., \blacklozenge_k$.

*Individual constants* are 0-place function symbols. *Brackets* are treated as punctuation marks. Typically, the variables and the *non-logical constants* (predicates and functions) are symbols from the Roman alphabet, with or without subscripts, and the logical constants are special conventional symbols (e.g., the constants $\top$ and $\bot$, denoting truth and falsity or absurdity, respectively).

In the field of logic, an *object language* is a logical language within which "things" are proved and/or interpreted, whereas a *metalanguage*

## 2. Logical languages

is the language in which the study of an object language is conducted. In this book, the metalanguages are all English supplemented with logical jargon (e.g., operator) and symbols (e.g., ⊢) that are not part of the object languages to be approached. In the metalanguage, arbitrary formulae are denoted by the Greek lowercase letters $\phi, \chi, \psi, ...$ (but also by $A, B, C, ...$ or $P, Q, R, ...$) and arbitrary sets of formulae are denoted by the Greek uppercase letters $\Gamma, X, ...$ Given an object language F, $\mu^{\mathsf{F}}$ denotes in this text the metalanguage of F.[1]

The following definitions frame in a most general way the technical study of our object languages.

**2.1.3. (Def.)** For an alphabet $\Sigma_{\mathsf{F}}$ and a finite set of rules $\Omega_{\mathsf{F}}$ of a logical language F, we define the *expressions* of F as follows: An *atom* $P$ is an expression of the form $P(t_1, ..., t_n)$, $n \geq 0$, where $P$ is an $n$-place predicate symbol and $t_1, ..., t_n$, the *arguments* of $P$ (denoted by $arg(P)$), are *terms*, expressions built up from the variables and function–and therefore constant–symbols of $\Sigma$. An atom $P$ is a *formula* $A$. If $\heartsuit_i^n$ is an $n$-place operator and $P_1, ..., P_n$ atoms, then $\heartsuit_i(P_1, ..., P_n)$ is a formula, too. $\blacklozenge_i \rho(A)$, where $\rho$ is a symbol from Def. 2.1.2.1, is also a formula.

More generally, we represent a formula $\phi$ with at most $p_1, ..., p_n$ (propositional) variables as $\phi(p_1, ..., p_n)$.

Now follow some specifications for formulae of a logical language F.

**2.1.4. (Prop.)** *Unicity of decomposition* – For each and every formula $\phi \in \mathsf{F}$ one and only one of the following conditions holds:

1. $\phi$ is an atom.

2. There is a *unique* formula $\psi$ and a *unique* unary connective $\heartsuit_i^1$ such that $\phi = \heartsuit_i(\psi)$.

3. There is a *unique* pair $\phi_1, \phi_2$ and a *unique* binary connective $\heartsuit_j^2$ such that $\phi = \heartsuit_j(\phi_1, \phi_2)$ ($\phi = (\phi_1 \heartsuit_j \phi_2)$ in infix notation).

4. There is a *unique* quantifier $\blacklozenge_i$, a *unique* symbol $\rho$, and a *unique* formula $\chi$ such that $\phi = \blacklozenge_i \rho(\chi)$.

**2.1.5. (Def.)** For $\phi \in \mathsf{F}$, some unary connective $\heartsuit_i^1$, some binary connective $\heartsuit_j^2$, and some quantifier $\blacklozenge_i$, an *immediate subformula* is defined as:

---

[1] We shall use $\phi, \chi, \psi, ...$ for utmost generality, and $A, B, C, ...$ (or $P, Q, R, ...$) for the specific object languages.

## 2.1. Formal languages and logical languages

1. An atom $\phi$ has no subformulae.

2. $\phi = \heartsuit_i(\psi)$ has as immediate subformula $\psi$.

3. $\phi = \heartsuit_j(\phi_1, \phi_2)$ has as immediate subformulae $\phi_1$ and $\phi_2$.

4. $\phi = \blacklozenge_i \rho(\chi)$ has as immediate subformula $\chi$.

**2.1.6. (Def.)** A *substitution* of a formula $\phi(p_1, ..., p_n)$ with at most $p_1, ..., p_n$ variables is any formula $\sigma\phi = \phi(p_1/\sigma p_1, ..., p_n/\sigma p_n)$ obtained from $\phi$ by simultaneously substituting $p_1, ..., p_n$ by some expressions $\sigma p_1, ..., \sigma p_n$.

**2.1.7. (Def.)** Let F be a logical language. If F is provided with a grammar (a syntax) specifying the rules according to which expressions are well formed and *derivable* from well-formed expressions by means of logical constants, then we have a *logical calculus* $\mathcal{F}$.

**2.1.8. (Def.)** In a logical language F, with each atomic formula $\phi$, each operator $\heartsuit_i^n$ with arity $n$, and each quantifier $\blacklozenge_i$ there can be associated

1. a *valuation function* $val : \phi \longrightarrow W$,

2. a *truth table* $\widetilde{\heartsuit_i^n} : W^n \longrightarrow W$, and

3. a *truth* (or *distribution*) *function* $\widetilde{\blacklozenge}_i : (2^W - \emptyset) \longrightarrow W$

where $W = \{v_0, v_1, ..., v_{n-1}\}$ is the set of *logical truth values*.

**2.1.9.** Given a truth-value set $W = \{v_0, v_1, ..., v_{n-1}\}$, typically, $v_0 = $ f (`false`) and $v_{n-1} = $ t (`true`).[2]

As said above, 2.1.2-7 constitute the syntax of F; 2.1.8-9 constitute the corresponding basic *semantics*. It can be argued that a logical calculus $\mathcal{F}$ with a semantics inaugurates the field of logic proper, as a *logical*

---
[2] We write the symbols designating logical truth values in a different font from that of the main text, in order to emphasize that they are not to be understood as their natural language correlates. That is, `true`, abbreviated t, does not mean *true* as in, for example, "It is true that snow is white." (The reader can benefit here from Tarski, 1935.) `True` of a formula or statement is a logical value that can be assigned to this formula or statement interchangeably with the distinguished element 1; the same holds for `false` and the distinguished element 0. The mathematical explanation for this can be found in a complete lattice in which 0 is the lower bound and 1 is the upper bound. This said, we do not wish to remove from `true`, `false`, etc. (see Chapter 5) the meanings they have in English. Not entirely, that is. And this mostly for the sake of the intuitive semantics mentioned above (Section 1.3).

## 2. Logical languages

*system* is typically required to be both a *system of proof*, or a calculus, and a *system of entailment*, with respect to (sentences of) a theory.

The function $\widetilde{\bigcirc}_i^n$ is called a *truth table* because it can easily be depicted by means of a table whose every cell contains *at least* one truth value. Note how in Example 2.1.1, for instance, $\widetilde{\vee}$ corresponds to:

$$\widetilde{\vee}(\mathtt{t},\mathtt{t}) = \mathtt{t}$$
$$\widetilde{\vee}(\mathtt{t},\mathtt{f}) = \mathtt{t}$$
$$\widetilde{\vee}(\mathtt{f},\mathtt{t}) = \mathtt{t}$$
$$\widetilde{\vee}(\mathtt{f},\mathtt{f}) = \mathtt{f}$$

**Example 2.1.1.** It will be useful to present straightaway the standard truth tables for the connectives $O = \{\neg, \wedge, \vee, \rightarrow, \leftrightarrow\}$ and $W_2$:

| $A$ | $\widetilde{\neg}A$ |
|---|---|
| t | f |
| f | t |

| $A$ | $B$ | $A \widetilde{\wedge} B$ | $A \widetilde{\vee} B$ | $A \widetilde{\rightarrow} B$ | $A \widetilde{\leftrightarrow} B$ |
|---|---|---|---|---|---|
| t | t | t | t | t | t |
| t | f | f | t | f | f |
| f | t | f | t | t | f |
| f | f | f | f | t | t |

The tabular "abbreviation" allows for a commodious way to represent all the possible valuations of formulae, especially when these have many subformulae: every row of the table corresponds to a distinct valuation.

**Example 2.1.2.** The truth table for the formula $(\neg P \wedge (\neg Q \vee P)) \wedge (\neg Q \vee \neg R)$ is:

| $P$ | $Q$ | $R$ | $(\neg P$ | $\wedge$ | $(\neg Q$ | $\vee$ | $P))$ | $\wedge$ | $(\neg Q$ | $\vee$ | $\neg R)$ |
|---|---|---|---|---|---|---|---|---|---|---|---|
| t | t | t | f | f | f | t | t | f | f | f | f |
| t | t | f | f | f | f | t | t | f | f | t | t |
| t | f | t | f | f | t | t | t | f | t | t | f |
| t | f | f | f | f | t | t | t | f | t | t | t |
| f | t | t | t | f | f | f | f | f | f | f | f |
| f | t | f | t | f | f | f | f | f | f | t | t |
| f | f | t | t | t | t | t | f | t | t | t | f |
| f | f | f | t | t | t | t | f | t | t | t | t |
|   |   |   |   | 2 |   | 1 |   | 4 |   | 3 |   |

The digits 1 to 4 below the binary connectives indicate the order by which they were evaluated by means of the corresponding truth tables $\widetilde{\heartsuit}_i$. But note the truth tables for $\widetilde{\neg}$ for the atomic formulae.

**2.1.10. (Def.)** *Logical equivalence* – Two formulae $\phi, \psi \in \mathsf{F}$ are said to be (logically) equivalent, denoted by $\phi \equiv_{(\mathsf{F})} \psi$, if and only if (henceforth abbreviated as "iff") both receive the same truth value under every assignment of truth values.

**Example 2.1.3.** We show that

$$(P \to Q) \equiv (\neg P \vee Q)$$

by means of the truth tables of both formulae. As it can be seen, the valuations for the main connectives are exactly the same in the four rows of both truth tables.

| P | Q | ¬P | ∨ | Q | ≡ | P | → | Q |
|---|---|---|---|---|---|---|---|---|
| t | t | f | t | t |   | t | t | t |
| t | f | f | f | f |   | t | f | f |
| f | t | t | t | t |   | f | t | t |
| f | f | t | t | f |   | f | t | f |

Below, we give examples of the distribution function.

## 2.2. Propositional and first-order languages

Logic comes in *orders* according to the following definitions:

**2.2.1. (Def.)** A predicate has order 1 if all of its arguments are terms; otherwise, it has order $n + 1$, for $n$ the highest order of its argument that is not a term. A quantifier has order 1 if it quantifies an individual variable; otherwise, it has order $n + 1$, for $n$ the order of the predicate (or function) symbol quantified. The *order of a formula* is the highest order of any of its quantifiers and predicates. An *n-th order logic* is a logic whose formulae have order $n$ or less: We speak of a zeroth-order logic when $n = 0$, of a first-order (second-order) logic for $n = 1$ ($n = 2$, respectively), and of a higher-order logic for $n > 2$.

**2.2.2.** When the order is 0, i.e. when there are only 0-ary predicates (propositional variables), we speak of a *propositional language* and of a

## 2. Logical languages

*propositional logic/calculus*; when the order is 1, we have a *first-order* (abbreviated: *FO*) *language* and a *(FO) predicate logic/calculus*.

We begin by defining the propositional language L, and then augment it in order to obtain the FO language L*. These will be our object languages of reference.

**2.2.3. (Def.)** A *propositional language* L is a pair $(V, O)$ where $V = \{p, q, r, p_1, q_1, r_1, ...\}$ is a denumerable (or countably infinite) set of *propositional variables* and $O = \{\heartsuit_1^n, ..., \heartsuit_r^n\}$ is a finite set of operators with arity $n \geq 1$ for finite $n$ called *(logical) connectives*.[3]

We can also define a propositional language as the pair $L = (F_L, O)$, where $F$ denotes a set of formulae, and $F_L$ the set of formulae of the propositional language L (often simply $F$ if L is understood). In fact, a propositional language L can, in very general terms, just be identified with its set of formulae $F_L$, because $F_L \subseteq L$.

**2.2.4. (Def.)** A *well-formed formula* of $F_L$ is defined inductively as:[4]

1. Every propositional atom is a well-formed formula: $V \subseteq F$.

2. A formula $\phi$ is a well-formed formula iff it has a finite number $k$ of propositional variables (or atoms), i.e. iff $\phi = \phi(p_1, ..., p_k)$.

3. If $\phi_1, ..., \phi_n$ are well-formed formulae and $\heartsuit_i^n$ is a logical connective with arity $n$, then $\heartsuit_i(\phi_1, ..., \phi_n)$ is a well-formed formula.

**2.2.5. (Def.)** A *FO language* L* is the language L augmented with the following countable sets:

1. $Q = \{\blacklozenge_1, ..., \blacklozenge_k\}$ of quantifiers.

2. $Vi = \{x, y, z, x_1, y_1, z_1, ...\}$ of nominal or individual variables.

3. $Cons = \{a, b, c, a_1, b_1, c_1, ...\}$ of individual constants.

4. $Pred = \{P, Q, R, P_1, Q_1, R_1, ...\}$ of predicate symbols.

5. $Fun = \{f, g, h, f_1, g_1, h_1, ...\}$ of function symbols.

**2.2.6. (Def.)** Given the FO language L*, the triple

$$\Upsilon = (Pred_{L^*}, Fun_{L^*}, ar)$$

---

[3] Actually, we often have $n = 0$, for example when we consider $\top$ and $\bot$ as operators rather than propositional constants.

[4] We shall write "formula" to abbreviate "well-formed formula," as nothing else is here a formula.

## 2.2. Propositional and first-order languages

where $ar : (Pred_{L^*} \cup Fun_{L^*}) \longrightarrow \mathbb{N}$ is the arity of the predicate and function symbols of $L^*$, is called a *signature (for $L^*$)*.[5] Over the signature for $L^*$, the expressions of $L^*$ are inductively defined as follows:[6]

$$
\begin{array}{llll}
\textit{Terms} & t & ::= & x \mid a \mid f(t_1, ..., t_n) \\
\textit{Atoms} & P\,(Q, ...) & ::= & p \mid P(t_1, ..., t_n) \\
\textit{Formulae} & A\,(B, ...) & ::= & P \mid \heartsuit_i^1 A \mid A \heartsuit_i^2 B \mid \blacklozenge_i x\,(A)
\end{array}
$$

**2.2.7. (Def.)** The set $F \subseteq L^*$ is defined as in L; atomic formulae are treated similarly as propositional variables and the following additional conditions hold:

1. A formula $\phi \in F$ is well-formed iff it has a finite number $k$ of terms, i.e. iff $\phi = \phi(t_1, ..., t_k)$.

2. Given a substitution $\sigma$, we have

   a) $\sigma c = c$, for $c \in Cons$;

   b) $\sigma(\diamond(t_1, ..., t_n)) = \diamond(\sigma t_1, ..., \sigma t_n)$, for $\diamond \in (\Upsilon - Cons)$.

**2.2.8. (Def.)** In particular, if $\phi$ belongs to $F$ and $x$ is a variable, then $\blacklozenge_i x\,(\phi) \in F \subseteq L^*$. The expression $\psi = \blacklozenge_i x\,(\phi)$ is a *quantified formula* where $(\phi)$ is the *scope* of the quantifier $\blacklozenge_i$ and $\blacklozenge_i$ *quantifies* $x$. If $x \notin Vi\,(\phi)$, then $\phi$ is said to be *trivially quantified*. Every occurrence of a variable $x$ in the scope of a quantifier $\blacklozenge_i$ is said to be *bound*; *free* otherwise. A formula is *closed* (or a *sentence*) if it has no free variables; otherwise, it is *open*. An open formula $\phi(x)$ can be closed by means of *quantifier closure* as $\blacklozenge_i x \phi\,(x)$.

We often write simply $\blacklozenge_i x \phi$, but to distinguish a *quantified* from a *trivially quantified* formula we shall often write the former as $\blacklozenge_i x \phi\,(x)$. This notation is also used to suggest that $x$ may (also) be free in $\psi$.

---

[5] A 0-ary function symbol is considered a constant. As a matter of fact, an alternative equivalent definition of a signature for $L^*$ is

$$\Upsilon_{L^*} := Pred_{L^*} \cup Fun_{L^*} \cup Cons_{L^*}.$$

[6] This is a specification of Definition 2.1.3; below (cf. Def. 2.3.1), we shall further specify the expressions of classical logic.

## 2. Logical languages

No serious ambiguity arises from this, as every $\blacklozenge_i x \phi(x)$ can equally be turned into (its subformula) $\phi(x)$, as we shall see.

We give some examples of open and/or closed formulae, as there are several aspects to be considered.

**Example 2.2.1.** The following are examples of FO formulae:

- The formula $\forall x \, (R(x, y))$ is open, as the variable $y$ is not bound by any quantifier, but the formula $\forall x \, (R(x, a))$ is closed, as the only variable $x$ in $R(x, a)$ is bound by $\forall$. More strictly, $\forall x \, (R(x, y))$ is *neither* closed *nor* open, as $x$ is a bound variable, so it is not open, but $y$ is free, so it is not closed.

- The formula $A = P(x) \to Q$ is both open and closed, as the variable $x$ in $P(x)$ is not bound by a quantifier, and every propositional formula is both open and closed, open because there is no quantifier and closed because there is no free variable.

- Let now $B = (P(x) \land R(x, y)) \to Q$. The *universal closure* of $B$ is $\forall x \forall y \, ((P(x) \land R(x, y)) \to Q)$ and its *existential closure* is $\exists x \exists y \, ((P(x) \land R(x, y)) \to Q)$.

- Let $B$ be given as above and let $\sigma$ and $\theta$ be the substitutions $\sigma x = a$ and $\theta y = g(x)$. Then, we have (cf. Def.s 2.2.7.2.b. and 2.6.3) $\sigma \circ \theta (B) = (P(a) \land R(a, g(a))) \to Q$.

- In the closed formula $\forall x \, (P(x) \land \exists x \, (Q(x)))$, the scope of $\forall$ is $P(x) \land \exists x \, (Q(x))$, but $\forall$ only binds the variable $x$ in $P(x)$, as the variable $x$ in $\exists x \, (Q(x))$ is already bound by $\exists$. In order to disambiguate we *rename* either one variable (e.g., $\forall y \, (P(y) \land \exists x \, (Q(x)))$) or both (e.g., $\forall x_1 \, (P(x_1) \land \exists x_2 \, (Q(x_2)))$).

We now introduce some important semantical notions for $\mathsf{L}^*$.

**2.2.9. (Def.)** A *frame* for an object language $\mathsf{L}^*$ with an alphabet $\Sigma$ and truth-value set $W$ is a pair $(\mathscr{D}, \Theta)$ where $\mathscr{D}$ is a non-empty *domain of discourse* and $\Theta$ is a *signature interpretation*, a mapping assigning the functions $\mathscr{D}^n \longrightarrow \mathscr{D}$ and $\mathscr{D}^n \longrightarrow W$ to each $n$-place function symbol and to each $n$-place predicate symbol of $\Sigma_{\mathsf{L}^*}$, respectively.

**2.2.10. (Def.)** An *interpretation* $\mathcal{I}$ for $\mathsf{L}^*$ is a triple $(\mathscr{D}, \Theta, \varpi)$ where $(\mathscr{D}, \Theta)$ is a frame and $\varpi$ is a *variable assignment* $\varpi : Vi \longrightarrow \mathscr{D}$. We say that $\mathcal{I}$ is based on the frame $(\mathscr{D}, \Theta)$.

**2.2.11. (Def.)** For $\mathcal{I} = (\mathscr{D}, \Theta, \varpi)$ an interpretation for $\mathsf{L}^*$, there is a corresponding (e)valuation function $val_{\mathcal{I}}$ defined inductively as:

## 2.2. Propositional and first-order languages

1. $val_\mathcal{I}(x) = \varpi(x)$ for all $x$ in $\Sigma$.

2. $val_\mathcal{I}(f(t_1, ..., t_n)) = \Theta(f)(val_\mathcal{I}(t_1), ..., val_\mathcal{I}(t_n))$ for all $n$-place function symbols $f$, $n \geq 0$, in $\Sigma$.

3. $val_\mathcal{I}(P(t_1, ..., t_n)) = \Theta(P)(val_\mathcal{I}(t_1), ..., val_\mathcal{I}(t_n))$ for all $n$-place predicate symbols $P$, $n \geq 0$, in $\Sigma$.

4. $val_\mathcal{I}(\heartsuit_i(\phi_1, ..., \phi_n)) = \widetilde{\heartsuit}_i(val_\mathcal{I}(\phi_1), ..., val_\mathcal{I}(\phi_n))$ for all logical operators $\heartsuit_i$, $n \geq 0$, in $\Sigma$.

5. $val_\mathcal{I}(\blacklozenge_i x(\phi)) = \widetilde{\blacklozenge}_i(\Delta_{\mathcal{I},x}(\phi))$ for all quantifiers $\blacklozenge_i$ in $\Sigma$, where $\Delta_{\mathcal{I},x}(\phi) = \{val_{\mathcal{I}_a^x}(\phi) \,|\, a \in \mathscr{D}\}$ is the distribution of $\phi$ in $\mathcal{I}$ with respect to $x$, and $\mathcal{I}_a^x$ is the interpretation identical to $\mathcal{I}$ when setting $\varpi(x) = a$.

**Example 2.2.2.** Let there be given the FO formula

$$\phi = \forall x \forall y \exists z \, (L(f(x, y), z) \rightarrow L(x, y)).$$

We consider the interpretation $\mathcal{I} = (\mathscr{D}, \Theta, \varpi)$ such that $\mathscr{D} = \mathbb{Z}$, $\varpi(z) = 0 \in \mathbb{Z}$, $f(x, y) = x - y$, and $L(x, y) = \{(x, y) \in \mathbb{Z} \times \mathbb{Z} | x < y\}$. Then, we have the formula:

$$\phi^\mu = \forall x \forall y \, ((x - y < 0) \rightarrow x < y)$$

In order to apply the evaluation function $val_{\mathcal{I}_a^x}$, we are required to substitute $x$, $y$, and $z$ by elements in the set $\mathbb{Z}$, which is the domain of the given interpretation $\mathcal{I}$. In particular, we have to replace $x$ and $y$ by every $a \in \mathbb{Z}$, as they are both bound by the *universally quantified*:

$$val_\mathcal{I}(\forall x, y(\phi)) = \widetilde{\forall}(\Delta_{\mathcal{I},x,y}(\phi)) = v_i \text{ for } \Delta_{\mathcal{I},x,y}(\phi) = \{val_{\mathcal{I}_a^{x,y}}(\phi) \,|\, a \in \mathbb{Z}\}$$

As for $z$, it is to be replaced by *some* $a \in \mathbb{Z}$, as it is *existentially quantified*. Additionally, we are already given the variable assignment $\varpi(z) = 0$:

$$val_\mathcal{I}(\exists z(\phi)) = \widetilde{\exists}(\Delta_{\mathcal{I},z}(\phi)) = v_i \text{ for } \Delta_{\mathcal{I},z}(\phi) = \{val_{\mathcal{I}_0^z}(\phi) \,|\, 0 \in \mathbb{Z}\}$$

Although it is unfeasible to replace every $x$ and $y$, as $\mathbb{Z}$ is an infinite set, it should be easy to see that in both cases we have $v_i = $ **true**, and $val_\mathcal{I}(\phi) = $ **true**, as given any two integers $x$ and $y$, if $x - y < 0$, then $x < y$.

This example illustrates an interpretation for a FO formula from a very general viewpoint. The substitution of the bound variables actually follows strict rules, which are introduced below in Section 2.3, and are further elaborated on in Section 2.6, as well as in Chapter 4 for the specific calculi. For now, it suffices to say that $x$ and $y$ should not be substituted simultaneously by the same integer, as they are meant to be denoting different individuals; we write $\{val_{\mathcal{I}_a^{x,y}}(\phi) \,|\, a \in \mathbb{Z}\}$ as an abbreviation for $\{val_{\mathcal{I}_{a,b}^{x,y}}(\phi) \,|\, a, b \in \mathbb{Z}\}$.

## 2.3. The language of classical logic

Classical logic (CL) is fundamental in the sense that the many and diverse logics today available commonly either are extensions or generalizations of it, or diverge from it in several ways (see Fig. 1.1.1). These extensions or generalizations and diversions reflect on the consequence operation/relation, thus affecting the sets of classical tautologies and contradictions, as well as the sets of satisfiable formulae (see Augusto, 2017). This said, their languages are either extensions or reducts of L*. We now introduce some *elementary* aspects of the language L* of CL that will be required for a good grasping of the many-valued logics to be discussed below.[7] In effect, though many of the many-valued logics have additional connectives, the set of connectives of CL features in all of them. As for the quantifiers, we shall not study many-valued systems with quantifiers other than those of CFOL.

**2.3.1. (Def.)** *Expressions of CL* – If given L* we define terms, atoms and formulae (whose corresponding sets are $T$, $At$, and $F$, respectively) inductively in the following way:

| | | | |
|---|---|---|---|
| *Terms* | $t$ | ::= | $x \,|\, a \,|\, f(t_1, ..., t_n)$ |
| *Atoms* | $P\,(Q,...)$ | ::= | $p \,|\, P(t_1, ..., t_n)$ |
| *Formulae* | $A\,(B,...)$ | ::= | $P \,|\, \neg A \,|\, A \wedge B \,|\, A \vee B \,|\, A \to B \,|\, A \leftrightarrow B \,|$ |
| | | | $\forall x\,(A) \,|\, \exists x\,(A)$ |

then we have the CL syntax of the logical connectives ¬ (*negation*), ∧ (*conjunction*), ∨ (*disjunction*), → (*material implication*, or *conditional*), and ↔ (*material equivalence*, or *biconditional*), and of the quantifiers ∀ (*universal quantifier*, read "for all") and ∃ (*existential quantifier*, read "there is a" or "for some").

---

[7]We encourage a good foundational knowledge of CL. A comprehensive treatment of CL can be found in, for instance, Mendelson (2009); see also Augusto (2019).

## 2.3. The language of classical logic

**2.3.2.** We shall refer to the propositional language of CL as L, and the first-order language of CL will be referred to as L*, simply. L provides the language for the *classical propositional logic* (CPL), and L* does so for the *classical first-order logic* (CFOL).

**2.3.3. (Def.)** The language L of CPL is said to be *functionally complete*, because it has at least one set of connectives that is functionally complete, i.e. a subset $O'$ of the set $O$ of connectives by means of which all the remaining connectives of $O$ can be defined.

**Example 2.3.1.** The set $O'_{\neg,\to} = \{\neg, \to\}$ of L is functionally complete. In effect, we have the following inductive definitions of the remaining connectives of $O_L$:

1. $(\vee_{df})$      $(A \vee B) := (\neg A \to B)$
2. $(\wedge_{df})$      $(A \wedge B) := \neg (A \to \neg B)$
3. $(\leftrightarrow_{df})$      $(A \leftrightarrow B) := (A \to B) \wedge (B \to A)$

**Example 2.3.2.** As a matter of fact, the sets $O'_{\neg,\vee}$ and $O'_{\neg,\wedge}$ of L are also functionally complete. Obviously, $O' = \{\neg, \wedge, \vee\}$ is also functionally complete, as we have:

1. $(\vee_{df})$      $(A \vee B) := \neg(\neg A \wedge \neg B)$
2. $(\to_{df})$      $(A \to B) := (\neg A \vee B)$
3. $(\leftrightarrow_{df})$      $(A \leftrightarrow B) := (A \wedge B) \vee (\neg A \wedge \neg B)$

**2.3.4. (Prop.)** The language of CL is functionally complete.
*Proof:* Trivial from the above.

Below (Section 3.2.1), we show why *functional completeness* is actually an abbreviation of *truth-functional completeness*, i.e. functional completeness is a semantical notion.

Another important aspect of the language of CL that holds generally in the many-valued logics is *substitution*.

**2.3.5. (Def.)** For the unary connective $\neg$ and the binary connectives $\dot{\heartsuit} = \wedge, \vee$ and $\mathring{\heartsuit} = \to, \leftrightarrow$, the following additional facts concerning a substitution $\sigma$ hold in CL (cf. Def. 2.1.6):

1. $\sigma(\neg A) = \neg(\sigma A)$.

2. $\sigma\left(A_1 \dot{\diamond} ... \dot{\diamond} A_n\right) = \sigma A_1 \dot{\diamond} ... \dot{\diamond} \sigma A_n$.

3. $\sigma\left(A \dot{\diamond} B\right) = \sigma A \dot{\diamond} \sigma B$.

Recall now Definition 2.2.7.2. The following definition specifies the conditions for a substitution over L* in CFOL.

**2.3.6. (Def.)** Let $A = \blacklozenge x A(x)$. Then, $t$ is said to be *substitutable* (or *free*) *for* $x$ *in* $A$, and we write $\sigma(A) = A_t^x = A[x/t]$, if (i) $Vi(t) = \emptyset$, (ii) $t \neq y$ or $y \notin Vi(t)$ if $y$ is bound in $A$, or (iii) $t = x$.

Note with respect to Definition 2.3.6 that $x$ must be free in $A$ for the application of $A_t^x$. For instance,
$$[\forall x \, (P(x)) \wedge Q(x)]_a^x = \forall x \, (P(x)) \wedge Q(a).$$
But
$$[\forall x \, (P(x)) \wedge Q(x)]_x^x = P(x) \wedge Q(x).$$
This shows that a bound variable $x$ can become free by a substitution $\sigma x = x$.

**Example 2.3.2.** Let $A = \forall x \, (P(x) \wedge \exists y \, (Q(y,z)))$. Then,
$$A_x^x = P(x) \wedge \exists y \, (Q(y,z)),$$
$$A_c^z = \forall x \, (P(x) \wedge \exists y \, (Q(y,c))),$$
and
$$A_y^y = \forall x \, (P(x) \wedge Q(y,z)).$$
$A_y^z$, $A_x^z$, and $A_{f(y)}^z$, for instance, are not allowed.

The substitutions in quantified formulae in CFOL are governed by the specific rules of inference of the diverse available calculi for this logical system, but the above Definition 2.3.6 holds generally. We refer here mostly to axiom systems, natural deduction, the sequent calculi, and also the analytic tableaux calculus (see Section 4.4). A specific kind of substitution, called *ground substitution*, and associated *unification* procedures that are essential for both the analytic tableaux and the refutation calculi, require a more extensive elaboration to be carried out in Section 2.6.

## 2.4. Clausal and normal forms

In order to computerize logical theories–for example, with automated theorem proving in mind–certain specific *normal forms* are often required for the formulae of a logical language. These normal forms, in turn, are associated with further important notions such as *literal* and *clause*. We begin with these.

### 2.4.1. Literals and clauses

**2.4.1. (Def.)** We define a *literal*, denoted by $L$, to be an atom (e.g., $P$) or the negation of an atom (e.g., $\neg P$). We say that the literals $L$ and $\neg L$ are *complementary*. A literal $L$ is said to be *ground* if it has no individual variables.

**2.4.2. (Def.)** A *clause* $\mathcal{C}$ is a finite disjunction of literals: $\mathcal{C} = L_1 \vee ... \vee L_n = \|L_1, ..., L_n\|$. A one-literal clause is a *unit* clause. The empty clause $\| \|$, denoted by $\square$, is a clause that contains no literals.

We shall further denote a finite set of clauses $\{\|\cdot\|_1, ..., \|\cdot\|_n\}$ by $\mathcal{C}$.[8]

**2.4.3. (Prop.)** The empty clause $\square$ is always false.

*Proof:* The falsity of the empty clause derives from the fact that it has no literal that can be satisfied by an interpretation. **QED**

**2.4.4. (Def.)** A clause is called *ground* if no individual variables occur in it.

**2.4.5. (Def.)** $\mathcal{C}$ is a *Horn clause* if it contains at most one positive literal. A Horn clause with exactly one positive literal is a *definite clause*. $\mathcal{C}$ is a *dual-Horn clause* if it has at most one negative literal.

**Example 2.4.1.** $P(a, f(b))$ and $Q$ are ground literals. $P(x, a)$ is not a ground literal. The clause

$$\mathcal{C} = \neg P(a, f(b)) \vee \neg Q \vee P(b, g(a)) \vee \neg R(c) =$$

$$= \|\neg P(a, f(b)), \neg Q, P(b, g(a)), \neg R(c)\|$$

is a definite ground clause.

---

[8] Note that $\|\cdot\|$ just is another way to represent the *set* of literals of a clause $\mathcal{C}$. It is a convenient way, as the common practice of representing a set of clauses as $\mathcal{C} = \{\{\cdot\}, ..., \{\cdot\}\}$ can be confusing.

## 2. Logical languages

### 2.4.2. Negation normal form

**2.4.6. (Def.)** A formula $A \in \mathsf{L}^*$ is said to be in *negation normal form* (NNF) iff the negation connective $\neg$ is applied only to atoms and the only other connectives are conjunction and disjunction.

**2.4.7 (Prop.)** Any formula $A \in \mathsf{L}^*$ is equivalent to a formula $A' \in \mathsf{L}^*$ in NNF.

*Proof:* Left as an exercise. (Hint: Algorithm 2.1.)

In effect, any formula $A \in \mathsf{L}^*$ can be transformed into NNF by applying specific rewriting rules.

**Example 2.4.2.** Let there be given the formula

$$A = \exists x \, (P(x) \wedge \forall y Q(x,y)) \rightarrow \forall x \neg (P(x) \rightarrow \exists y Q(x,y)).$$

By applying 1.b ($\rightarrow_{df}$) of Algorithm 2.1 both to the whole formula and the consequent, we obtain the equivalent formula

$$\neg \exists x \, (P(x) \wedge \forall y Q(x,y)) \vee \forall x \neg (\neg P(x) \vee \exists y Q(x,y)).$$

Applying $DM_\vee$ to the second disjunct gives the equivalent formula

$$\neg \exists x \, (P(x) \wedge \forall y Q(x,y)) \vee \forall x (\neg \neg P(x) \wedge \neg \exists y Q(x,y)).$$

An application of LDN gives

$$\neg \exists x \, (P(x) \wedge \forall y Q(x,y)) \vee \forall x \left( P(x) \wedge \neg \exists y Q(x,y) \right).$$

We next apply $QN_\exists$ to obtain

$$\forall x \neg (P(x) \wedge \forall y Q(x,y)) \vee \forall x \left( P(x) \wedge \forall y (\neg Q(x,y)) \right).$$

And a final application of $DM_\wedge$ gives

$$\forall x (\neg P(x) \vee \neg \forall y Q(x,y)) \vee \forall x \, (P(x) \wedge \forall y (\neg Q(x,y))).$$

## Algorithm 2.1 NNF transformation

**Input**: A formula $A \in L^*$

**Output**: A formula $A' \in L^*$ in NNF such that $A \equiv A'$

Apply the following rewriting rules until $A$ is in NNF (cf. Def. 2.4.6):

1. *Connective inter-definitions:*

   a) $(\leftrightarrow_{df})$ $\quad (A \leftrightarrow B) \implies \begin{cases} (A \to B) \wedge (B \to A) \\ \\ (A \wedge B) \vee (\neg A \wedge \neg B) \end{cases}$

   b) $(\to_{df})$ $\quad A \to B \implies \neg A \vee B$

2. *De Morgan's laws:*

   a) (DM$_\vee$) $\quad \neg(A \vee B) \implies \neg A \wedge \neg B$
   b) (DM$_\wedge$) $\quad \neg(A \wedge B) \implies \neg A \vee \neg B$

3. *Double negation, or involution, law:*

   $$(\text{LDN}) \quad \neg\neg A \implies A$$

4. *Quantifier duality:*

   a) (QN$_\exists$) $\quad \neg \exists x\,(A) \implies \forall x\,(\neg A)$
   b) (QN$_\forall$) $\quad \neg \forall x\,(A) \implies \exists x\,(\neg A)$

## 2. Logical languages

Finally, application of QN$_\forall$ produces

$$\forall x \left( \neg P(x) \vee \underbrace{\exists y (\neg Q(x,y))}_{\text{QN}_\forall} \right) \vee \forall x (P(x) \wedge \forall y (\neg Q(x,y))).$$

The application of Algorithm 2.1 outputs the formula:

$$A' = \forall x (\neg P(x) \vee \exists y (\neg Q(x,y))) \vee \forall x (P(x) \wedge \forall y (\neg Q(x,y)))$$

$A'$ is in NNF and $A \equiv A'$.

### 2.4.3. Prenex normal form

**2.4.8. (Def.)** A formula $A \in F_{L^*}$ is said to be in *prenex normal form* (PNF) iff if it is in the form

$$(\dagger) \qquad \blacklozenge_1 x_1, ..., \blacklozenge_n x_n (M)$$

where every $\blacklozenge_i x_i$, $i = 1, ..., n$ is either $\forall x_i$ or $\exists x_i$, and $M$ is a formula containing no quantifiers. $\blacklozenge_1 x_1, ..., \blacklozenge_n x_n$ is the *prefix* and $(M)$ is the *matrix* of $A$.[9]

**2.4.9. (Prop.)** Every formula $A \in L^*$ is equivalent to a formula $A' \in L^*$ in PNF.

*Proof:* Left as an exercise. (Hint: Algorithm 2.2.)

In effect, every formula $A \in L^*$ can be transformed into a PNF by an algorithmic process specified in Algorithm 2.2. The transformation process is speeded if $A$ is already in NNF.[10]

**Example 2.4.3.** We begin with the formula in NNF obtained in

---

[9] Not to be confused with the *logical matrix* $\mathfrak{M}$; see below Section 3.2.3.

[10] Although computationally advantageous, the matrix of a formula $A$ in PNF is not required to be in NNF. For instance, the formula $\exists x \forall y \forall z ((P(x) \rightarrow Q(y)) \vee R(z,x,y))$ is in PNF. If $A$ is not in NNF, to convert it to PNF apply $\leftrightarrow_{df}$, 2-4 of Algorithm 2.1, and the following equivalences can be applied together with 2.a-d in Algorithm 2.2:

1. $\forall x A(x) \rightarrow B \equiv \exists x (A(x) \rightarrow B)$
2. $\exists x A(x) \rightarrow B \equiv \forall x (A(x) \rightarrow B)$
3. $B \rightarrow \forall x A(x) \equiv \forall x (B \rightarrow A(x))$
4. $B \rightarrow \exists x A(x) \equiv \exists x (B \rightarrow A(x))$

## Algorithm 2.2 PNF transformation

**Input:** A formula $A \in L^*$ in NNF

**Output:** A formula $A' \in L^*$ in PNF such that $A \equiv A'$

1. Rename any bound variables so that each variable occurs only once in $A$.

2. Push the quantifiers outwards by means of the following equivalences *when $x$ does not appear as a free variable in $B$*:

    a) $\forall x A(x) \land B \equiv \forall x (A(x) \land B)$
    b) $\forall x A(x) \lor B \equiv \forall x (A(x) \lor B)$
    c) $\exists x A(x) \land B \equiv \exists x (A(x) \land B)$
    d) $\exists x A(x) \lor B \equiv \exists x (A(x) \lor B)$

3. If $x$ appears as a free variable in $B$ (i.e, if we have $B(x)$), then rename the bound variable $x$ in $\blacklozenge x A(x)$ as $y$, obtaining the equivalent $\blacklozenge y (A[x/y]) = \blacklozenge y A(y)$.

## 2. Logical languages

Example 2.4.2:

$$A = \forall x \, (\neg P(x) \vee \exists y \, (\neg Q(x,y))) \vee \forall x \, (P(x) \wedge \forall y \, (\neg Q(x,y))).$$

Renaming of the variables (Step 1 of Algorithm 2.2) in the second disjunct gives

$$\forall x \, (\neg P(x) \vee \exists y \, (\neg Q(x,y))) \vee \forall u \, (P(u) \wedge \forall z \, (\neg Q(u,z))).$$

Repeated application of Step 2 produces first

$$\forall x \exists y \, (\neg P(x) \vee \neg Q(x,y)) \vee \forall u \forall z \, (P(u) \wedge \neg Q(u,z))$$

and then

$$A' = \forall x \exists y \forall u \forall z \, ((\neg P(x) \vee \neg Q(x,y)) \vee (P(u) \wedge \neg Q(u,z))).$$

$A'$ is in PNF and $A \equiv A'$.

Finally, it should be remarked that any formula with no quantifiers is said to be *trivially* in PNF.

### 2.4.4. Skolem normal form

Skolemization – an expression coined after the Norwegian mathematician T. Skolem, namely after Skolem (1920) – is a further transformation required for automated deduction. Contrarily to the NNF and PNF transformations, this procedure (see Algorithm 2.3) does not guarantee equivalence of formulae, namely as far as validity is concerned. In particular, we shall need the following new notion of *equisatisfiability*:

**2.4.10. (Def.)** Two formulae $\phi$ and $\psi$ are *satisfiability-equivalent*, or *equisatisfiable*, denoted by $\phi \equiv_{sat} \psi$, iff they are both satisfiable or they are both unsatisfiable.

**2.4.11. (Def.)** $A_{Sk}$ in Algorithm 2.3 is said to be in *Skolem normal form* (SNF). The constant $c$ and the function $f(x_{s_1}, ..., x_{s_m})$ used to replace the existential variables of $A$ are called *Skolem constant* and *Skolem function*, respectively.

**Example 2.4.4.** From the formula obtained in Example 2.4.3

$$A = \forall x \exists y \forall u \forall z \, ((\neg P(x) \vee \neg Q(x,y)) \vee (P(u) \wedge \neg Q(u,z)))$$

by applying Algorithm 2.3 above we obtain the formula

$$A_{Sk} = \forall x \forall u \forall z \, ((\neg P(x) \vee \neg Q(x, f(x))) \vee (P(u) \wedge \neg Q(u,z))).$$

## 2.4. Clausal and normal forms

**Algorithm 2.3** Skolemization

**Input:** A formula $A \in L^*$ in PNF

**Output:** A formula $A' \in L^*$ in SNF such that $A' \equiv_{sat} A$

Let $A$ be a formula in the PNF †. Let $\blacklozenge_r$, $1 \leq r \leq n$, be an existential quantifier in the prefix of $A$. Then:

1. If there is no universal quantifier before $\blacklozenge_r$, we choose a new constant $c$ that does not occur in $M$, replace all $x_r$ in $M$ with $c$, and delete $\blacklozenge_r x_r$ from the prefix.

2. If before $\blacklozenge_r$ there are the universal quantifiers $\blacklozenge_{s_1}, ..., \blacklozenge_{s_m}$, i.e. $1 \leq s_1 < s_2 < ... < s_m < r$, we select a new $m$-place function symbol $f$ which does not occur in $M$, replace all $\blacklozenge_r$ in $M$ by $f(x_{s_1}, ..., x_{s_m})$ and delete $\blacklozenge_r x_r$ from the prefix.

---

The aim of the Skolemization algorithm is the elimination of the existential closure of a formula, but the (remaining) universal closure can be simply removed from $A_{Sk}$ (cf. remark to Def. 2.3.6). In this book, we shall be mostly interested in obtaining sets of clauses in SNF. We can obtain a set of clauses by the end of Skolemization if the matrix $M$ of $A$ is already in a *conjunctive* normal form (CNF; see next Section), a normal form central for some calculi. For example, resolution, one of the main calculi elaborated on in this book, requires formulae to be in CNF.

**Example 2.4.5.** Let there be given the formula in PNF

$$A = \exists x \forall y \forall z \forall u \exists v \left( (P(x) \lor Q(y)) \land R(z, u, v) \right).$$

By applying the Skolemization algorithm we obtain the formula

$$A_{Sk} = \forall y \forall z \forall u \left( (P(a) \lor Q(y)) \land R(z, u, f(y, z, u)) \right).$$

We can now represent $A_{Sk}$ as the set of clauses (understood to be universally quantified):

$$C_{A_{Sk}} = \{ P(a) \lor Q(y), R(z, u, f(y, z, u)) \} =$$

$$= \{\|P(a), Q(y)\|, \|R(z, u, f(y, z, u))\|\}$$

From the above examples, it should be easy to see that, for the sake of simplicity, whenever possible we should obtain a PNF with all the existential quantifiers before the universal ones. Importantly, CFOL over $\mathsf{L}^*$ is typically *not* decidable (see Def. 4.1.1) when $Fun \subseteq \mathsf{L}^*$, an aspect we need not worry about when working with Herbrand semantics, but which is relevant for the analytic tableaux calculus.[11]

As said above, the process of transforming a formula $A$ into a formula in SNF does not necessarily preserve equivalence, but only satisfiability.

**2.4.12. (Prop.)** Every formula $A \in \mathsf{L}^*$ can be transformed into a formula $A' = A_{Sk} \in \mathsf{L}^*$ such that $A_{Sk} \equiv_{sat} A$.
*Proof:* Left as an exercise.

Preservation of satisfiability is, however, all that is required in refutation-based proof procedures, such as resolution and analytic tableaux.

### 2.4.5. Conjunctive and disjunctive normal forms

**2.4.13. (Def.)** A formula $A$ is said to be in a *conjunctive normal form* (CNF) iff $A$ has the form $A = C_1 \wedge ... \wedge C_n$, $n \geq 1$, where each of $C_1, ..., C_n$ is a clause, i.e.

$$A = \bigwedge_{i=1}^{n} \left( \bigvee_{j=1}^{m_i} L_{i,j} \right)$$

and is in NNF.

**2.4.14. (Def.)** A formula $A$ is said to be in a *disjunctive normal form* (DNF) iff $A$ has the form $A = \mathcal{E}_1 \vee ... \vee \mathcal{E}_n$, $n \geq 1$, where each of $\mathcal{E}_1, ..., \mathcal{E}_n$ is a conjunction of literals, i.e.

$$A = \bigvee_{i=1}^{n} \left( \bigwedge_{j=1}^{m_i} L_{i,j} \right)$$

and is in NNF.

**2.4.15. (Prop.)** Let $\{A_1, ..., A_n\}$ be a finite set of formulae. Then

$$\neg \left( \bigwedge_{i=1}^{n} A_i \right) \equiv \left( \bigvee_{i=1}^{n} \neg A_i \right)$$

---

[11] See Augusto (2019; 2020a) for a comprehensive discussion of this important topic.

and
$$\neg\left(\bigvee_{i=1}^{n} A_i\right) \equiv \left(\bigwedge_{i=1}^{n} \neg A_i\right).$$

*Proof:* (Sketch) We have it that $\neg(A) \equiv (\neg A)$. Then, obviously the proposition is proved for $n = 1$ by the equivalence $\neg\left(\bigwedge_{i=1}^{1} A_i\right) \equiv \left(\bigvee_{i=1}^{1} \neg A_i\right)$. The proof then follows by induction on $n$. **QED**

**2.4.16. (Prop.)** Let $A$ be a formula in CNF and $B$ be a formula in DNF. Then $\neg A$ is equivalent to a formula in DNF, and $\neg B$ is equivalent to a formula in CNF.

*Proof:* If $A$ is in CNF, then $A$ is the formula $\bigwedge_{i=1}^{n}\left(\bigvee_{j=1}^{m_i} L_{i,j}\right)$. By Proposition 2.4.15, we have

$$\neg A = \neg \bigwedge_{i=1}^{n}\left(\bigvee_{j=1}^{m_i} L_{i,j}\right) \equiv \bigvee_{i=1}^{n} \neg\left(\bigvee_{j=1}^{m_i} L_{i,j}\right) \equiv \bigvee_{i=1}^{n}\left(\bigwedge_{j=1}^{m_i} \neg L_{i,j}\right).$$

The proof runs similarly for $\neg B$ being equivalent to a formula in CNF. **QED**

Just as in the case of formulae in NNF, PNF, and SNF, there is an algorithm to convert formulae into CNF/DNF. We provide here Algorithm 2.4.

**2.4.17. (Prop.)** Any formula can be transformed into a CNF/DNF in three main steps involving the properties of a Boolean algebra.

*Proof:* Left as an exercise. (Hint: See Algorithm 2.4 and Def. 9.3.3).

**Theorem 2.4.1.** *Every formula $A$ is equivalent to any formula $A_1$ in DNF and some formula $A_2$ in CNF.*

*Proof:* Follows immediately from the above. The proof is by induction on the complexity of $A$. **QED**

An alternative method to Algorithm 2.4 is by truth-table computations. We begin with a truth table for $A$. By considering the rows in which $A$ is true, we obtain a formula in DNF equivalent to $A$. Likewise, by negatively considering the rows in which $A$ is false, i.e. by considering the DNF for $\neg A$ and subsequently negating it, we obtain a formula in CNF equivalent to $A$.

**Example 2.4.6.** Let $A = (B \vee C) \wedge ((\neg B \wedge C) \vee D)$. We leave the complete truth table for $A$ as an exercise. The abbreviated truth table for $A$ and $\neg A$ is as follows:

## 2. Logical languages

**Algorithm 2.4** CNF/DNF transformation

**Input:** A formula $A \in L^*$
**Output:** A formula $A' \in L^*$ in CNF/DNF such that $A \equiv A'$

1. Apply $\rightarrow_{def}$ and/or $\leftrightarrow_{def}$.

2. Apply $DM_{\wedge,\vee}$ and LDN.

3. Apply the Boolean properties for $\wedge$ and $\vee$, in particular the *distributive laws*:

$$(D_\wedge) \quad A \wedge (B \vee C) \equiv (A \wedge B) \vee (A \wedge C)$$

$$(D_\vee) \quad A \vee (B \wedge C) \equiv (A \vee B) \wedge (A \vee C)$$

| B | C | D | A | ¬A |
|---|---|---|---|----|
| t | t | t | t | f |
| t | t | f | f | t |
| t | f | t | t | f |
| t | f | f | f | t |
| f | t | t | t | f |
| f | t | f | t | f |
| f | f | t | f | t |
| f | f | f | f | t |

The DNF of $A$ is:

$$(B \wedge C \wedge D) \vee (B \wedge \neg C \wedge D) \vee (\neg B \wedge C \wedge D) \vee (\neg B \wedge C \wedge \neg D)$$

The DNF of $\neg A$ is:

$$(B \wedge C \wedge \neg D) \vee (B \wedge \neg C \wedge \neg D) \vee (\neg B \wedge \neg C \wedge D) \vee (\neg B \wedge \neg C \wedge \neg D)$$

By Proposition 2.4.16, the negation of this DNF produces the CNF:

$$(\neg B \vee \neg C \vee D) \wedge (\neg B \vee C \vee D) \wedge (B \vee C \vee \neg D) \wedge (B \vee C \vee D)$$

Although Algorithm 2.4 and the truth-table procedure in Example 2.4.6 guarantee logical equivalence of a formula $A$ with its DNF or CNF,

## 2.4. Clausal and normal forms

---
**Algorithm 2.5** Tseitin CNF transformation

**Input:** A formula $A \in \mathsf{L}^*$

**Output:** A formula $A' \in \mathsf{L}^*$ in CNF such that $A \equiv A'$

---

1. Introduce a new atom, say $p_1$, not occurring anywhere else in $A$, to abbreviate the innermost subformula and conjoin the abbreviated formula with the definition of $p_1$.

2. Proceed as in 1., introducing new atoms $p_2, p_3, ..., p_n$ as required in the direction of the main connective in $A$.

3. Put each of the conjuncts in CNF by applying the rules in Figure 2.4.1.

---

they can lead to an exponential growth (a.k.a. "explosion"). If one wishes to restrict this to a linear growth, then one may have to give up on equivalence in favor of equisatisfiability. This can be done by applying the *Tseitin algorithm* for CNF formulae (Tseitin, 1968). We give it as Algorithm 2.5 and refer the reader also to Figure 2.4.1.

Note in Figure 2.4.1 the following equivalences:

$$(\circledast) \quad p \to (q \land r) \equiv (p \land q) \to (p \land r)$$

$$(\odot) \quad ((p \land q) \lor (\neg p \land \neg q)) \to r \equiv ((p \land q) \to r) \land ((\neg p \land \neg q) \to r)$$

**Example 2.4.7.** Let $A = (p \lor (q \land \neg r)) \to s$. Then we have the following definitions:

$p_1 \leftrightarrow (q \land \neg r)$
$p_2 \leftrightarrow (p \lor p_1)$
$p_3 \leftrightarrow (p_2 \to s) \equiv p_3 \leftrightarrow (\neg p_2 \lor s) \equiv p_3 \leftrightarrow (p_2 \land \neg s)$

We have the conjuncts:

$$(p_1 \leftrightarrow (q \land \neg r)) \land (p_2 \leftrightarrow (p \lor p_1)) \land (p_3 \leftrightarrow (p_2 \land \neg s)) \land p_3$$

By rewriting the conjuncts into CNF (see Fig. 2.4.1), we obtain the formula:

| ($\heartsuit$) | | | Rule |
|---|---|---|---|
| $x \leftrightarrow (\neg p)$ | $\equiv$ | $(x \to \neg p) \land (\neg p \to x)$ | $\leftrightarrow_{df}$ |
| $x \leftrightarrow (p \land q)$ | $\equiv$ | $(x \to p) \land (x \to q) \land ((p \land q) \to x)$ | $\leftrightarrow_{df}, \circledast$ |
| | $\equiv$ | $(\neg x \lor p) \land (\neg x \lor q) \land (\neg (p \land q) \lor x)$ | $\to_{df}$ |
| | $\equiv$ | $(\neg x \lor p) \land (\neg x \lor q) \land (\neg p \lor \neg q \lor x)$ | $DM_\land$ |
| $x \leftrightarrow (p \lor q)$ | $\equiv$ | $(p \to x) \land (q \to x) \land (x \to (p \lor q))$ | $\leftrightarrow_{df}, \circledast$ |
| | $\equiv$ | $(\neg p \lor x) \land (\neg q \lor x) \land (\neg x \lor p \lor q)$ | $\to_{df}$ |
| $x \leftrightarrow (p \leftrightarrow q)$ | $\equiv$ | $(x \to (p \leftrightarrow q)) \land ((p \leftrightarrow q) \to x)$ | $\leftrightarrow_{df}$ |
| | $\equiv$ | $(x \to ((p \to q) \land (q \to p))) \land$ $((p \leftrightarrow q) \to x)$ | $\leftrightarrow_{df}$ |
| | $\equiv$ | $(\neg x \lor ((\neg p \lor q) \land (\neg q \lor p))) \land$ $\left(((p \land q) \lor (\neg p \land \neg q)) \to x\right)$ | $\to_{df}$ $\leftrightarrow_{df}$ |
| | $\equiv$ | $\left(\neg x \lor (\neg(\neg p \lor q) \lor \neg(\neg q \lor p))\right) \land$ $((p \land q) \to x) \land ((\neg p \land \neg q) \to x)$ | $DM_\land$ $\odot$ |
| | $\equiv$ | $\left(\neg x \lor ((p \land \neg q) \lor (q \land \neg p))\right) \land$ $\left((\neg p \lor \neg q) \to x\right) \land \left((p \lor q) \to x\right)$ | $DM_\lor$ $DM_\land$ |
| | $\equiv$ | $\left(\neg x \lor (\neg(p \land \neg q) \land \neg(q \land \neg p))\right) \land$ $(\neg(\neg p \lor \neg q) \lor x) \land (\neg(p \lor q) \lor x)$ | $DM_\lor$ $\to_{df}$ |
| | $\equiv$ | $\neg x \lor \left((\neg p \lor q) \land (\neg q \lor p)\right) \land$ $\left((p \land q) \lor x\right) \land \left((\neg p \land \neg q) \lor x\right)$ | $DM_\land$ $DM_\lor$ |
| | $\equiv$ | $(\neg x \lor \neg p \lor q) \land (\neg x \lor \neg q \lor p) \land$ $(\neg p \lor \neg q \lor x) \land (q \lor p \lor x)$ | $D_\lor$ $DM_\land$ |

Figure 2.4.1.: Tseitin transformations for the connectives of L.

$$(\neg p_1 \vee q) \wedge (\neg p_1 \vee r) \wedge (\neg q \vee \neg r \vee p_1) \wedge$$
$$(\neg p_2 \vee p \vee p_1) \wedge (p_2 \vee \neg p) \wedge (p_2 \vee \neg p_1) \wedge$$
$$(\neg p_3 \vee p_2) \wedge (\neg p_3 \vee \neg s) \wedge (\neg p_2 \vee s \vee p_3) \wedge p_3$$

**2.4.18. (Prop.)** Given a formula $A$, we can obtain a formula $A'$ in CNF by means of the Tseitin algorithm such that $A \equiv A'$.
*Proof:* Left as an exercise.

## 2.5. Signed formalisms

### 2.5.1. Signed logic

In *signed logic*, a set $S \subseteq W_n = \{v_0, ..., v_{n-1}\}$, $n \geq 2$, (or a truth value $v_i \in S$) is associated with each formula $\phi$, and then pairs $(S, \phi)((v_i, \phi)$, respectively) are considered as atoms. Importantly, if these are considered as Boolean atoms, then signed logic is comparable to CL in terms of computational complexity (cf. Beckert, Hähnle, & Manyà, 2000), constituting thus an interesting trade-off between expressivity and complexity.

**2.5.1. (Def.)** Let $S$ be a subset of the truth-value set $W_n = \{v_0, ..., v_{n-1}\}$, $n \geq 2$, and let $\phi$ be a formula. An expression of the form $S[\phi]$, where $S$ is the *sign* of $\phi$, is called a *signed formula*.[12] If $S$ is a sign, then $(W \setminus S) = \overline{S}$ is also a sign. The signed formulae $S[\phi]$ and $\overline{S}[\phi]$ are said to be *complementary*.

1. $\top$ and $\bot$ are signed formulae.

2. If $\chi$ and $\psi$ are signed formulae, then $(\chi \wedge \psi)$, $(\chi \vee \psi)$, $(\chi \rightarrow \psi)$, $(\chi \leftrightarrow \psi)$, and $\neg \chi$ are signed formulae.

3. If $x$ is a variable and $\phi$ is a signed formula, then so are $\forall x \phi$ and $\exists x \phi$.

**2.5.2. (Def.)** For $v_i \in S$ an arbitrary truth value, we abbreviate the signed formula $\{v_i\}[\phi]$ as $i[\phi]$. If $S$ has two or more members,

---

[12] More precisely, a signed formula is of the form $S\phi$; the square brackets around $\phi$ are just for readability purposes, much in the way of, for instance, $-(1) = -1$. However, if there are square brackets around $S$, then this indicates an interval. In the literature, $S\phi$ is often written $S : \phi$.

## 2. Logical languages

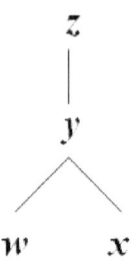

Figure 2.5.1.: A partially ordered set.

we represent the signed formula $S\,[\phi]$ as $\{v_1,..,v_n\}\,[\phi]$. $\{v_i\}\,[\phi]$ and $\{v_1,...,v_n\}\,[\phi]$ are read as "$\phi$ takes the value $v_i$" and "$\phi$ takes one of the values $v_i$ in the set $S = \{v_1,...,v_n\}$," respectively.

1. A formula is *mono-signed* if its sign $S$ is a singleton.

2. A *regular sign* is either of the form $\{v_j \in W | j \geq i\}$, denoted by $\uparrow v_i$, or of the form $\{v_j \in W | j \leq i\}$, denoted by $\downarrow v_i$, for $v_i, v_j \in W$ and $\leq$ a partial ordering on $W$.

We often abbreviate $\uparrow \{v_i\}$ and $\downarrow \{v_i\}$ as $\uparrow i$ and $\downarrow i$, respectively.

**Example 2.5.1.** Let $W = \{w, x, y, z\}$ and let $W$ be partially ordered as shown in Figure 2.5.1. Then, $\uparrow w = \{w, y, z\}$, $\downarrow w = \{w\}$, $\overline{\uparrow w} = \{x\}$, and $\overline{\uparrow y} = \{w, x\}$. The former two signs are regular and the latter two are complements of regular signs, whereas the signs $\{y\}$ and $\{w, z\}$ are neither, because there is no $i$ in $W$ such that $\uparrow i = \{y\}$ or $\uparrow i = \{w, z\}$, $\downarrow i = \{y\}$ or $\downarrow i = \{w, z\}$, as well as there is no $i$ in $W$ such that $\overline{\uparrow i} = \{y\}$, etc. Note also that $\overline{\uparrow y}$ is not regular, though it is the complement of a regular sign.

**2.5.3. (Def.)** Let $P$ be an atom. An expression of the form $S\,[P]$ or $i\,[P]$ ($\uparrow i\,[P]$ or $\downarrow i\,[P]$), where $S$ or $i$ ($\uparrow i$ or $\downarrow i$) is the sign of $P$ (the regular sign of $P$), is a *signed literal* (*regular signed literal*, respectively). The *complement* of the signed literal $L = S\,[P]$ is the signed literal $\overline{L} = \overline{S}\,[P]$. The complement of the signed literal $L = \{v_i\}\,[P]$ is the signed literal $\overline{L} = \{v_j\}\,[P]$ for $i \neq j$. The complement of the regular signed literal $L = \uparrow i\,[P]$ ($\downarrow i\,[P]$) is the regular signed literal $\overline{L} = \overline{\uparrow i}\,[P]$ (respectively, $\overline{\downarrow i}\,[P]$).

**2.5.4. (Def.)** *Subsumption* – A signed literal $S\,[P]$ *subsumes* a signed literal $S'\,[P']$, denoted by $S\,[P] \sqsubseteq S'\,[P']$, iff $P = P'$ and $S \subseteq S'$.

**2.5.5. (Def.)** The signed formula $S\,[\phi]$ (or $\overline{S}\,[\phi]$) is a *signed formula expression* (SFE) with sign $S$ (or $\overline{S}$, respectively) if it is built up from signed formulae using only $\wedge$ and $\vee$.[13]

### 2.5.2. Signed clause logic

Signed logic allows us to create signed formulae that can be used to translate any finitely-valued formula into an equisatisfiable signed CNF formula (signed CNF formulae are a subclass of signed formulae; cf. Chapter 6).

**2.5.6. (Def.)** A *signed clause* is a finite set of signed literals. A signed clause is called a *signed unit clause* if it contains exactly one literal, and it is called a *signed binary clause* if it contains exactly two literals. The *empty signed clause*, denoted by $\square$, is the clause whose literals have all the *empty sign* ($\{\}$ or $\emptyset$).

**2.5.7. (Def.)** A *signed CNF formula* (SCNFF) is a finite set of signed clauses. A SCNFF constituted by binary clauses is a *signed 2-CNF formula*. A SCNFF whose signs of the literals are regular is called a *regular CNF formula*. A signed CNF formula whose signs are singletons is a *mono-signed CNF formula*.

When using signed clauses and/or SCNFFs, regularity (as defined above) provides a means to generalize straightforwardly the notion of Horn clauses (Def. 2.4.5); in effect, we can speak of regular Horn clauses and regular Horn formulae by attributing to a regular sign of the form $\uparrow i$ a *positive polarity*, and to a sign of the form $\downarrow i$ a *negative polarity*.

**2.5.8. (Def.)** A regular clause is a *regular Horn clause* iff it contains at most one literal of positive polarity and the signs of all its other literals are complements of signs with positive polarity. A regular CNF formula is a *regular Horn formula* iff all its clauses are regular Horn clauses.

**Example 2.5.2.** Using the set $W$ of Example 2.5.1 and its associated

---

[13] By "signed formula expression" we mean simply "signed CNF/DNF representation" of a formula. The distinction is simply intended to emphasize the fact that a signed formula and its corresponding SFE are not logically equivalent, but solely satisfiability-equivalent, or equisatisfiable (see Chapter 6), which is all it takes when checking for satisfiability (cf. Chapters 7 and 8). In particular, a SFE functions very much like a schema of which particular signed CNFs/DNFs are instances (e.g., Zach, 1993).

ordering, the following is a set of regular Horn clauses:

$$\{\uparrow w\,[P], \uparrow x\,[P] \vee \overline{\uparrow y}\,[Q], \uparrow z\,[Q]\}$$

The clause $\downarrow w\,[P]$ is a regular Horn clause, as $\downarrow w = \overline{\uparrow x}$, whereas $\downarrow z\,[P]$ is not a regular Horn clause ($\downarrow z \neq \overline{\uparrow i}$ for all $i \in W$).

It is implicitly established that clauses of a SCNFF are conjunctively connected and the literals in a signed clause are disjunctively connected; thus we represent the signed clause $\|S_1\,[P_1], ..., S_k\,[P_k]\|$ as $S_1\,[P_1] \vee ... \vee S_k\,[P_k]$.

**2.5.9 (Prop.)** For all signs $S_1, ..., S_k \subseteq W_n, k \in \mathbb{N}$, and every propositional variable $P$,

$$S_1\,[P] \vee ... \vee S_k\,[P] \vee D \equiv (S_1 \cup ... \cup S_k)\,[P] \vee D.$$

*Proof:* Left as an exercise.

However, this should be applied with care, as it is true iff it does not destroy the regularity of a clause (cf. Def. 2.5.7).

## 2.6. Substitutions and unification for FOL

In Section 2.3, we set the basic rule for substitutions in FOL, namely from the viewpoint of the notions of permissible substitutions (*substitutable* or *free for*) and with a view to turning bound variables into free ones (cf. Def. 2.3.6). In this Section, with automated deduction in mind we concentrate on ground substitutions and associated unification techniques. In effect, these are essential for the main calculi to be elaborated on in this book, to wit, resolution and analytic tableaux, both classical and many-valued.

**2.6.1. (Def.)** Given a set $Vi = \{x_1, ..., x_n\}$ of variables and a set $T = \{t_1, ..., t_n\}$ of terms, a *substitution* is a mapping $\sigma : Vi \longrightarrow T$ such that $\sigma x_i = t_i$ almost everywhere, and where typically $t_i \neq x_i$. We represent a substitution $\sigma$ as a finite set of expressions of the form $x_i \mapsto t_i$, where no two different terms substitute the same variable:

$$\sigma = \{x_1 \mapsto t_1, ..., x_n \mapsto t_n\}$$

We define the domain of $\sigma$ as $dom\,(\sigma) = \{x | \sigma x \neq x\}$ and its range as $rg\,(\sigma) = \{\sigma x | x \in dom\,(\sigma)\}$.

## 2.6. Substitutions and unification for FOL

1. We say that $\sigma$ is a *ground substitution* when $Vi\,(rg\,(\sigma)) = \emptyset$, i.e. when $t_1, ... t_n$ are ground terms.

2. For $\sigma = \emptyset$, we speak of the *empty substitution*, and we denote it by $\epsilon$.

3. An injective substitution $\sigma$ such that $rg\,(\sigma) \subseteq Vi$ is called a *(variable) renaming*.

**2.6.2. (Def.)** For a substitution $\theta = \{x_1 \mapsto t_1, ..., x_n \mapsto t_n\}$ and an expression $E$, $E\theta = \theta E$ is an expression obtained from $E$ by replacing simultaneously each occurrence of the variable $x_i$, $1 \leq i \leq n$, in $E$ by the term $t_i$. We say that $E\theta$ is an *instance* of $E$. If $Vi\,(E\theta) = \emptyset$, then $E\theta$ is called a *ground instance*.

**Example 2.6.1.** Let $\theta = \{x \mapsto a, y \mapsto f(b), z \mapsto c\}$, $E = P(x, y, z)$. Then $E\theta = P(a, f(b), c)$.

**2.6.3. (Def.)** Given two substitutions $\theta = \{x_1 \mapsto t_1, ..., x_n \mapsto t_n\}$ and $\lambda = \{y_1 \mapsto u_1, ..., y_n \mapsto u_n\}$, their *composition*, denoted by $\theta \circ \lambda$, is obtained from the set $\{x_1 \mapsto t_1\lambda, ..., x_n \mapsto t_n\lambda, y_1 \mapsto u_1, ..., y_n \mapsto u_n\}$ by deleting any element $x_j \mapsto t_j\lambda$ for which $t_j\lambda = x_j$, and any element $y_i \mapsto u_i$ such that $y_i$ is among $\{x_1, ..., x_n\}$.

**Example 2.6.2.** Given the substitution sets $\theta = \{x \mapsto f(y), y \mapsto z\}$ and $\lambda = \{x \mapsto a, y \mapsto b, z \mapsto y\}$, then $\theta \circ \lambda = \{x \mapsto f(b), z \mapsto y\}$.

**2.6.4. (Def.)** *Unification* – We say that a set of expressions $E = \{E_1, ..., E_n\}$ is *unifiable* by a substitution $\sigma$ (called the *unifier* for $E$) if $E_i\sigma = E_j\sigma$ for all $E_i, E_j \in E$. A unifier $\sigma$ for the set $E$ is a *most general unifier (MGU)* iff for each unifier $\theta$ for the set there is a substitution $\lambda$ such that $\theta = \sigma \circ \lambda$. (Equivalently, let us say that $\lambda$ is more general than $\theta$, denoted $\lambda \leq_s \theta$, if there is a substitution $\sigma$ such that $\sigma \circ \lambda = \theta$. Then, $\sigma$ is the MGU of the set $E$ if for every other unifier $\lambda$ it is the case that $\sigma \leq_s \lambda$.)

**Example 2.6.3.** For substitutions $\theta = \{x \mapsto z, y \mapsto z, u \mapsto f(z, z)\}$, $\lambda = \{y \mapsto z\}$, and $\sigma = \{x \mapsto y, u \mapsto f(y, z)\}$, $\sigma$ is the MGU, because $\theta = \sigma \circ \lambda$.

**Example 2.6.4.** $\theta = \{x \mapsto f(b), y \mapsto b, z \mapsto u\}$ is a unifier of the expressions $E_1 = f(x, b, g(z))$ and $E_2 = f(f(y), y, g(u))$, because $E_1\theta = E_2\theta = f(f(b), b, g(u))$.

## 2. Logical languages

**Algorithm 2.6** Robinson unification procedure

**Input:** A set *Lit* of literals
**Output:** A set $\sigma$ such that $\sigma$ is a unifier for a disagreement set $D(E)$

I. Set $\sigma := \epsilon$

II. Select a disagreement pair in $D(E)$ in $\sigma(Lit)$

  1. If $Vi(D(E)) = \emptyset$, then stop and return "not unifiable."
  2. If $(x, t) \in D(E)$:
     a) If $x$ occurs in $t$, then "*occurs check*": return "not unifiable."
     b) Otherwise, let $\sigma := \sigma \circ (x \mapsto t)$ and return $\sigma$.

III. Repeat while $|\sigma(Lit)| > 1$.

---

**2.6.5. (Def.)** Let $C_1$ and $C_2$ be two clauses. Then, $C_1 \leq_{ss} C_2$ if there is a substitution $\sigma$ such that $C_1 \sigma \subseteq C_2$, and we say that $C_1$ *subsumes* $C_2$.

**2.6.6. (Def.)** For a non-empty set of expressions $E$, the *disagreement set* $D(E)$ of $E$ is the set of sub-expressions of $E$ obtained by locating the leftmost symbol at which not all expressions in $E$ have exactly the same symbol, and then extracting from each expression in $E$ the sub-expression that begins with the symbol occupying that position.

**Example 2.6.5.** The disagreement set of

$$E = \{P(x, g(z)), P(x, a), P(x, f(h(z, u)))\}$$

is

$$D(E) = \{g(z), a, f(h(z, u))\}.$$

The *unification problem* is that of finding a MGU of two given terms. This is a purely mechanical procedure, for which there is more than one algorithm, of which the *Robinson algorithm* (Robinson, 1965) appears to be the most efficient (cf. Hoder & Voronkov, 2009). We give it as Algorithm 2.6.

**2.6.7. (Prop.)** The Robinson algorithm is an efficient procedure for the unification problem.

*Proof:* Left as an exercise.

The application of unification rules is an alternative (more efficient) procedure (see, e.g., Baader & Snyder, 2001).

**2.6.8. (Def.)** Let $\sigma$ and $\theta$ stand for substitutions and denote failure in the application of a rule by $\bot$. For $P$ and $Q$ pairs of expressions $\{\langle E_1, F_1\rangle, ..., \langle E_n, F_n\rangle\}$, a rule **r** is said to be a *unification rule* for $P$ and $Q$ if it has the general form

$$P; \sigma \Longrightarrow Q; \theta$$

or

$$P; \sigma \Longrightarrow \bot.$$

The (successive) application of the unification rules ends either in success, denoted by $\emptyset$, or in failure. The order of application of the rules is non-deterministic.

**2.6.9.** The unification rules are as follows:

1. *Trivial:*
$$\{\langle s, s\rangle\} \cup P'; \sigma \Longrightarrow P'; \sigma$$

2. *Decomposition:*
$$\{\langle f(s_1, ..., s_n), f(t_1, ..., t_n)\rangle\} \cup P'; \sigma \Longrightarrow \{\langle s_1, t_1\rangle, ..., \langle s_n, t_n\rangle\} \cup P'; \sigma$$
if $f(s_1, ..., s_n) \neq f(t_1, ..., t_n)$.

3. *Orient:*
$$\{\langle t, x\rangle\} \cup P'; \sigma \Longrightarrow \{\langle x, t\rangle\} \cup P'; \sigma$$
if $t$ is not a variable.

4. *Variable elimination:*
$$\{\langle x, t\rangle\} \cup P'; \sigma \Longrightarrow P'\theta; \sigma\theta$$
given that $x$ does not occur in $t$ and $\theta = \{x \to t\}$.

5. *Symbol clash:*
$$\{\langle f(s_1, ..., s_n), g(t_1, ..., t_m)\rangle\} \cup P'; \sigma \Longrightarrow \bot$$
if $f \neq g$.

## 2. Logical languages

6. *Occurs check:*
$$\{\langle x, t \rangle\} \cup P'; \sigma \Longrightarrow \bot$$
if $x$ occurs in $t$ but $x \neq t$.

**Example 2.6.6.** We provide examples of the application of each of the above rules:

1. For the pair of expressions $\langle P(a), P(a) \rangle$ we have

$\{\langle P(a), P(a) \rangle\}; \epsilon \Longrightarrow_{Tr}$
$\emptyset; \epsilon$

2. Given the pair of expressions $P(f(a), g(x))$ and $P(y, z)$ we have

$\{\langle P(f(a), g(x)), P(y, z) \rangle\}; \epsilon \Longrightarrow_{Dec}$
$\{\langle f(a), y \rangle, \langle g(x), z \rangle\}; \epsilon$

3. Applying this rule to the example immediately above, we have

$\{\langle f(a), y \rangle, \langle g(x), z \rangle\}; \epsilon \Longrightarrow_{Or}$
$\{\langle \underline{y, f(a)} \rangle, \langle g(x), z \rangle\}; \epsilon$

4. Given the pair of expressions $P(x, f(a))$ and $P(g(y), x)$ we have

$\{\langle P(x, f(a)), P(g(y), x) \rangle\}; \epsilon \Longrightarrow_{VE}$
$\{\langle g(y), f(a) \rangle\}; \{x \to f(a)\}$

5. Applying this rule to the result obtained immediately above, we have

$\{\langle g(y), f(a) \rangle\}; \epsilon \Longrightarrow_{SCl}$
$\bot$

6. For the pair of expressions $P(x)$ and $P(f(x))$ we have

$\{\langle x, f(x) \rangle\}; \epsilon \Longrightarrow_{OCh}$
$\bot$

## 2.6. Substitutions and unification for FOL

$$\{\langle P(a,x,h(g(z))), P(z,h(y),h(y))\rangle\}; \epsilon \Longrightarrow_{Dec}$$
$$\{\langle a,z\rangle, \langle x,h(y)\rangle, \langle h(g(z)),h(y)\rangle\}; \epsilon \Longrightarrow_{Or}$$
$$\{\langle z,a\rangle, \langle x,h(y)\rangle, \langle h(g(z)),h(y)\rangle\}; \epsilon \Longrightarrow_{VE}$$
$$\{\langle x,h(y)\rangle, \langle h(g(a)),h(y)\rangle\}; \{z \mapsto a\} \Longrightarrow_{VE}$$
$$\{\langle h(g(a)),h(y)\rangle\}; \{z \mapsto a, x \mapsto h(y)\} \Longrightarrow_{Dec}$$
$$\{\langle g(a),y\rangle\}; \{z \mapsto a, x \mapsto h(y)\} \Longrightarrow_{Or}$$
$$\{\langle y,g(a)\rangle\}; \{z \mapsto a, x \mapsto h(y)\} \Longrightarrow_{VE}$$
$$\emptyset; \{z \mapsto a, x \mapsto h(y), y \mapsto g(a)\}$$

Figure 2.6.1.: Unifying the pair $\langle P(a,x,h(g(z))), P(z,h(y),h(y))\rangle$.

**Example 2.6.7.** The pair $P(a,x,h(g(z)))$ and $P(z,h(y),h(y))$ is unifiable (see Fig. 2.6.1).

## 2. Logical languages

# Exercises

**Exercise 2.1.** For each formula, indicate the order of its constituents and then the order of the formula:

1. $P(x)$

2. $P(x) \wedge Q(P)$

3. $\forall x P(x)$

4. $P(x, f, f(x))$

5. $\forall x \exists S \exists T \exists q \, (S(x, q(x)) \wedge T(S))$

**Exercise 2.2.** Formalize the following proposition in 2nd-order logic: "There is a set $N$ of natural numbers $\mathbb{N}$ that does not contain 5."

**Exercise 2.3.** Write the following statements in $\mathsf{L}^*$:

1. Everybody loves something.

2. There is something that everybody loves.

3. No one loves everything.

4. Not everything is loved by everybody.

5. Not everybody loves everything.

6. Everything that is loved is loved by someone.

7. Joe likes everybody.

8. Joe likes everybody who likes him.

9. Not one single person does Joe like.

10. Not one single person who likes him does Joe like.

11. Joe likes nothing that is liked by everybody.

12. If someone likes Joe, Joe likes him or her.

13. Everybody who likes Joe is a duck.

14. If you schmoo a schmink, then some schmink is schmooed.

15. All schmoos are stinks if one schmoo is a stink.

**Exercise 2.4.** Translate the following sentences in L* into English by paying attention to the domains and the defined predicates.

1. Let the domain $\mathscr{D}$ be the set of all the people and let $L(x, y)$ stand for "$x$ loves $y$":

$$(\forall x \exists y \, (L(x,y)) \wedge \neg \exists x \forall y \, (L(x,y))) \vee (\exists x \forall y \, (L(x,y)) \wedge \exists x \forall y \, (\neg L(x,y)))$$

2. Let the domain $\mathscr{D}$ be the set of all the people and of all the times, and define $F(x, y, t)$ as "$x$ can fool $y$ at time $t$":

$$\forall x \, (\exists y \forall t \, (F(x,y,t)) \wedge \forall y \exists t \, (F(x,y,t)) \wedge \exists y \exists t \, (\neg F(x,y,t)))$$

**Exercise 2.5.** Transform each of the following formulae into a set of ground clauses:

1. $(A \wedge B) \vee \neg (C \vee D)$
2. $(P \wedge Q) \to ((R \wedge S) \to P)$
3. $\exists y \forall x \, (P(x, y) \to Q(x))$
4. $\exists y \forall x \, (P(x, y)) \to Q(x)$
5. $\forall x \forall y \, (P(x, y) \vee \exists z Q(x, y, z))$
6. $\forall w \exists u \exists v \, ((\neg P(w, u) \wedge Q(w, v)) \to R(w, u, v))$
7. $\forall y \neg \exists z \, (P(y) \to Q(y, z)) \wedge \forall x \forall u \exists w \, (R(x, w) \leftrightarrow S(x, u, w))$
8. $\forall x \, ((P(x) \leftrightarrow Q(x)) \wedge \forall z \neg \exists y \, (R(x, z) \to S(x, y, z)))$

**Exercise 2.6.** Rewrite the following formulae in CNF clearly showing every step of the transformation:

1. $\forall x \neg \exists y \, ((P(x) \to S(y)) \wedge \forall z \, (Q(x, y, z) \leftrightarrow R(x, z)))$
2. $\forall x ((P(x) \leftrightarrow Q(x)) \wedge \forall z \neg \exists y (R(x, z) \to S(x, y, z)))$
3. $\forall y \neg \exists z \, (P(y) \to Q(y, z)) \wedge \forall x \forall u \exists w \, (R(x, w) \leftrightarrow S(x, u, w))$

## 2. Logical languages

**Exercise 2.7.** Given the set (of truth values) and associated ordering of Figure 9.2.1, find the following signs and say of each whether it is (or not) a regular sign or the complement of a regular sign:

1. ↑□
2. ↓□
3. $\overline{↑□}$
4. ↓◇
5. $\overline{↓◇}$
6. $\overline{↓@}$
7. $\overline{↑@}$
8. ↓◆
9. $\overline{↑◆}$
10. $\overline{↑△}$

**Exercise 2.8.** For the given pairs of atoms, determine whether $\sigma$ is (i) a unifier and (ii) a MGU:

1. $P(a, f(y), z)\,;\, S(x, f(f(b)), b)$

    $\sigma = \{x \mapsto a, y \mapsto f(b), z \to b\}$

2. $P(x, h(a, z), f(x))\,;\, P(g(g(v)), y, f(w))$

    $\sigma = \{x \mapsto g(g(v)), y \mapsto h(a, z), w \mapsto x\}$

3. $Q(f(x), g(y))\,;\, Q(z, g(v))$

    $\sigma = \{x \mapsto a, z \mapsto f(a), y \mapsto v\}$

4. $\begin{array}{l} Knows\,(boss\,(jane)\,,trainee\,(x,at\,(y)));\\ Knows\,(boss\,(jane)\,,trainee\,(john,at\,(party\,(xmas))))\end{array}$

$$\sigma = \{x \mapsto john, y \mapsto party\,(spring)\}$$

**Exercise 2.9.** Determine whether you can unify the following pairs of atoms:

1. $Q(x,y,z)$ and $Q(u,h(v,v),u)$
2. $P(x,f(x))$ and $P(y,y)$
3. $P(f(x))$ and $P(a)$
4. $P(x,y)$ and $P(y,f(y))$
5. $S\,(f\,(a)\,,g\,(x))$ and $S\,(u,u)$
6. $R\,(a,x,h\,(g\,(z)))$ and $R\,(z,h\,(y)\,,h\,(y))$
7. $P\,(f\,(a)\,,g\,(x))$ and $P\,(x,x)$
8. $R\,(a,x,(f\,(x)))$ and $R\,(a,y,y)$

**Exercise 2.10.** Prove (Complete the proof of) the propositions and theorems in this Chapter that were left without a proof (with a sketchy proof, respectively).

# 3. Logical systems

As said in the Introduction, logic is the science of reasoning, and formal logic is the science of formal(ized) reasoning. Once we have a logical language, we can start doing "logical things" with its formulae, i.e. we can start *reasoning formally*. In other words, we can start making (and analyzing) *logical arguments*. This is done within a *logical system*. For this, we require a notion of *logical consequence*. We elaborate here–Sections 3.1-3–on the very basic aspects of logical consequence, namely with a view to *adequateness* of a logical system (Section 3.4).[1] We end this Chapter with some aspects of the logical system known as *classical logic* that are relevant for the study of many-valued logics (Section 3.5).

## 3.1. Logical consequence and inference

**3.1.1. (Def.)** In very general terms, given a logical language L and formulae $A_1, ..., A_n, B \in F_L$, a *(logical) argument* is a construction of the form

$$(\alpha) \quad \frac{A_1, ..., A_n}{B}$$

where $A_1, ..., A_n$ are *premises* and $B$ is the *conclusion*.[2]

$\alpha$ is a construct of the object language L (abbreviating L*), but there are several other ways to represent $\alpha$ in the metalanguage $\mu^L$ (abbreviating $\mu^{L^*}$), of which we first introduce the following two

---

[1] For a comprehensive elaboration on the central topic of logical consequence, see Augusto (2017).

[2] The following are alternative ways to represent $\alpha$:

- $$A_1, ..., A_n / B$$

- $$\begin{array}{c} A_1 \\ \vdots \\ A_n \\ \hline B \end{array}$$

## 3. Logical systems

$$(Cn) \quad B \in Cn(\{A_1, ..., A_n\})$$

$$(\Vdash) \quad \{A_1, ..., A_n\} \Vdash B$$

to express the fact that $B \in F_L$ is a conclusion of the set of premises $\{A_1, ..., A_n\} \subseteq F_L$.

From the viewpoint of the notion of logical consequence, denoted above by $Cn$ or $\Vdash$, a logical system S is the theory of what formulae (more generally: assertions or sentences) *follow from (are consequences of)* which other formulae.[3]

**3.1.2. (Def.)** A *logical system* S is a pair $(L, Cn)$, where L (abbreviating L*) is a logical language and $Cn$ a *consequence operation*. Equivalently, $S = (L, \Vdash)$, where $\Vdash$ is a (syntactical or semantical) *consequence relation*.[4]

**3.1.3. (Def.)** Given a logical system $S = (L, Cn)$, for every $X \subseteq F_L$ the set $Cn_S(X)$ is the set of all the consequences of $X$ in S. For a formula $\phi \in F_L$, $\phi \in Cn_S(X)$ denotes the fact that $\phi$ is a consequence of $X$ in S, or, which is the same, that $\phi$ can be inferred from $X$ in S.

We shall more often than not omit the subscript in $Cn$, because either the logical system is left unspecified, or we know what the logical system being studied is.

**3.1.4. (Def.)** For a logical system $S = (L, Cn)$, we define a *(logical) consequence operation* $Cn$ as a mapping $Cn : 2^F \longrightarrow 2^F$ satisfying the following conditions for any $X, Y \subseteq F_L$:

| | | |
|---|---|---|
| (C1) | $X \subseteq Cn(X)$ | *Inclusion* |
| (C2) | $Cn(Cn(X)) = Cn(X)$ | *Idempotency* |
| (C3) | $Cn(X) \subseteq Cn(Y)$ whenever $X \subseteq Y$ | *Monotonicity* |

Basic properties of $Cn$ are as follows:

- 
$$A_1$$
$$\vdots$$
$$A_n$$
$$\therefore B$$

---

[3] Note that we do not say "follow *necessarily* from," as this is a feature of deductive logic alone. In the non-deductive logics, conclusions do also follow from premises, but do so with a certain degree of plausibility or probability (inductive and probabilistic logics), or as an explanandum (abductive logic). See Augusto (2017) for a discussion on this topic.

[4] Other, more *ad-hoc*, definitions of a logical system are possible (e.g., Def. 7.4.7; see also Section 1.3).

## 3.1. Logical consequence and inference

**3.1.5. (Def.)** Let $Cn$ be a consequence operation on a set of formulae $F_L$. Let $X \subseteq F_L$ be given.

1. If for all substitutions $\sigma$ it is the case that

$$\sigma Cn(X) \subseteq Cn(\sigma X)$$

then $Cn$ is said to be *structural*.

2. $Cn$ is *finitary* if it is the case that

$$Cn(X) = \bigcup \left\{ Cn(X') \mid X' \text{ is a finite subset of } X \right\}.$$

Otherwise, it is *infinitary*.

3. $Cn$ is *standard* if it is both finitary and structural.

4. The strongest consequence operation on L is the operation $Cn$ such that $Cn(X) = F = L$. It is called the *inconsistent* or *trivial consequence operation* on $F$.

5. The weakest consequence operation on L is the operation $Cn$ defined by $Cn(X) = X$. This is called the *idle consequence operation* on $F$.

With the consequence operation $Cn$ in a logical system S there is naturally associated a consequence relation $\Vdash$:

**3.1.6. (Prop.)** The consequence operation $Cn$ induces a *consequence relation*:

$$(\Vdash_{Cn}) \qquad (X, \phi) \in \Vdash \text{ iff } \phi \in Cn(X)$$

And $\Vdash$, in turn, induces the *consequence operation*:

$$(Cn_{\Vdash}) \qquad Cn(X) = \{\phi \in F_L \mid (X, \phi) \in \Vdash\}$$

*Proof:* Left as an exercise.

**3.1.7. (Def.)** For a logical system $S = (L, \Vdash)$, we define a *(logical) consequence relation* $\Vdash$ as a relation $\Vdash \subseteq 2^F \times F$ satisfying the following conditions for any sets $X, Y \subseteq F_L$ and for arbitrary formulae $\phi, \chi \in F_L$:

(R1)     If $\phi \in X$, then $(X, \phi) \in \Vdash$
(R2)     If $(X, \phi) \in \Vdash$ and $X \subseteq Y$, then $(Y, \phi) \in \Vdash$
(R3)     If $(X, \phi) \in \Vdash$ and $(Y, \chi) \in \Vdash$ for every $\chi \in X$, then $(Y, \phi) \in \Vdash$

## 3. Logical systems

R1-3 can be reformulated so that they more clearly convey three important properties of a logical system, to wit, reflexivity (R), monotonicity (M), and transitivity (T).

**3.1.8. (Prop.)** We rewrite R1-R3 as follows:[5]

(R)  $\phi \Vdash \phi$
(M)  If $X \Vdash \phi$, then $X, Y \Vdash \phi$
(T)  If $X \Vdash \phi$ and $X, \phi \Vdash \psi$, then $X \Vdash \psi$

*Proof:* Left as an exercise.

*Reflexivity* is the case when $X = \{\phi\}$, and it expresses the fact that every formula is a logical consequence of itself. *Monotonicity* is an important property of some logical systems (e.g., CL; see Section 3.5), and it conveys the fact that the addition of formulae to a theory does not change its set of consequences. *Transitivity* expresses the fact that if $\phi$ is a lemma in the proof of a theorem, then we are allowed to "cut" $\phi$, i.e. substitute it by its proof.[6]

The notion of consequence operation (or relation) is essential for defining a logical inference and an inference system:

**3.1.9. (Def.)** Let L (henceforth abbreviating L*) be a logical language. An *inference* is a couple $(X, \psi)$ such that $X \subseteq F \subseteq \mathsf{L}$ and $\psi \in F$. An alternative notation is $\psi \in Cn(X)$ or $X \Vdash \psi$. In effect,

1. a consequence operation $Cn$ (relation $\Vdash$) on $F_\mathsf{L}$ is called an *inference operation (relation)* on L if it satisfies C1 and C2 (R1 and R2, respectively) for every $X \subseteq F \subseteq \mathsf{L}$ and $\psi \in F$.

2. an *inference system* on L is a pair $(\mathsf{L}, Cn)$ (a pair $(\mathsf{L}, \Vdash)$) where $Cn$ ($\Vdash$, respectively) is an inference operation (relation) on L.

Clearly, a logical system is an inference system.

---

[5]Note the following abbreviations: we write $X, \phi \Vdash \psi$ for $X \cup \{\phi\} \Vdash \{\psi\}$. In particular, we write $\Vdash \phi$ for $\emptyset \Vdash \{\phi\}$, $\phi$ is a theorem or a tautology.

[6]Reason why this is also known as *cut*. In particular, R3 expresses the fact that if $X \Vdash \phi$ and $Y \Vdash \chi$ for every $\chi \in X$, then $\chi$ can be used (as a lemma) to obtain $\phi$ from $Y$. Other, not necessarily equivalent, formalizations of T/R3 are

(T')  If $X \Vdash \phi$ and $\phi \Vdash \psi$, then $X \Vdash \psi$
(T'')  If $X \Vdash \phi$ and $Y, \phi \Vdash \psi$, then $X \cup Y \Vdash \psi$
(T''')  If $X \cup \{\phi_1, ..., \phi_n\} \Vdash \psi$ and $X \Vdash \phi_i$ for $i = 1, ..., n$, then $X \Vdash \psi$

## 3.2. Semantics and model theory

Recall from 2.2.10-11 the definition of *interpretation*. An interpretation is, in very general terms, an assignment of meaning to the symbols of a logical language. This assignment of meaning is, in formal terms, an attribution of a truth value (a valuation) to a formula: we speak of a formula $\phi$ being true or false under an interpretation $\mathcal{I}$. This is the fundamental notion of *model theory*, the study of the interpretation of formal languages by means of set-theoretical structures.

### 3.2.1. Truth-functionality and truth-functional completeness

The following "axiom" makes this notion formally clear for a propositional logical system; generalization to the case of FO systems is then feasible, though not without limitations and restrictions (see, e.g., Wójcicki, 1988).

**3.2.1. (Prop.)** *The Fregean Axiom* – Let $\phi$ be a formula, $\phi \in F \subseteq \mathsf{L}$; in order to interpret $\mathsf{L}$, $\phi$ is provided with a *meaning*. The meaning of $\phi$ is its *semantical correlate*. Let $G$ be the range of all the semantical correlates; a mapping $r : F \to G$ requires two conditions (Frege, 1892):

1. With each $\phi \in F$, exactly one semantical correlate is associated (i.e. $r$ is a function).

2. *Principle of extensionality* – Two formulae $\chi, \psi \in F$ are interchangeable in any propositional context $\phi \in F$ whenever $r(\chi) = r(\psi)$ or, in other words, when for any $\phi \in F$, $p \in V$ we have it that
$$r(\phi[p/\chi]) = r(\phi[p/\psi]) \quad \text{iff} \quad r(\chi) = r(\psi)$$
where $\phi[p/\varsigma]$ stands for the formula that results from $\phi$ after the substitution $\varsigma$ instead of $p$.

Proposition 3.2.1.2 states that the *denotation* of a proposition is a function of the denotation of its components. This is known as *truth-functionality*.

Although it is a property that one must be prepared to dispense with under certain circumstances (e.g., in the case of infinitely many-valued logics), the truth-functional completeness of $n$-valued algebras is highly relevant in that the propositional logics that are founded on them are logics of all possible extensional $n$-valued connectives.

## 3. Logical systems

We next present some important notions and results with respect to (truth-)functional completeness. For the sake of brevity, we shall abbreviate and speak of "functional completeness," because it is implicit that this is a semantical notion. We show how (truth-)functional completeness is related to truth-functionality.

**3.2.2. (Def.)** Let $n \geq 2 \in \mathbb{N}$; we denote by $G_n$ the set $\{0, 1, ..., n-1\}$ and by $\mathfrak{U}_n$ any algebra of the form $(G_n, f_1, ..., f_m)$. In particular, we denote the set of all $m$-ary mappings defined on $G_n$ with values in the same set by

$$Z_n^m = \{f | f : G_n^m \longrightarrow G_n\}, \quad m \geq 0, m \text{ finite}$$

and we denote the set of all mappings defined on $G_n$ with values in $Z_n^m$ by

$$Z_n = \bigcup_{m \in \omega} Z_n^m$$

for $\omega$ the set of all natural numbers.

**Example 3.2.1.** Let $G_2 = \{0, 1\}$. Then, we have the functions $f_1 : G_2^1 \longrightarrow G_2$ and $f_2 : G_2^2 \longrightarrow G_2$ for unary and binary mappings respectively defined on $G_2$. The functions $f_1$ and $f_2$ are *Boolean functions*, i.e. functions defined on $G_2 = W_2$.

**3.2.3. (Def.)** Let now $\mathfrak{L} = (F, o_1, ..., o_m)$ be an algebra of formulae freely generated by the set of generators $V \subseteq F_\mathsf{L}$. Then $\mathfrak{L}$ is a propositional language. Given an algebra $\mathfrak{U} = (G_n, f_1, ..., f_m)$ similar to $\mathfrak{L}$, it is easy to see that a homomorphism $hom(\mathfrak{L}, \mathfrak{U})$ can give rise to a function $h : V_\mathfrak{L} \longrightarrow G_\mathfrak{U}$, which in turn can be extended to the function $h : T_\mathfrak{L} \longrightarrow G_\mathfrak{U}$ such that for each operation $o_i$ with arity $m$ and given terms $t_i \in T$ we have

$$h(o_{i_\mathfrak{L}}(t_1, ..., t_m)) = f_{i_\mathfrak{U}}(h(t_1), ..., h(t_m))$$

Then, $h$ is an assignment of truth values, or a valuation, into $\mathfrak{L}$.

More specifically, the algebra $\mathfrak{U}$ is similar to the set of all formulae of $\mathfrak{L}$ formed by means of the $m$ operations of $\mathfrak{L}$ (the connectives), in other words, the algebra of $F_\mathfrak{L}$. From the semantical viewpoint, the fundamental importance of the homomorphism $hom(\mathfrak{L}, \mathfrak{U})$ is that it is an embedding of $F_\mathfrak{L}$ in $\mathfrak{U}$ that is in fact an interpretation for the formulae of $\mathfrak{L}$ (in $\mathfrak{U}$).

**3.2.4. (Def.)** A set of functions $X \subseteq Z_n$ is said to be *functionally complete* iff every function $f \in Z_n$ can be defined by means of functions

## 3.2. Semantics and model theory

in the set $X$.

In other words, the finite mapping $f : G_n^m \longrightarrow G_n$ can be represented as a composition of the operations $f_1, ..., f_m$. Then, we say that an algebra $\mathfrak{U}_n = (G_n, f_1, ..., f_m)$ is functionally complete (Post, 1921).

**Theorem 3.2.1.** *(Post, 1921). If $\mathfrak{U}_n$ is functionally complete for $m$ variables, $m \geq 2$, then it is also functionally complete for $m+1$ variables and hence also functionally complete.*

*Proof:* Left as a research exercise.

Theorem 3.2.1 reduces the problem of functional completeness of many-valued algebras to that of the definability of all unary and binary connectives:

**3.2.5. (Prop.)** Given $n \in \mathbb{N}$, in an $n$-valued logic any given place in the truth table can be occupied by $n$ truth values. For a $k$-place connective, the truth table has room for entries

$$\underbrace{n \times n \times ... \times n}_{k \text{ times}} = n^k$$

and for each there will be any of $n$ possibilities, so that we can have

$$\underbrace{n \times n \times ... \times n}_{n^k \text{ times}} = n^{n^k}$$

possible $k$-place truth tables for a given $n$-valued logic. We thus have it that the number of unary and binary connectives equals $n^n$ and $n^{n^2}$, respectively.
*Proof:* Trivial. **QED**

Further, "handier," criteria of (truth-)functional completeness are available. For instance, and considering $G$ as a set of semantical correlates or, more simply, truth values:

**Theorem 3.2.2.** *(Słupecki, 1939). For finite $n \geq 2$, an $n$-valued algebra $\mathfrak{U}_n$ is functionally complete iff in it one can define*

1. *all one-argument operations on $G_n$;*

2. *at least one two-argument operation $f(x, y)$ whose range consists of all the values $i$ for $1 \leq i \leq n$.*

## 3. Logical systems

*Proof:* Left as a research exercise.

**Example 3.2.2.** Post (1921) provided the first functionally complete $n$-valued logic algebra for $n \geq 2$ by showing that in order to establish functional completeness of $\mathfrak{U}_n$ the generation of two functions suffice. Given the linearly ordered set $P_n = \{t_1, ..., t_n\}$, where $t_i$ is less true than $t_j$ just in case $i < j$, these are the one-argument *cyclic rotation function* and the two-argument *maximum function*,

$$\neg t_i = \begin{cases} t_i & \text{if } i = n \\ t_{i+1} & \text{if } i \neq n \end{cases}$$

and

$$t_i \vee t_j = t_{max(i,j)}$$

for negation and disjunction, respectively. Thus, for $n = 4$, we have the following tables:

| $p$ | $\neg p$ |
|---|---|
| $t_1$ | $t_2$ |
| $t_2$ | $t_3$ |
| $t_3$ | $t_4$ |
| $t_4$ | $t_1$ |

| $\vee$ | $t_1$ | $t_2$ | $t_3$ | $t_4$ |
|---|---|---|---|---|
| $t_1$ | $t_1$ | $t_2$ | $t_3$ | $t_4$ |
| $t_2$ | $t_2$ | $t_2$ | $t_3$ | $t_4$ |
| $t_3$ | $t_3$ | $t_3$ | $t_3$ | $t_4$ |
| $t_4$ | $t_4$ | $t_4$ | $t_4$ | $t_4$ |

**Theorem 3.2.3.** *(Post, 1921) Every n-valued Post algebra*

$$\mathfrak{P}_n = (\{t_1, ..., t_n\}, \neg, \vee)$$

*is functionally complete.*

**Theorem 3.2.4.** *The two-element Boolean algebra*

$$\mathfrak{B}_2 = (\{0, 1\}, \vee, \wedge, \neg, 1, 0)$$

*of which* $\mathfrak{P}_2 = (G_2, \neg, \vee)$ *is a reduct, is functionally complete.*

The proofs of the above theorems are left as exercises, for which the respective cited literature should be helpful.

## 3.2. Semantics and model theory

### 3.2.2. Semantics and deduction

For a logical system L, the *semantical consequence relation*, denoted by $\models_L$, specifies the class of *valid inferences* in L, or, in other terms, which inferences in L *preserve truth*, or are *deductively valid*.[7] We say that an inference is *deductive* (vs. inductive or abductive) if the conclusion follows necessarily from a set of premises. We next provide definitions and theorems that specify the deductive consequence relation from the semantical viewpoint.

The fundamental concepts of semantics *validity* and *satisfiability* depend on the notion of *interpretation* (cf. Def.s 2.2.10-11).

**3.2.6. (Def.)** *Satisfiability* – An interpretation $\mathcal{I}$ is said to *satisfy*

1. a formula $\psi$, and $\psi$ is said to be *satisfiable*, iff $val_\mathcal{I}(\psi) = \mathsf{t}$ and we denote this satisfiability relation by $\models_\mathcal{I} \psi$ (abbreviating $\models_{val_\mathcal{I}} \psi$). Otherwise, we write $\not\models_\mathcal{I} \psi$. If there is no interpretation $\mathcal{I}$ that satisfies $\psi$, then $\psi$ is *unsatisfiable*, and we write $\not\models \psi$.

2. a set of formulae (a theory) $X = \{\chi_1, ..., \chi_n\}$, and $X$ is said to be *satisfiable*, iff for all $\chi_i \in X$, $val_\mathcal{I}(\chi_i) = \mathsf{t}$, and we denote this satisfiability relation by $\models_\mathcal{I} X$.[8] Otherwise, we write $\not\models_\mathcal{I} X$. If there is no interpretation $\mathcal{I}$ that satisfies $X$, then $X$ is *unsatisfiable*, and we write $\not\models X$.

**3.2.7. (Def.)** *Model* – We say that $\mathcal{I}$ is a *model* of

1. $\psi$, and we write $\models_\mathcal{M} \psi$, iff $\models_\mathcal{I} \psi$. Thus, a formula $\psi$ is said to be satisfied iff there is an interpretation that is a model of $\psi$. Otherwise, we write $\not\models \psi$ when there is no model of $\psi$, or $\not\models_\mathcal{I} \psi$ for $\mathcal{I} = \overline{\mathcal{M}}$ a *countermodel* of $\psi$.

2. $X$, and we write $\models_\mathcal{M} X$, iff $\models_\mathcal{I} X$. Thus, a set of formulae (a theory) $X$ is satisfiable iff all its members have a (common) model. Otherwise, we write $\not\models X$ when there is no model of $X$, or $\not\models_\mathcal{I} X$ for $\mathcal{I} = \overline{\mathcal{M}}$ a *countermodel* of $X$.

**3.2.8. (Def.)** A *semantics* $\mathfrak{S}$ for a logical language L (abbreviating L*) is an infinite set of (classes of) models.

---

[7]This is a specification of $\Vdash$ (cf. Section 3.1). We omit the subscript in $\models$ when we know the logical system at hand. In particular, if no ambiguity arises, we write simply $\models$ for $\models_{CL}$ (or, alternatively, $\models_K$).

[8]This implies that $X$ can be viewed as a conjunction of the formulae $\chi_i$: $X \equiv \bigwedge_{i=1}^{n} \chi_i$.

## 3. Logical systems

**3.2.9. (Def.)** *Validity* – Let $X$ be a (possibly empty) set of formulae and $\psi$ a formula *entailed*–i.e. following semantically–from $X$. We say that $\psi$ is *valid* iff there is no interpretation assigning the value $\mathbf{t}$ to all the members of $X$ and $\mathbf{f}$ to $\psi$, and we write $X \models \psi$ (abbreviating $X \models_{\mathfrak{S}} \psi$, or $X \models_{\mathcal{M}_i} \psi$ for all $\mathcal{M}_i \subseteq \mathfrak{S}$). A formula is said to be *invalid*, written $X \not\models \psi$, iff it is not valid.

**3.2.10. (Def.)** Let $\mathfrak{S} = \{\mathcal{M}_1, \mathcal{M}_2, ...\}$ where the $\mathcal{M}_i$ are models. Then, we can define the *semantical consequence relation* $\models$ as the relation $\models \subseteq \mathfrak{S} \times \mathsf{L}$ ($\models \subseteq \mathcal{M}_i \times \mathsf{L}$) such that for an arbitrary formula $\psi \in F_{\mathsf{L}}$ and a (possibly empty) set of formulae $X \subseteq F_{\mathsf{L}}$ we have, for $\mathfrak{S}$ ($\mathcal{M}_i$):

$$(\models) \qquad X \models \psi \quad \text{iff} \quad X \models_{\mathfrak{S}} \psi \ (X \models_{\mathcal{M}_i} \psi, \text{ respectively})$$

This is more precisely the *semantical consequence relation induced by the semantics* $\mathfrak{S}$ *(the model* $\mathcal{M}_i$*)* as $X \models \psi$ means that for $\mathfrak{S}$ ($\mathcal{M}_i$) we have:

$$\text{If } \models_{\mathfrak{S}(\mathcal{M}_i)} X, \text{ then } \models_{\mathfrak{S}(\mathcal{M}_i)} \psi.$$

Obviously, Definition 3.2.10 means that $\psi$ is a semantical consequence of $X$ iff every semantics (model) that satisfies $X$ also satisfies $\psi$. Thus, we see the semantical consequence relation as coinciding with the satisfiability relation.

**3.2.11. (Def.)** A formula $\phi$ is said to be

1. a *tautology* iff every interpretation $\mathcal{I}$ of $\phi$ is also a model of $\phi$. In other words, $\phi$ is a tautology if it uniformly takes the truth value $\mathbf{t}$ for any and every assignment of truth values to its variables. We consequently have the set $\top$ (L) of all the tautologies in a logical system L with a language L:

$$\top(\mathsf{L}) = \{\phi \in F_{\mathsf{L}} | val(\phi) = \mathbf{t} \text{ for every } val : F_{\mathsf{L}} \longrightarrow W_n\}$$

2. a *contradiction* iff there is no $\mathcal{I}$ of $\phi$ that is a model of $\phi$. In other words, $\phi$ is said to be a contradiction if it uniformly takes the truth value $\mathbf{f}$ for any and every assignment of truth values to its variables, i.e. for the same system we have:

$$\bot(\mathsf{L}) = \{\phi \in F_{\mathsf{L}} | val(\phi) = \mathbf{f} \text{ for every } val : F_{\mathsf{L}} \longrightarrow W_n\}$$

3. *contingent* iff it is neither a tautology nor a contradiction.

We shall often denote an arbitrary tautology by $\top$ and an arbitrary contradiction by $\bot$.

## 3.2. Semantics and model theory

It is now obvious that:

**3.2.12. (Prop.)** A formula $\phi$ is valid iff

1. every interpretation of $\phi$ is also a model of $\phi$, and

2. $\phi$ is a tautology.

*Proof:* Left as an exercise.

The valid formulae of a logical system L are fundamental to the definition of the logic **L**:[9]

**3.2.13. (Def.)** The *logic* of L, denoted by **L**, is the set of tautologies generated by L, i.e.
$$\mathbf{L} := \{\phi \in \mathsf{L} | \models_L \phi\}.$$

It is also obvious that:

**3.2.14. (Prop.)** A formula $\phi$ is

1. valid iff its negation ($\neg\phi$) is unsatisfiable.

2. unsatisfiable iff its negation is valid.

3. invalid iff there is at least one interpretation that falsifies it.

4. satisfiable iff there is at least one interpretation that makes it true.

*Proof:* Left as an exercise.

**3.2.15. (Prop.)** If a formula is valid, then it is satisfiable (but not vice-versa). If a formula is unsatisfiable, then it is invalid (but not vice-versa).
*Proof:* Trivial. **QED**

**Theorem 3.2.5.** ($\models$-*Deduction theorem 1)* $X \models \psi$ iff $X \cup \{\neg\psi\}$ is unsatisfiable.

*Proof:* ($\Rightarrow$) For any interpretation $\mathcal{I}$, either $\mathcal{I}(\chi_i) = \mathsf{t}$ (abbreviating $val_{\mathcal{I}}(\chi_i) = \mathsf{t}$) for all $\chi_i \in X$ and $\mathcal{I}(\psi) = \mathsf{t}$ (hence, $\mathcal{I}(\neg\psi) = \mathsf{f}$), or $\mathcal{I}(\chi_i) = \mathsf{f}$ for some $\chi_i \in X$. Either way, $\mathcal{I}(X \cup \{\neg\psi\}) = \mathsf{f}$.

---
[9] See also Def. 3.3.9. See Augusto (2017), p. 64-5, for some critical remarks on this definition.

## 3. Logical systems

($\Leftarrow$) For any interpretation $\mathcal{I}$, either $\mathcal{I}(\chi_i) = \mathtt{t}$ for all $\chi_i \in X$ and $\mathcal{I}(\neg\psi) = \mathtt{f}$ (hence, $\mathcal{I}(\psi) = \mathtt{t}$), or $\mathcal{I}(\chi_i) = \mathtt{f}$ for some $\chi_i \in X$. Therefore, $\mathcal{I}(\psi) = \mathtt{t}$ whenever $\mathcal{I}(X) = \mathtt{t}$, and thus $X \models \psi$. **QED**

It will be useful to provide an equivalent formulation of the deduction theorem:

**Theorem 3.2.6.** ($\models$-*Deduction theorem 2*) *Given a set of formulae* $X = \{\chi_1, ..., \chi_n\}$ *and a formula* $\psi$, $\psi$ *is a logical consequence of* $X$ *iff the formula* $((\chi_1 \wedge ... \wedge \chi_n) \rightarrow \psi)$ *is valid, i.e. iff*

$$\text{if } (\chi_1 \wedge ... \wedge \chi_n) \models \psi, \text{ then } \models (\chi_1 \wedge ... \wedge \chi_n) \rightarrow \psi.$$

*Proof:* The proof follows immediately from the above. **QED**

The above allows for a reformulation of the definition of logical equivalence (cf. Def. 2.1.10):

**3.2.16. (Def.)** Two formulae $\phi$ and $\psi$ are said to be *equivalent* iff:

1. They have exactly the same truth value in every interpretation.

2. $\phi \models \psi$ and $\psi \models \phi$.

3. $\phi \leftrightarrow \psi$ is a tautology, i.e. $\models (\phi \leftrightarrow \psi)$.

**Theorem 3.2.7.** *A formula* $\psi$ *is a logical consequence of a set of formulae* $X = \{\chi_1, ..., \chi_n\}$ *iff the formula* $(\chi_1 \wedge ... \wedge \chi_n \wedge \neg\psi)$ *is unsatisfiable.*

*Proof:* It follows from Theorem 3.2.6 that $\psi$ is a logical consequence of $X = \{\chi_1, ..., \chi_n\}$ iff the negation of $((\chi_1 \wedge ... \wedge \chi_n) \rightarrow \psi)$ is unsatisfiable. In effect, $\neg((\chi_1 \wedge ... \wedge \chi_n) \rightarrow \psi) \equiv \chi_1 \wedge ... \wedge \chi_n \wedge \neg\psi$ by $\rightarrow_{df}$, $DM_\vee$, and the property of associativity of $\wedge$. **QED**

**Theorem 3.2.8.** *A formula* $\phi$ *is unsatisfiable iff it is possible to derive a contradiction from* $\phi$, *i.e. iff we have*

$$\phi \models \psi \wedge \neg\psi.$$

*Proof:* By the $\models$-deduction theorem, we have (i) $\phi \models \psi \wedge \neg\psi$ iff we have (ii) $\models \phi \rightarrow (\psi \wedge \neg\psi)$. In turn, we have (ii) iff, for every interpretation $\mathcal{I}$, (a) $val_\mathcal{I}(\phi) = \mathtt{f}$ or (b) $val_\mathcal{I}(\phi) = \mathtt{t}$ and $val_\mathcal{I}(\psi \wedge \neg\psi) = \mathtt{t}$. But the truth table of $\psi \wedge \neg\psi$ shows that it is a contradiction, and thus for every interpretation $\mathcal{I}$ we have $val_\mathcal{I}(\psi \wedge \neg\psi) = \mathtt{f}$. Therefore, we must have (a). Thus, $\models \phi \to (\psi \wedge \neg\psi)$ iff $\phi$ is unsatisfiable, and $\phi \models \psi \wedge \neg\psi$ iff $\phi$ is unsatisfiable. **QED**

### 3.2.3. Matrix semantics

Algebras provide many-valued logics with adequate semantics, and especially so logical matrices, algebras equipped with a distinguished set of truth values. In effect, a logical matrix is an interpretation structure with a distinguished subset of elements corresponding to sentences of a specified type such as, for example, tautologies.

**3.2.17. (Def.)** A *(logical) matrix* $\mathfrak{M}$ for a language $\mathsf{L} = (F, O)$ is a triple $\mathfrak{M} = (W, Fop, D)$, where

1. $|W| \geq 2$, $W$ is a set of truth values;

2. $Fop = \{f_{o_1}, ..., f_{o_n}\}$ is the set of *interpretation functions*, i.e. the functions corresponding to each operator in $O = \{o_1, ..., o_n\}$ such that $f_{o_i} : W^{n_{o_i}} \longrightarrow W$, $n_{o_i}$ is the arity of $o_i$;

3. $D \subset W$, $D \neq \emptyset$ is the set of *designated truth values*.

It is often highly fruitful (e.g., Rasiowa, 1974; Stachniak, 1996) to see a propositional language $\mathsf{L} = (F, o_1, ..., o_m)$ as an *algebra of formulae*

$$\mathfrak{L} = (F, o_1, ..., o_m)$$

where $F$ is a set of formulae, $o_1, ..., o_n$ are finitary operations on $F$, and the algebra $\mathfrak{L}$ is freely generated by the set $V$ of propositional variables (cf. Def. 3.2.3).

**Example 3.2.3.** For the specific case of CPL, which can be denoted by K, we have it that

$$\mathfrak{L}_\mathsf{K} = (F, \neg, \to, \vee, \wedge, \leftrightarrow)$$

is an abstract algebra of type $(1, 2, 2, 2, 2)$.

**3.2.18.** As seen in Definition 3.2.3, an interpretation for $\mathfrak{L}$ can be provided by an interpretation structure

$$\mathfrak{A} = (G, f_1, ..., f_m)$$

## 3. Logical systems

that is an algebra similar to $\mathfrak{L}$. This follows immediately from the following theorem:

**Theorem 3.2.9.** *(Suszko, 1957). When $G$ is the set of all the semantical correlates of the language $\mathfrak{L} = (F, o_1, ..., o_m)$ then for each $i = 1, ..., m$, $r$ is defined as in Proposition 3.2.1, the formula*

$$r(o_i(\phi_1, ..., \phi_m)) = f_i(r(\phi_1), ..., r(\phi_m))$$

*uniquely defines a function $f_i$ on $G$ of the same arity as $o_i$.*

*Proof:* Left as a research exercise.

Any mapping $s: V \longrightarrow G$ can be uniquely extended to the homomorphism $h_s : \mathfrak{L} \longrightarrow \mathfrak{A}$ (i.e. $h_s \in Hom(\mathfrak{L}, \mathfrak{A})$), and $\mathfrak{L}$ is an absolutely free algebra, freely generated by $V \subseteq F$, in its similarity class.

Given the above, we can now provide a reformulation of the definition of a logical matrix $\mathfrak{M}$:

**3.2.19. (Def.)** A *(logical) matrix* $\mathfrak{M}$ is a pair $(\mathfrak{A}, D)$ where $\mathfrak{A} = (G, f_1, ..., f_m)$ is an algebra similar to a propositional language $\mathfrak{L}$ and $D \subseteq G$ is a non-empty subset of the universe of $\mathfrak{A}$; more specifically, $D$ is called the set of designated values of $\mathfrak{M}$.

**3.2.20. (Def.)** With each matrix $\mathfrak{M}$ there is associated a set of formulae

$$E(\mathfrak{M}) = \{\phi \in F_\mathsf{L} | h(\phi) \in D \text{ for any } h \in Hom(\mathfrak{L}, \mathfrak{A})\}$$

called the *content* of $\mathfrak{M}$, and for any such matrix $\mathfrak{M}$ we define the relation $\models_\mathfrak{M}$ for every $X \cup \{\psi\}$ for which $X \subseteq F_\mathsf{L}, \psi \in F_\mathsf{L}$, as:

$$X \models_\mathfrak{M} \psi \text{ iff } \forall h \in Hom(\mathfrak{L}, \mathfrak{A}), \, h(\psi) \in D \text{ whenever } h(X) \subseteq D.$$

**3.2.21. (Def.)** With any relation $\models_\mathfrak{M}$ there may be uniquely associated an operation $Cn_\mathfrak{M} : 2^F \longrightarrow 2^F$ such that

$$\psi \subset Cn_\mathfrak{M}(X) \quad \text{iff} \quad X \models_\mathfrak{M} \psi.$$

We say that $Cn_\mathfrak{M}$ is a *matrix consequence operation* and the relation $\models_\mathfrak{M}$ is a *matrix consequence relation* $\models_\mathfrak{M} \subseteq 2^F \times F$ for $\mathfrak{M}$, it being the case that $\models_\mathfrak{M} \subseteq 2^F \times F$ is a natural generalization of the relation of classical consequence relation $\models$.

**Example 3.2.4.** Illustrating the above, for the propositional language L, the propositional language of CPL, we then have it that the

## 3.2. Semantics and model theory

*2-element algebra of classical logic* is of the form

$$\mathfrak{K} = (\{0,1\}, \neg, \rightarrow, \vee, \wedge, \leftrightarrow)$$

while the *classical matrix* has the form

$$\mathfrak{M}_2 = (\{0,1\}, \neg, \rightarrow, \vee, \wedge, \leftrightarrow, \{1\}) = (\mathfrak{K}, \{1\})$$

and the *classical matrix consequence relation* is characterized in the following way:

$$X \models_{\mathfrak{M}_2} \psi \text{ iff, } \forall h \in Hom(\mathfrak{L}, \mathfrak{K}), h(\psi) = 1 \text{ whenever } h(X) \subseteq \{1\}.$$

Thus, the content of $\mathfrak{M}_2$ is the set consisting of formulae that are consequences of the empty set,

$$\top(K) = E(\mathfrak{M}_2) = \{\phi | \models \phi\}$$

for $\top(K)$ the set of all classical tautologies.

**3.2.22.** The distinguishing property of the logic generated by a matrix $\mathfrak{M}$ is its *structurality*. In effect, for any substitution $\sigma \in End(\mathfrak{L})$, $X \models_{\mathfrak{M}} \psi$ implies $\sigma X \models_{\mathfrak{M}} \sigma \psi$, or equivalently $\psi \in Cn_{\mathfrak{M}}(X)$ implies $\sigma \psi \in Cn_{\mathfrak{M}}(\sigma X)$.

The most important property of structural logics is their representation by means of a matrix.

**3.2.23. (Prop.)** *Matrix representation* – For every consequence operation $Cn$ there is a class of matrices $\mathcal{M}$ such that for any $X \in F_L$:

$$Cn(X) = \bigcap \{Cn_{\mathfrak{M}}(X) | \mathfrak{M} \in \mathcal{M}\}$$

This can be restated as a *completeness theorem* by proceeding in the following way:

**Theorem 3.2.10.** *(Malinowski, 1989). Let any class $\mathcal{M}$ of matrices for a language* L *be called a* matrix semantics *for* L*; then every semantics $\mathfrak{S}_{\mathcal{M}}$ for* L *defines a function* $Cn_{\mathcal{M}} : L \supseteq X \longrightarrow Cn_{\mathcal{M}}(X) \subseteq L$ *such that for any* $X \subseteq L$ *and for* $\psi \in L, \psi \in Cn_{\mathcal{M}}(X)$ *iff for every matrix* $\mathfrak{M} = (\mathfrak{A}, D) \in \mathcal{M}$ *and every valuation* $h \in Hom(\mathfrak{L}, \mathfrak{A})$, *if* $h(X) \subseteq D$, *then* $h(\psi) \in D$. *It follows that if* $\mathfrak{S}_{\mathcal{M}}$ *is a semantics for* L, *then the function* $Cn_{\mathcal{M}}$ *is a structural consequence operation on* L, *and we say that a logical system* L $= (L, Cn)$ *is strongly complete relative to* $\mathcal{M}$ *for* L *if* $Cn = Cn_{\mathcal{M}}$.

**3.2.24. (Prop.)** It can be shown that each matrix consequence operation $Cn_\mathfrak{M}$ is structural, and conversely, each structural consequence $Cn$ on L and any set of formulae $X \subseteq F_L$ together determine a matrix $\mathfrak{M}_X = (L, Cn(X))$ called a *Lindenbaum matrix* for $Cn$. A class of matrices involving the same algebra $\mathscr{L}_{Cn} = \{\mathfrak{M}_X | X \subseteq L\}$ is called a *Lindenbaum bundle* for $Cn$.

**Theorem 3.2.11.** *(Lindenbaum; Wójcicki, 1969). Every logic is strongly complete with respect to a Lindenbaum bundle $\mathscr{L}_{Cn}$.*

*Proof:* A proof can be found in Malinowski (1989). **QED**

The importance of Theorem 3.2.11 lies in the fact that it establishes that every logic has a truth-functional semantics, namely a matrix semantics.

## 3.3. Syntax and proof theory

Recall from Definition 2.1.7 that a proof calculus is a logical language equipped with a set of rules specifying derivations from (sets of) formulae to (sets of) formulae. When proving that a formula or a set of formulae is a derivation from a set of formulae, we can apply these same rules–we, so to say, reconstruct the derivation by listing all the rules in it. But logical proofs can be carried out by applying other kinds of formal structures, namely by means of axiom schemata, alone or together with rules. In any case, we can speak of proof systems. *Proof theory* is the study of the *form* (the syntax) of formal derivations and proofs, and of the diverse proof systems. In effect, whereas model theory revolves around *meaning* (cf. Section 3.2), proof theory does so around *form*. This is to say that we can approach the study of deduction in a logical system by focusing on form alone, with no appeal to meanings.

### 3.3.1. Inference rules and proof systems

The central notion of proof theory is that of *inference rule*. This, in turn, is related to other fundamental proof-theoretical notions such as *proof*, and *derivation* or *inference*. We provide the most important aspects of these notions in the definitions that follow. Further material on these and other topics on proof theory can be found in Troelstra & Schwichtenberg (2000) and, more specialized, Rybakov (1997).

**3.3.1. (Def.)** Given an inference system $(\mathsf{L}, Cn)$ that is also a logical system, an *inference rule* $\mathbf{r}$ on a set of formulae $F \subseteq \mathsf{L}$ is a mapping assigning to some finite sequence $\chi_1, ..., \chi_n \in X$, $n \geq 0$, of formulae (the *premises*) a formula $\psi$ (the *conclusion*), i.e. $\mathbf{r} : X \longrightarrow F$ where $X \subseteq F^n$ for some $n = 1, 2, ....$ We write $\mathbf{r}(\chi_1, ..., \chi_n) = \psi$, $\mathbf{r}(\{\chi_1, ..., \chi_n\}, \psi)$, or more commonly,

$$(\mathbf{r}) \quad \frac{\chi_1, ..., \chi_n}{\psi}$$

or

$$(\mathbf{r}) \quad \chi_1, ..., \chi_n / \psi.$$

We denote by $RI$ the set of inference rules $\{\mathbf{r}_1, ..., \mathbf{r}_n\}$.

1. Given a substitution $\sigma$, $\mathbf{r}$ is called a *structural inference rule* of $\mathsf{L}$ if it has the form:

$$(\mathbf{r}) \quad \sigma\chi_1, ..., \sigma\chi_n / \sigma\psi$$

2. A rule $\mathbf{r}$ is said to *preserve* a set of formulae $X$, and $X$ is said to *be closed under* $\mathbf{r}$ iff for all $X' \subseteq X$ and for all formulae $\psi$, if $\mathbf{r}(X', \psi)$, then $\psi \in X$.

3. A subset $F_0 \subseteq F$ is said to be *closed under a rule of inference* $\mathbf{r}$ provided that $(\chi_1, ..., \chi_n) \in F_0^n \cap X$ implies that $\mathbf{r}(\chi_1, ..., \chi_n) \in F_0$.

**3.3.2. (Def.)** Given an inference of the form $X/\psi$,

1. we say that $\psi$ is *an axiom* when $X = \emptyset$ in (the application of) an inference rule $\mathbf{r}(X, \psi)$.

2. $X$ can comprise a set $AX$ of *axiom schemata*, formulae in the set $F \subseteq \mu^{\mathsf{L}}$ that can be replaced by formulae of $\mathsf{L}$.

**3.3.3. (Def.)** A *proof system* (or *calculus*) $\mathcal{P}$ for a logical language $\mathsf{L}$ is a pair $(RI, AX)$ where either $RI$ or $AX$–but not both–can be empty, i.e. a system of rules of inference and/or axiom schemata. The triple $(\mathsf{L}, RI, AX)$ is an instance of a *formal system*, namely an *inference system*.[10]

---

[10] Insofar as the pair $(RI, AX)$ gives rise to a consequence relation (see Def. 3.1.2). See also Def. 3.1.9.

## 3. Logical systems

**3.3.4. (Def.)** A *proof* of $\psi$ in $\mathcal{P}$ is a finite collection of rules of inference and/or axioms of $\mathcal{P}$ that leads to concluding that $\psi$ is a member of $F_\mathsf{L}$.

We shall denote a proof by the symbol ∎. We make Definition 3.3.4 more precise from the formal point of view by means of the notion of *provability* or *derivability*.

**3.3.5. (Def.)** A formula $\psi$ is *provable*, or *derivable*, *from* a (possibly empty) set of formulae $X$ by means of axioms in the set $AX$ and rules in the set $RI$ iff there is a finite sequence of formulae $\psi_1, ..., \psi_n$ that is a *proof* or a *derivation of* $\psi$ *from* $X$, i.e. there is a finite sequence $\psi_1, ..., \psi_n$ such that

1. $\psi_1 \in (X \cup AX \cup RI)$;

2. for every $1 < i \leq n$, either $\psi_i \in (X \cup AX \cup RI)$ or $\psi_i$ is the conclusion of one of the rules of inference $\mathbf{r}_j$, $j = 1, ..., k$ of which the premises are some of the $\psi_1, ..., \psi_{i-1}$;

3. $\psi_n = \psi$.

**3.3.6. (Def.)** Let $\psi$ be a conclusion in a proof (derivation). When proven (derived) by means of a rule of inference $\mathbf{r}$ from $X = \emptyset$, then $\psi$ is a *theorem*. If proven (derived) by means of a rule of inference $\mathbf{r}$ from the axioms of a theory, then $\psi$ is called a *theorem* of the theory. In any case, *theorems* are always provable (derivable) formulae.

**3.3.7. (Def.)** Two inference rules are particularly important:

1. *Modus ponens*:
$$(\text{MP}) \quad \frac{\phi, \phi \to \psi}{\psi}$$

2. The purely syntactical version of the principle of extensionality (cf. 3.2.1.2), known as *substitution rule*
$$(\text{SUB}) \quad \frac{\phi}{\sigma\phi}$$
which states that if a formula $\phi$ is a theorem, then any of its substitution instances (the extensionally equivalent formulae) is also a theorem.

### 3.3.2. Syntax and deduction

For a logical system L, the *syntactical consequence relation* $\vdash_L$ specifies what conclusions are derivable in L.[11] Just as in the case of the semantical consequence relation (cf. Section 3.2.2), we say that an inference is *deductive* (vs. inductive or abductive) when the conclusion follows necessarily from a set of premises. We now provide definitions and theorems that specify the consequence relation from the syntactical viewpoint.

For the language L (abbreviating L*), derivability and consistency are the fundamental concepts of the proof theory for L. Recall the above (3.3.5) definition of derivability. We now reformulate this definition from the viewpoint of the syntactical consequence relation.

**3.3.8. (Def.)** *Derivability and refutation* – Let $X$ be a (possibly empty) set of formulae and $\psi$ a formula. We say that $\psi$ is derivable from (i.e. is a conclusion of) the set of premises (or assumptions) $X$ iff there is a proof ■ such that we have the relation $X \vdash_■ \psi$. Otherwise, we write $X \nvdash_■ \psi$ and call ■ a *counter-proof* or *refutation*.

We often abbreviate this consequence relation as $X \vdash \psi$, especially when $\phi$ is an axiom, i.e. when we have $\vdash \phi$.

Obviously, every axiom is a theorem–is derivable (from $\emptyset$). Theorems are of fundamental importance for the definition, given a logical system L, of the logic **L**:[12]

**3.3.9. (Def.)** The *logic* of L, denoted as **L**, is the set of theorems generated by L, i.e.

$$\mathbf{L} := \{\phi \in \mathsf{L} | \vdash_L \phi\}.$$

**3.3.10. (Def.)** *Consistency* – We say that a set $X = \{\chi_1, ..., \chi_n\}$ is *inconsistent* iff we can derive from it a formula together with its negation (in semantical terms, a contradiction), i.e. iff we have

$$X \vdash (\psi \wedge \neg\psi).$$

Otherwise, the set $X$ is *consistent*.

**3.3.11. (Def.)** Let the proof system $\mathcal{P}$ be given for L, and let $F \subseteq \mathsf{L}$. Then, we can define the *syntactical consequence relation* $\vdash$ as the relation $\vdash \subseteq \mathcal{P} \times \mathsf{L}$ such that for an arbitrary formula $\psi \in F$ and a (possibly empty) set of formulae $X \subseteq F$ we have

$$(\vdash) \qquad X \vdash \psi \quad \text{iff} \quad X \vdash_\mathcal{P} \psi.$$

---
[11] This is a specification of $\Vdash$ (cf. Section 3.1). Just as in the case of $\models$, we omit subscripts whenever no ambiguity arises.
[12] See also Def. 3.2.13. See Augusto (2017), p. 64-5, for some critical remarks.

## 3. Logical systems

This is more precisely the *syntactical consequence relation induced by* $\mathcal{P}$ as, given $\mathcal{P}$, we have

$$X \vdash_{\mathcal{P}} \psi \quad \text{iff} \quad \{\mapsto \chi | \chi \in X\} \vdash_{\mathcal{P}, \zeta} \mapsto \psi$$

where $\zeta \in (RI_{\mathcal{P}} \cup AX_{\mathcal{P}})$ and $\mapsto$ denotes derivability.

Thus, we see the syntactical consequence relation as coinciding with the derivability (or provability) relation.

**Theorem 3.3.1.** *($\vdash$-Deduction theorem 1) $X \vdash \psi$ iff $X \cup \{\neg \psi\}$ is inconsistent.*

**Theorem 3.3.2.** *($\vdash$-Deduction theorem 2) Given a set of formulae $X = \{\chi_1, ..., \chi_n\}$ and a formula $\psi$, $\psi$ is a logical consequence of $X$ iff the formula $((\chi_1 \wedge ... \wedge \chi_n) \rightarrow \psi)$ is a theorem, i.e. iff*

$$\text{if } (\chi_1 \wedge ... \wedge \chi_n) \vdash \psi, \text{ then } \vdash (\chi_1 \wedge ... \wedge \chi_n) \rightarrow \psi.$$

**Theorem 3.3.3.** *A formula $\psi$ is a logical consequence of a set of formulae $X = \{\chi_1, ..., \chi_n\}$ iff the formula $(\chi_1 \wedge ... \wedge \chi_n \wedge \neg \psi)$ is inconsistent.*

The proofs of the above theorems are left as exercises.

**Theorem 3.3.4.** *(Łoś-Suszko) Any structural logical consequence relation can be generated by a proof system $\mathcal{P}$.*

The proof is left as a research exercise. See Augusto (2017), p. 61f, for some remarks on this theorem.

## 3.4. Adequateness of a deductive system

Both notions of logical consequence, syntactical and semantical, are essential for a logical system in that they express the fundamental metatheoretical results of *soundness* and *completeness*.

Let $L_1 = (L, \vdash)$ and $L_2 = (L, \models)$ be respectively syntactically and semantically defined logical systems over the same language L. As seen above (cf. Def. 3.3.3), $L_1$ can be the inference system $L_1 = (L, RI, AX)$, where the pair $(RI, AX)$ is a proof system $\mathcal{P}$ for the language L. By appealing to Definition 3.3.5, we can make the notion of an inference more precise from the logical viewpoint, namely with respect to *deduction*:

## 3.4. Adequateness of a deductive system

**3.4.1. (Def.)** A *deductive system* L *over a language* L is a pair L = (L, (RI ∪ AX)) where either RI or AX–but not both–can be empty, and $RI \cap AX = \emptyset$.

As a matter of fact, any set of sentences $X$ of a logical language (a theory) that contains all its consequences ($X \supseteq Cn(X)$) can be seen as a deductive system (Tarski, 1930). In particular:

**3.4.2. (Def.)** A logical system whose consequence relation (operation) satisfies R1-3 (C1-3, respectively) and in addition satisfies the following conditions for finite $X' \subseteq X \subseteq F_L$ and $\psi \in F_L$,

$$(\Vdash') \qquad \text{If } X \Vdash \psi, \text{ then } X' \Vdash \psi$$

and

$$(Cn') \qquad Cn(X) \subseteq \bigcup \{Cn(X') \mid X' \subseteq X\}$$

is a deductive system.

**3.4.3. (Def.)** Conditions $\Vdash'$ and $Cn'$ define the property of *compactness*.[13]

Recall from Sections 3.2.2 and 3.3.2 that deduction can be both semantically and syntactically defined; for this reason, we started above by letting $L_1 = (L, \vdash)$ and $L_2 = (L, \models)$ be syntactically and semantically, respectively, defined logical systems over the same language L. Now suppose that our logical system L at hand is equipped with both a syntactical and a semantical consequence relation such that we have the deductive system L = (L, $\Vdash$) satisfying the following theorem:[14]

**Theorem 3.4.2.** *(Deduction theorem)* For a (possibly empty) set $X \subseteq F_L$ and for any formulae $\phi, \psi \in F_L$,

$$(DT) \qquad \text{If } X, \phi \Vdash \psi, \text{ then } X \Vdash \phi \to \psi.$$

---

[13] This property can be expressed either syntactically or semantically in terms of the consequence relation:

**Theorem 3.4.1.** *(The compactness theorem; Gödel, 1930)* Given an infinite set of formulae $X$ and finite subsets $X'_i$ thereof, (1) if $X \vdash \psi$, then $X' \vdash \psi$; (2) if every $X'_i$ has a model, then $X$ has a model.

Note that this theorem is equivalent to Gödel's completeness theorem for classical FOL (Theorem 3.5.1 below). The equivalence lies mostly in the fact that any inference entails a finite number of steps in the application of rules of inference and/or axiom schemata.

[14] Our aim is to prove soundness and completeness of L as a single deductive system. However, we could leave $L_1$ and $L_2$ as defined above, and then prove these properties by relating them, i.e. we would prove the soundness and completeness of $L_1$ with respect to $L_2$, and vice-versa, an unnecessary complication.

## 3. Logical systems

*Proof:* A proof of this theorem can be found in, for example, Mendelson (2009).[15] **QED**

Any logical system $L = (L, \Vdash)$ in which DT holds is indeed a deductive system.

Then, we have the following fundamental theorems for the deductive system L:

**Theorem 3.4.3.** L *is* sound *if, if* $X \vdash_L \phi$, *then* $X \models_L \phi$.

*Proof:* (Sketch) We first sketch a proof when $X = \emptyset$. Assume that $\vdash_L \phi$. Then, there is a sequence $\phi_1, ..., \phi_n$ that is a proof of $\phi$ in L in which every step is either an axiom or derived from previous steps by means of rules of inference (cf. Def. 3.3.5). The idea is to show, by induction, that every step of the proof is a tautology: assume that all steps up to $\phi_i$ are tautologies; it is now necessary to show that $\phi_i$ itself is a tautology. But $\phi_i$ is either an axiom, in which case it is a tautology (a truth-table verification of the axioms of the deductive system at hand may elucidate this), or is derived from a previous result by means of a rule of inference, which assures us that it is also a tautology. Therefore, $\models_L \phi$ by proof induction up to $\phi_n = \phi$. For $X = \{\chi_1, ..., \chi_n\}$, we have $\vdash (\chi_1 \to (\chi_2 \to ... (\chi_n \to \phi)))$ by multiple applications of DT from $\{\chi_1, ..., \chi_n\} \vdash \phi$; by soundness we conclude that $\models (\chi_1 \to (\chi_2 \to ... (\chi_n \to \phi)))$, and hence $\{\chi_1, ..., \chi_n\} \models \phi$.[16] **QED**

**Theorem 3.4.4.** L *is* strongly complete *if, if* $X \models_L \phi$, *then* $X \vdash_L \phi$.

*Proof:* The proof of this theorem is left as an exercise.

Just as in the case of Theorem 3.4.3, this is actually the corollary to the *completeness theorem*

$$\models_L \phi \quad \Rightarrow \quad \vdash_L \phi.$$

The proof of Theorem 3.4.4 is more complex and we skip it here, leaving it as an exercise for which the reader will possibly require some research work (e.g., Boolos, Burgess, & Jeffrey, 2007; Mendelson, 2009).

Theorem 3.4.3 expresses the fact that everything that is provable in L (every theorem of L) is also logically true: L does not prove falsities.

---

[15] Mendelson (2009) uses $\mathscr{A}, \mathscr{B}, \mathscr{C}, ...$ where we use lowercase letters from the Greek alphabet.

[16] Theorem 3.4.3 is actually a corollary of the case when $X = \emptyset$. See, e.g., Boolos, Burgess, & Jeffrey (2007), Chapter 14, for a complete proof of the soundness of the sequent calculus (see below for this topic).

## 3.4. Adequateness of a deductive system

As for Theorem 3.4.4, it expresses the fact that every logical truth in L is provable in L (is a theorem of L); that is to say that L requires no additional inference rules to prove every logical truth in L, being thus *complete* in this sense.[17] Together, soundness and completeness express the fact that in a deductive system L one can derive everything one should (*the system is complete*), and nothing one should not (*the system is sound*). However, it is obvious that if one has to choose, then one should choose soundness; completeness is a nice property of a logical system, but one has to make sure first and foremost that one does not prove falsities.

Given Theorems 3.4.3-4, the following proposition is obvious.

**3.4.4. (Prop.)** L is sound if $\vdash_L \;\subseteq\; \models_L$ and complete[18] if $\models_L \;\subseteq\; \vdash_L$.
*Proof:* Left as an exercise.

**3.4.5. (Def.)** *Adequateness* – Given the deductive system $L = (L, \Vdash)$, we say that the axiomatization $\mathcal{P}$ of L is adequate whenever the set of theorems of L coincides with the set of tautologies of L, i.e.

$$\vdash_{L,\mathcal{P}} \psi \quad \Leftrightarrow \quad \models_L \psi.$$

We say that a semantics $\mathfrak{S}$ is adequate for a deductive system L whenever the set of tautologies of L coincides with the set of theorems of L, i.e.

$$\models_{L,\mathfrak{S}} \psi \quad \Leftrightarrow \quad \vdash_L \psi.$$

**3.4.6. (Prop.)** A deductive system $L = (L, \Vdash)$ is *adequate* iff, given a (possibly empty) set of formulae $X$ and a formula $\psi$, $X \subseteq F_L$ and $\psi \in F_L$, we have

$$X \vdash_{L,\mathcal{P}} \psi \quad \Leftrightarrow \quad X \models_{L,\mathfrak{S}} \psi.$$

*Proof:* Left as an exercise.

Note that the property of adequateness is especially desirable when talking about theories. We thus say that a deductive system L is adequate for, say, Peano arithmetic. However, adequateness is also a central property for (automated) theorem proving, as we can answer questions related to validity by ultimately checking for derivability (cf. Fig. 3.4.1). In particular, Herbrand's theorem (see Section 4.3) relates validity with

---

[17] Some authors (e.g., Chiswell & Hodges, 2007) use the label "completeness" to actually refer to what we call "adequateness," and vice-versa. However, with Gödel (1930), we say that a logical system or a theory is *complete* (*incomplete*) if every logical truth of the system/theory is (not, respectively) a theorem thereof.
[18] Actually, strongly complete.

## 3. Logical systems

| MODEL THEORY | PROOF THEORY |
|---|---|
| $\xrightarrow{completeness}$ | $\xleftarrow{soundness}$ |
| *Validity* <br><br> Given a semantics $\mathfrak{S}$, we have $\chi_1, ..., \chi_n \models \psi$ iff, if $\models_{\mathfrak{S}} \chi_1, ..., \chi_n$, then $\models_{\mathfrak{S}} \psi$. | *Derivability* <br><br> Given a proof ■, we have $\chi_1, ..., \chi_n \vdash \psi$ iff $\chi_1, ..., \chi_n \vdash_{\blacksquare} \psi$. |
| *Satisfiability* <br><br> Given a model $\mathcal{M}$, $\{\chi_1, ..., \chi_n\}$ is satisfiable iff $\models_{\mathcal{M}} \chi_1$ ... and ... and $\models_{\mathcal{M}} \chi_n$. | *Consistency* <br><br> $\{\chi_1, ..., \chi_n\}$ is consistent iff $\{\chi_1, ..., \chi_n\} \nvdash (\psi \wedge \neg \psi)$. |
| $\Downarrow$ <br> **Theorem 3.2.6** | $\Downarrow$ <br> **Theorem 3.3.3** |

Figure 3.4.1.: Adequateness of a deductive system L = (L, ⊩).

satisfiability, and relates the latter with consistency by appealing to computable functions. In effect, we have it that:

**Theorem 3.4.5.** $\chi_1, ..., \chi_n \Vdash \psi$ iff $\{\chi_1, ..., \chi_n, \neg \psi\}$ *is unsatisfiable.*

Importantly, if a deductive system L for a specific logic, say, CFOL, is adequate, then it is equivalent to any other adequate deductive system S for the same logic, i.e. any (classical) FOL proof in L can be converted to a (classical) FOL proof in S.

Still with regard to classical FOL, we now present two results that have to do with the proper use of variables (and substitutions) in the adequate system CFOL. Recall the deduction theorem above. We have the following important result known as *substitution lemma*:

**Lemma 3.4.6.** *If* $\Vdash \phi \to \psi$, *then* $\Vdash \forall x \phi(x) \to \forall x \psi(x)$.

*Proof:* We present the semantical version of the proof. Assume that $\models \phi \to \psi$. Let $\mathcal{I}$ be an interpretation and let $\varpi \in \mathcal{I}$ be a variable assignment with $val_{\varpi}(\forall x \phi(x)) = \mathsf{t}$. Then, for all $x$-variants $\varpi'$ of $\varpi$ (i.e., assignments that differ from $\varpi$ at most in what is assigned to $x$)

we have $val_{\varpi'}(\phi) = \text{t}$. By assumption, we have $val_{\varpi'}(\psi) = \text{t}$, and consequently $val_{\varpi}(\forall x \psi(x)) = \text{t}$. **QED**

An equally important result–whose proof is left as an exercise–with respect to the adequateness of CFOL is:

**3.4.7. (Prop.)** For a formula $\phi(x)$ with $x$ as a free variable and its corresponding universal closure, we have it that

$$\Vdash \phi(x) \quad \Leftrightarrow \quad \Vdash \forall x \phi(x).$$

*Proof:* Left as an exercise.

We next give further important aspects of CL.

## 3.5. The system of classical logic

If, given L*, we define expressions as in Definition 2.3.1 and we restrict the set of truth values $W$ so that we have $W_2 = \{\text{f}, \text{t}\}$ (also: $W_2 = \{0, 1\}$), then we have the logical system known as *classical logic* if the connectives $O = \{\neg, \rightarrow, \wedge, \vee, \leftrightarrow\}$ and the quantifiers $Q = \{\forall, \exists\}$ satisfy the Tarski-style conditions for classical deduction with respect to the consequence relation $\Vdash$ and the consequence operation $Cn$ (see Augusto, 2017, p. 83ff; see also Exercises below).

The most widely used interpretation for CPL is provided by the truth tables for $A, B \in F_L$ (see Example 2.1.1), corresponding to the following conditions for the semantical consequence relation (we leave the connective $\leftrightarrow$ as an exercise).

**3.5.1. (Prop.)** For the set $O_L$ and any two formulae $A, B \in F_L$:

1. $\models \neg A$ is classical iff $\not\models A$

2. $\models A \wedge B$ is classical iff $\models A$ and $\models B$

3. $\models A \vee B$ is classical iff $\models A$ or $\models B$

4. $\models A \rightarrow B$ is classical iff $\not\models A$ or $\models B$

*Proof:* Left as an exercise.

Given these truth tables and the definitions of the connectives of L, it is obvious that:

## 3. Logical systems

**3.5.2.** (**Prop.**) The set $O'_L = \{\neg, \rightarrow\}$ makes of CPL a *truth-functionally complete* logic.

*Proof:* Left as an exercise. (Hint: cf. Example 2.3.1.)

**3.5.3.** (**Prop.**) The following are well known tautologies of CPL:[19]

| (T1) | $A \vee \neg A$ | PEM |
| --- | --- | --- |
| (T2) | $\neg(A \wedge \neg A)$ | PNC |
| (T3) | $A \rightarrow A$ | Law of identity |
| (T4) | $\neg\neg A \leftrightarrow A$ | LDN |
| (T5) | $((A \rightarrow B) \wedge A) \rightarrow B$ | *Modus ponens* (MP) |
| (T6) | $((A \rightarrow B) \wedge \neg B) \rightarrow \neg A$ | *Modus tollens* |
| (T7) | $(A \rightarrow B) \leftrightarrow (\neg B \rightarrow \neg A)$ | Law of contraposition |
| (T8) | $((A \rightarrow B) \wedge (B \rightarrow C)) \rightarrow (A \rightarrow C)$ | Hypothetical syllogism |
| (T9) | $(\neg A \rightarrow (B \wedge \neg B)) \rightarrow A$ | *Reductio ad absurdum* |

*Proof:* The proof is by examination of the truth tables. **QED**

**3.5.4.** Clearly, given the Boolean character of CPL, the negation of any tautology of CPL is a contradiction, as are the formulae $A \leftrightarrow \neg A$, $(A \wedge B) \wedge \neg(A \vee B)$, and $(\neg A \vee \neg B) \leftrightarrow (A \wedge B)$, for example. In particular, and denoting a contradiction or falsity by the symbol $\Box$, we have in CPL

$$(\Box_{df}) \qquad A \wedge \neg A$$

and

$$A \wedge \Box = \Box.$$

**3.5.5.** The definitions of tautologousness and contradictoriness in CPL are generalizable to CFOL.

CFOL is an adequate logical system for most mathematical theories. In particular, we have the following fundamental theorem in which $\Vdash$ without any subscripts denotes the classical FOL consequence relation:

**Theorem 3.5.1.** *(Gödel's completeness theorem) Standard FOL has a complete proof system, i.e. $X \models \psi$ iff $X \vdash \psi$.*

*Proof:* The proof is left as a research exercise (cf. Gödel, 1930).

First and foremost, CL is characterized by the *principle of bivalence*, stating that a proposition $P$ (or a formula $\phi$) has exactly one truth

---

[19] PEM, PNC, and LDN are, in full, principle of excluded middle (also: *tertium non datur*), principle of non-contradiction, and law of double negation (also: involution law), respectively.

## 3.5. The system of classical logic

value, either `true` or `false`. Indeed, C(FO)L is the bivalent logic of reference, and we say of a logic that is bivalent that it behaves classically. The logics that constitute the main subject of this book are essentially *not* bivalent, as their truth-value sets have more than two elements, but they either exhibit a quasi-classical behavior, or can be "made" to behave (quasi-)classically. From a theoretical viewpoint, this means that we can approach logical consequence in the many-valued logics from the perspective of classical deduction. This is the approach taken in this book, reason why the summary contents above–plus the exercises below–in this Section should, so to say, be readily at hand.

We leave the definitions of the classical conditions for the quantifiers $Q = \{\forall, \exists\}$ as an exercise.

## 3. Logical systems

## Exercises

**Exercise 3.1.** The Sheffer stroke, whose symbol is $|$, is defined as $(\phi|\psi) := \neg(\phi \wedge \psi)$. Show that the connectives of $O_L$ can all be expressed by means of this connective.

**Exercise 3.2.** Let the logical system $L = (L, Cn)$ be given, $X \subseteq F_L$, and $\phi, \psi \in F_L$. Let further L be the classical logic system. Provide informal, intuitive, accounts of the following Tarski-style conditions for $Cn$ (conditions imposed on $Cn$ involving exactly one connective of the language of $Cn$):

1. $(\neg_{Cn})$ $\quad \phi \in Cn(X)$ iff $Cn(X, \neg \phi) = F_L$
2. $(\rightarrow_{Cn})$ $\quad \phi \rightarrow \psi \in Cn(X)$ iff $\psi \in Cn(X, \phi)$
3. $(\wedge_{Cn})$ $\quad Cn(\phi \wedge \psi) = Cn(\phi, \psi)$
4. $(\vee_{Cn})$ $\quad Cn(X, \phi \vee \psi) = Cn(X, \phi) \cap Cn(X, \psi)$

**Exercise 3.3.** With respect to the previous exercise, formalize $\leftrightarrow_{Cn}$.

**Exercise 3.4.** Formalize the Tarski-style conditions above in terms of the consequence relation $\Vdash$.

**Exercise 3.5.** The following can be said to specify the behavior of the classical connectives involved (in terms of the classical consequence relation). For each, provide an informal, intuitive, account, taking into consideration their traditional designations ($\Rightarrow$ is a metalanguage symbol denoting a conditional formulation):

1. *Redundancy:* $\Gamma, \phi \vdash \psi$ and $\Gamma, \neg \phi \vdash \psi \Rightarrow \Gamma \vdash \psi$
2. *Principle of explosion:* $\phi, \neg \phi \vdash \psi$
3. *Ex falso quodlibet (From a false proposition follows any proposition whatsoever):* $\bot \vdash \phi$
4. *(Weak) monotonicity:* $\Gamma, \top \vdash \phi \Rightarrow \Gamma \vdash \phi$
5. *Adjunction:* if $\Gamma \vdash \phi$ and $\Gamma \vdash \psi$, then $\Gamma \vdash \phi \wedge \psi$
6. *(Strong) reductio ad absurdum:* $\Gamma, \neg \phi \vdash \phi \Rightarrow \Gamma \vdash \phi$

**Exercise 3.6.** What objections can be made with respect to each of the conditions of the previous exercise? In which ways could these objections motivate new logical systems?

**Exercise 3.7.** Let $\mathcal{I} = (\mathscr{D}, \Psi, \varpi)$ be an interpretation for L*. Complete the following statements with respect to the classical quantifiers:

1. $\models_\mathcal{I} \forall x \phi(x)$ is classical iff... (Hint: Consider every $a \in \mathscr{D}$.)
2. $\models_\mathcal{I} \exists x \phi(x)$ is classical iff... (Hint: Consider some $a \in \mathscr{D}$.)

**Exercise 3.8.** Comment informally on the behavior of a binary connective $\wr$ defined as follows in terms of consequence relation:

$$\Vdash \phi \wr \psi \text{ iff } \Vdash \phi \text{ and } \nVdash (\phi \wedge \psi).$$

(Hint: Build its truth table.)

**Exercise 3.9.** Given the following truth table, characterize the involved connective in terms of the consequence relation $\models$:

| $P$ | $Q$ | $P \boxtimes Q$ |
|---|---|---|
| 1 | 1 | 0 |
| 1 | 0 | 0 |
| 0 | 1 | 1 |
| 0 | 0 | 0 |

**Exercise 3.10.** Let the following axiom schemata together with the rule MP constitute the proof system $\mathcal{L}$ over the language L (Mendelson, 2009):

($\mathcal{L}$1)    $A \to (B \to A)$
($\mathcal{L}$2)    $(A \to (B \to C)) \to ((A \to B) \to (A \to C))$
($\mathcal{L}$3)    $(\neg B \to \neg A) \to ((\neg B \to A) \to B)$

1. Show that $\mathcal{L}$1-3 are indeed axioms.

2. How would you proceed to prove the soundness of the deductive system $\mathsf{L} = (\mathcal{L}, \vdash_\mathrm{L})$?

3. Is L functionally complete for $O_\mathsf{L} = \{\neg, \to, \wedge, \vee, \leftrightarrow\}$?

## 3. Logical systems

**Exercise 3.11.** Provide informal, intuitive, accounts for each classical tautology in Proposition 3.5.3.

**Exercise 3.12.** Provide an informal, intuitive, account of the *deduction-detachment theorem*, whose formal formulation is as follows for $X \subseteq F_L$, $X$ may be empty, and $\phi, \psi \in F_L$ (note that there are two "directions" in this theorem, the $\Rightarrow$- and $\Leftarrow$-direction):

$$X, \phi \Vdash \psi \quad \text{iff} \quad X \Vdash \phi \to \psi.$$

**Exercise 3.13.** Prove (Complete the proof of) the propositions and theorems in this Chapter that were left without a proof (with a sketchy proof, respectively).

# 4. Logical decisions

In this Chapter, after briefly discussing the important topics of *decidability* and *computational complexity* from a very general perspective, we introduce the central problem in computability theory, to wit, the *Boolean satisfiability problem*, or *SAT* (Section 4.1).[1] We then elaborate at length on both important aspects of the *automation of deduction* (Sections 4.2-3) and *refutation calculi* that have been shown to be amenable to automation, to wit, *resolution* and *analytic tableaux* (Sections 4.5-6). Our focus is on both general aspects and refinements of these calculi that are required for a satisfactory understanding of the extensions or adaptations thereof to the many-valued logics, to be discussed at length in Part III. For historical and pedagogical reasons, it is advisable to have some knowledge of *calculi for the VAL*, and we give the basics of some notorious ones in Section 4.4.

## 4.1. Meeting the decision problem and the SAT

Given a theory $\Theta$–a set of formulae in an $n$-order logical language L closed under a specific notion of logical consequence–we are often faced with the need to prove that a certain formula $\phi$ is true in $\Theta$–whether ever (*the satisfiability problem*, or SAT) or always (*the validity problem*, or VAL). This is known as the *deduction problem*. Once we have specified the deduction problem with respect to a theory $\Theta$ and a formula $\phi$ as $\Theta \Vdash^? \phi$, we are faced with yet another problem: The *decision problem*–often referred to as the *Entscheidungsproblem* when L* is meant–which asks whether there is a *decision procedure* for the deduction problem, where by "decision procedure" an algorithmic procedure producing a "Yes/No" answer in useful time must be understood. In other words, we ask whether the deduction problem for the pair $(\Theta, \phi)$ is *decidable*.

**4.1.1. (Def.)** *Decidability* – A theory is said to be *decidable* if there is an *effective decision procedure* or, what is the same, an *algorithm*

---

[1] The specifications for many-valued logics will be given below, namely in Section 6.1.

## 4. Logical decisions

that solves the decision problem. By "solving the decision problem" it is meant that the algorithm (i) halts, and (ii) outputs "Yes" if $\Theta \Vdash \phi$ or "No" otherwise. A theory is *undecidable* if it is not decidable. If besides (i) and (ii) the algorithm (iii) does not halt if $\Theta \nVdash \phi$, then we speak of *semi-decidability*: the algorithm is only guaranteed to halt if $\Theta \Vdash \phi$; otherwise, it may or may not halt.

**4.1.2. (Def.)** *Tractability* – A decidable problem is said to be *tractable* if a decision can be found in polynomial time, i.e. if given an input of size $n$, the running time of the algorithm, denoted by $T(n)$, is upper-bounded by a polynomial expression in $n$: Given a constant $k$ and a notation (*big O*) characterizing functions with respect to their (asymptotic) growth rates, we have $T(n) = O(n^k)$. In other words, a problem is tractable if it belongs to the complexity class **P**. A problem that is not tractable is said to be *intractable*: It can be solved, in theory and given a large but finite time window, but not in polynomial time. This leaves typically exponential time. Let now **NP** be the complexity class of problems that can be *solved* by a non-deterministic Turing machine[2] (whereas the problems in **P** can be solved by a deterministic Turing machine), or whose solution can be *checked* in polynomial time by a deterministic Turing machine. Then, if–an unsolved problem in computability and complexity theory–$\mathbf{P} \neq \mathbf{NP}$, then **NP**-complete problems, or the "hardest problems in **NP**," are also intractable.

As seen above, while classical truth tables are effective decision procedures in CPL for formulae with a small $n$ of propositional variables, they are no longer so for a larger $n$ of variables, as the number of lines in a classical truth table is determined by $2^n$ (cf. Prop. 3.2.5). Moreover, this resource is not available for FOL no matter how small $n$ might be; other validity proof methods are usable in FOL, but only fragments of FOL are decidable: *FOL is undecidable*. This result is known as the *Church-Turing theorem* (Church, 1936a, 1936b; Turing, 1936-7).

We thus require more efficient proof procedures than those based on validity.

### 4.1.1. The Boolean satisfiability problem, or SAT

Approached from a more formal perspective than the one above (cf. Def. 4.1.1), the decision problem is formulated as some language $L$ conceived as a set of strings $w$ of terminal symbols such that for a formula

---

[2] A Turing machine is said to be deterministic if it prescribes a single action for each state on reading a tape symbol; otherwise, it is non-deterministic.

## 4.1. Meeting the decision problem and the SAT

$\varphi$ we have the language[3]

$$L = \{\varphi | \varphi \in L\}$$

where $\varphi \in L$ is an abbreviation for $P(\varphi)$ for $P(x)$ some property of interest. For example, if $\phi = p \wedge (q \vee \neg r)$ and $L = \{\psi | \underbrace{\psi \text{ is in CNF}}_{P}\}$, then $P(\phi) = \mathtt{t}$ and $\phi \in L$.[4] In particular, the satisfiability problem is typically formulated as the language

$$SAT = \{\langle \varphi \rangle \, | \models_\mathcal{M} \varphi\}$$

where $\langle \varphi \rangle$ denotes a Boolean encoding of $\varphi$–frequently a CNF. This explains why one often speaks of the SAT as *a language*, as well as why SAT is an abbreviation for *Boolean satisfiability problem*. Defined equivalently in the due logical detail, we have:

**4.1.4. (Def.)** *The Boolean satisfiability problem, or SAT* – Given a (propositional) formula $\phi(x_1, ..., x_n)$ with connectives $\heartsuit_j \in \{\neg, \wedge, \vee\}$, $j = 0, 1, ..., m$, it is asked if $\phi$ can be evaluated to $\mathtt{t}$ by some assignment of the truth-value set $W_2 = \{\mathtt{f}, \mathtt{t}\}$ to the $x_i$, $1 \leq i \leq n$. We say that a (propositional) formula $\phi(x_1, ..., x_n)$ is *satisfiable* if truth values can be assigned to its variables $x_i$ in such a way as to make $\phi$ true.

In terms of complexity classes, the SAT is **NP**-complete, whereas its dual, VAL (or $\overline{SAT}$), is **co-NP**-complete.[5] Although these two classes may be identical in terms of computational costs (an undecided question), as said in the Introduction validity-checking methods are "not easy to use." Recall the *deduction theorem* (DT): This theorem formulates the fact that satisfiability and validity are dual concepts in the sense that some formula $\phi$ is valid iff $\neg \phi$ is unsatisfiable (cf. Propositions 3.2.14-15). This entails the fundamental result:

**4.1.4. (Prop.)** The decision problem can be reduced to the Boolean satisfiability problem, or SAT.

---

[3] Recall that DT allows us to see the pair $(\Theta, \phi)$, $|\Theta| \geq 0$, as the single formula $\varphi = \Theta \rightarrow \phi$.

[4] Mathematically formulated, we say that a decision procedure computes the *characteristic function* of a language $L$ (with respect to $\phi$), denoted by $\chi_L(\phi)$, such that we have

$$\chi_L(\phi) = \begin{cases} 1 & \text{if } \phi \in L \\ 0 & \text{if } \phi \notin L \end{cases}.$$

Hence, we speak of *decidable* or *undecidable languages* (rather than theories).

[5] Note, however, that they coincide for quantified Boolean formulae. A problem is said to be **co-NP**-complete if its complement is in the class **NP**-complete.

*Proof:* Left as a research exercise. (See Augusto, 2019; 2020b.)

The practical importance of Proposition 4.1.4 is that it informs us as to the complexity of the decision problem. From the viewpoint of complexity theory, the SAT is a **NP**-complete problem (in fact, it was the first problem proved to be **NP**-complete; Cook, 1971). Translating the complexity jargon, this means that there is no known method to find a solution to every SAT problem, though, if it is known, a solution can be checked quickly, namely in polynomial time. However, **NP**-completeness characterizes only the worst-case instances in terms of running time, with many applications being solved much more quickly. On the other hand, a polynomial-time algorithm may be actually impractical or even useless (e.g., check the running time for an input as low as $n = 100$ and $k = 15!$). Also, a complexity issue may arise already at the transformation of a formula into a DNF or a CNF, as it may take exponential time (and space) to convert a SAT problem to a DNF-/CNF-SAT.

**4.1.5. (Def.)** The problem of determining the satisfiability of a formula in DNF (CNF) is called *DNF-SAT* (*CNF-SAT*, respectively). The SAT for formulae in CNF whose clauses have at most two literals (three literals) is known as the *2-SAT* (*3-CNF-SAT* or just *3-SAT*, respectively). This can be generalized to formulae with each clause containing up to (or exactly) $k$ literals, in which case we speak of the *k-CNF-SAT* or *k-SAT*. The problem of determining the satisfiability of a Horn formula is called *HORN-SAT*.

**4.1.6. (Prop.)** The 2-SAT and HORN-SAT are solvable in polynomial time. The $k$-SAT for $k \geq 3$ is **NP**-complete.

*Proof:* Left as a research exercise. (See Augusto, 2020b.)

### 4.1.2. Refutation proof procedures

In general terms, in theorem proving we are interested in finding out about *theoremhood*. In order to obtain this result a problem, or an argument, is transformed into a formula or set of formulae; the formula/set is then input, as is or also after undergoing some transformation or translation, and is operated upon by a proving method that is expected to produce an output, preferably in useful time. Satisfiability-based theorem proving methods operate upon (sets of) formulae by verifying whether they are satisfiable or unsatisfiable. The former test for satisfiability with the aim of producing a model for a formula/set in case it is verified to be satisfiable. The latter is a *refutation* method, i.e. it tests for unsatisfiability: if a set of formulae is unsatisfiable, then the corresponding argument is valid, and a proof can be produced. This is called a refutation method because the formula/conclusion to be proven is negated as input to the proof procedure (see Theorem 3.2.7).

Thus, testing for satisfiability does not prove validity, whereas if a negated formula is proven to be unsatisfiable, this is a proof of its validity (cf. Prop. 3.2.15). This shows that a refutation proof procedure is indeed a decision procedure if its algorithm is guaranteed to terminate.

Recall the ⊩-deduction theorem (Theorems 3.2.5 and 3.3.1) stating that a formula $\psi$ is a consequence of a set of formulae $X$ iff $X \cup \{\neg \psi\}$ is unsatisfiable (semantical version) or inconsistent (syntactical version). Recall also that the deduction theorem states that $X \Vdash \psi$ if $\Vdash (\bigwedge \chi_i \in X) \to \psi$. This allows us to check for the validity of $X \Vdash \psi$ by testing for the unsatisfiability of $\neg[(\bigwedge \chi_i \in X) \to \psi]$, which is equivalent to $(\bigwedge \chi_i \in X) \wedge \neg \psi$ (cf. proof of Theorem 3.2.7). All this means that in order to verify the validity of $\alpha = X/\psi$ we need only negate the conclusion and then check for the unsatisfiability of the argument with the negated conclusion, denoted by $\alpha_\neg$:

$$\alpha_\neg = \{\chi_i\}_1^n \cup \{\neg \psi\}$$

If $\alpha_\neg$ is unsatisfiable, then we conclude for the validity of $\alpha$.

Analytic tableaux and resolution are the refutation-based calculi of election for classical logic. Moreover, when in conjunction with signed formalisms, they are very efficient proof procedures for many-valued logics. We shall thus concentrate on these two refutation methods, after providing the reader with some basics on automated theorem proving and its history, on the general impact of the mathematician J. Herbrand on the field, and on the main validity- and satisfiability-based proof procedures.

## 4.2. Some historical notes on automated theorem proving

We require more efficient decision procedures for CPL, as well as procedures to reduce any statement in FOL to an equisatisfiable, or satisfiability-equivalent, statement in CPL. These methods do indeed exist, but issues of complexity might render them largely useless for the human computer, unless they can be implemented by machines that can cope with high complexity and can return a yes-or-no output in useful time. This is known as *theorem proving automation*, a field that has

## 4. Logical decisions

kept a lot of people busy for some decades now.

The ambition of automating deduction, i.e. finding algorithmic methods that decide the truth of propositions, is not recent; it predates not only the field of artificial intelligence (AI), which emerged in the 1950s, but also the formalization of logic with G. Boole and F. L. G. Frege in the late 19th century. As a matter of fact, it can be said to date at least as far back as to the 17th century, when the philosopher G. W. von Leibniz envisaged a *calculus ratiocinator*, a "reasoning calculus" capable of reducing the solution of an arbitrary, adequately formalized, problem to a mechanical procedure; the adequate formalization would be carried out in what he called the *lingua universalis*, a universal conceptual language finding an interpretation in a *characteristica universalis*, a general theory of signs.

Since the early 1900s, D. Hilbert's formalization program in mathematics and his formulation of the *Entscheidungsproblem* (e.g., Hilbert, 1900; Hilbert & Ackermann, 1928) had motivated a clear mathematical definition of algorithmic procedure. While A. Church and A. Turing solved–independently from each other–negatively the *Entscheidungsproblem* (i.e. there is no algorithmic procedure to decide if a given proposition in FOL is provable in FOL),[6] they came up with the required definition of an effectively calculable function, or algorithm (Church, 1936a, 1936b; Turing, 1936-7). Although the two are equivalent, it was Turing's solution, featuring the famous Turing machine, that provided a clearer and more intuitive definition of an algorithmic procedure.[7]

**4.2.1.** The way was paved for the first steps in the automation of deduction and these were taken in the 1950s with M. Davis programming a computer with the Presburger procedure in 1954 (Davis, 1957), and Newell and colleagues' Logic Theory Machine (Newell, Shaw, & Simon, 1957), which can be considered the first *de-facto* AI program. This emulated the reasoning of Whitehead and Russell's *Principia mathematica* (Whitehead & Russell, 1910) and it was mainly heuristic; the former, though algorithmic in nature, did not go beyond proving that the sum of two even numbers is even, mainly because the Presburger procedure has worse than exponential complexity. Around this time, various important results in proof procedures and simplified completeness proofs for FOL impacting on the design of the first theorem provers were published (e.g., Beth, 1955; Hintikka, 1955; Quine, 1955; Schütte, 1956).

**4.2.2.** As a matter of fact, the basic tools required for the automation of deduction for FOL had already been found in the 1920s, namely by

---

[6] As seen, this is known as the Church-Turing *theorem*.

[7] In effect, the Church-Turing *thesis* states that the class of the effectively calculable functions is precisely the class of functions computable by a Turing machine.

Skolem (e.g., 1920) (Section 2.4.4) and Herbrand (1930) (Section 4.3), and real progress took place in the late 1950s with Davis and Putnam (1958; 1960), who used CNFs for unsatisfiability testing. They applied four rules, to wit, the *tautology rule*, the *one-literal rule*, the *pure-literal* rule, and the *splitting rule*. Theirs was a more efficient method than Gilmore's (1960) multiplication method, which was also based on Herbrand's theorem (version 2, and thus suggesting a refutation procedure; see Section 4.3), and it was further improved by Davis, Logemann, and Loveland (1962) with a view to satisfiability testing. This is now known as the DPLL procedure (see below, Section 4.4.5).

**4.2.3.** We can say that a revolution took place when Robinson (1965) rediscovered the *resolution rule*, first discovered by Blake (1937), and *unification*, first discovered by Prawitz (1960). The group at the Argonne National Laboratory directed by L. Wos and G. Robinson improved resolution; in effect, it was necessary to reduce the search space of the generated resolvents. In order to do this they introduced *factoring*, *unit preference*, as well as the *set of support strategy*, and later on they worked on *equality*, a work that resulted in the techniques of *modulation* and *paramodulation*. All these were implemented in the prover Otter (see Kalman, 2001), whose present successor is Prover9-Mace4.

With regard to the DPLL procedure, major breakthroughs came in the late 1990s and early 2000s with such systems as SATO, MiniSAT, Chaff, and GRASP.

These latter developments are all based on satisfiability, being in one way or another connected to the SAT and the search for SAT solvers.

## 4.3. Herbrand semantics

Testing for SAT in a proof calculus was actually given a semantical foundation by the mathematician J. Herbrand. This, which we can call *Herbrand semantics*, exhibits many advantages over the *Tarskian semantics* elaborated on in (especially) Section 3.2.2, in particular with respect to automated deduction. This is so mostly because Herbrand semantics provides a fixed interpretation for FOL, thus allowing for a purely syntactical manipulation of symbols in formulae while at the same time providing these symbols with a bivalent semantics. Although our discussion of Herbrand semantics serves mainly our treatment of resolution, it is also relevant to the analytic tableaux calculus (see below).

We begin by presenting some fundamental theorems for testing for SAT. As our focus is on refutation procedures, we shall be more con-

## 4. Logical decisions

cerned with unsatisfiability.

**Theorem 4.3.1.** *Let $C_\phi$ be a set of clauses that represents a SNF of a formula $\phi$. Then $\phi$ is unsatisfiable iff $C_\phi$ is unsatisfiable.*

*Proof:* See, e.g., Chang & Lee (1973), p. 48-9. **QED**

This theorem requires that we specify when $C$ is unsatisfiable. This was done by J. Herbrand in a theorem that can be divided into two versions:[8]

**Theorem 4.3.2.** *(Herbrand, 1930 - version 2) A set $C$ of clauses is unsatisfiable iff there is a finite unsatisfiable set $C'$ of ground instances of $C$.*

Although not problem-free, Herbrand's results are fundamental in more than one way. Firstly, this theorem tells us that in order to verify whether $C$ is unsatisfiable, we need only focus on Herbrand interpretations, as a set of clauses $C$ is unsatisfiable iff $C$ is false under all Herbrand interpretations. This, in turn, means that we need only consider the Herbrand universe (vs. all possible domains). We next introduce the terminology to fully understand the above. Let $C$ be a set of clauses;[9] then the following definitions hold:

**4.3.1. (Def.)** The *Herbrand universe* of $C$, denoted by $H_C$, is the set of all ground terms built up from the constants and functions of $C$ in the way stipulated in Algorithm 4.1.[10]

$H_i$, for $1 \leq i \leq \infty$, is called the *i-level constant set of $C$*. Clearly, $H_C$ is infinite iff $Fun(C) \neq \emptyset$.

---

[8] We chose to present first version 2 of Herbrand's theorem; we prove it after stating and proving version 1 of the same theorem. These two versions are actually equivalent formulations of Herbrand's original theorem, which, in an already simplified formulation, runs as follows:

> Let $\Theta$ be a theory axiomatized by exclusively universal formulae. Suppose that $\Theta \models \forall x \, (\exists y_1, ..., y_k) \, (P(x, y))$ where $P(x, y)$ is a quantifier-free formula. Then, there is a finite sequence $t_{ij} = t_{ij}(x)$ of terms, $1 \leq i \leq r$ and $1 \leq j \leq k$, such that
> $$\Theta \vdash \forall x \left( \bigvee_{i=1}^{r} P(x, t_{i1}, ..., t_{ik}) \right)$$

Informally, and formulated for unsatisfiable formulae, this theorem states that a closed formula $\phi$ in SNF is unsatisfiable iff there is a finite number of clause instances of $\phi$ whose conjunction is unsatisfiable in terms of truth-functionality.

[9] Although we restrict our discussion to sets of clauses, here denoted by $C$, Herbrand's results generalize to any set $X$ of FOL formulae.

[10] In other words, the Herbrand universe of $C$ is the set of all ground terms definable over $\Upsilon_C$. See Definition 2.2.6.

## 4.3. Herbrand semantics

---
**Algorithm 4.1** Construction of the Herbrand universe
---
**Input:** A set of clauses $C$

**Output:** $H_C$, the Herbrand universe of $C$

---

1. $H_C = H_0$ if

   a) *$C$ contains no function*: then $H_C$ is the set of constants occurring in $C$; and/or

   b) *No constant occurs in $C$*: then $H_C$ consists of a single arbitrary constant, say, $H_C = \{a\}$. I.e.

   $$H_0 = \begin{cases} Cons\,(C) & \text{if } Cons\,(C) \neq \emptyset \\ \{a\} & \text{if } Cons\,(C) = \emptyset \end{cases}.$$

2. *$C$ contains a function*: then $H_C = \bigcup_{i=0}^{\infty} H_i$, $H_i = H_{i-1} \cup Fun$,

   $Fun = \{f(t_1, ..., t_n) \,|\, f \in Fun\,(C), (t_1, ..., t_n) \in H_{i-1}, n \in \mathbb{N}\}.$

---

## 4. Logical decisions

**Example 4.3.1.** We exemplify the above:

1. Let $C = \{P(b), \neg P(x) \vee Q(y)\}$. Then, $H_C = H_0 = \{b\}$.

2. Let $C = \{P(x) \vee Q(x), R(z), T(y) \vee \neg S(y)\}$. Then, we let $H_C = H_0 = \{a\}$.

3. Let $C = \{P(f(x)), a, g(y), b\}$. Then:

$H_0 = \{a, b\}$
$H_1 = \{a, b, f(a), f(b), g(a), g(b)\}$
$H_2 = \{a, b, f(a), f(b), g(a), g(b), f(f(a)), f(f(b)), f(g(a)), f(g(b)),$
$\quad g(f(a)), g(f(b)), g(g(a)), g(g(b))\}$
$\vdots$
$H_C = \{a, b, f(a), f(b), g(a), g(b), f(f(a)), f(f(b)), f(g(a)), f(g(b)),$
$\quad g(f(a)), g(f(b)), g(g(a)), g(g(b)), ...\}$

**4.3.2. (Def.)** A *ground instance* of a clause $\mathcal{C}$ of $C$ is a clause obtained by replacing variables in $\mathcal{C}$ by members of $H_C$. A *Herbrand instance* of $\mathcal{C}$ is a ground instance $\mathcal{C}\theta$ of $\mathcal{C}$ such that $\theta$ is based on $C$. The *Herbrand base* of $C$, denoted by $H(C)$, is the set of all Herbrand instances of atoms occurring in clauses of $C$.

Obviously, $H(C)$ is finite iff $H_C$ is finite.

**4.3.3. (Def.)** A *Herbrand interpretation (H-interpretation)* for $C$, denoted by $H\mathcal{I}_C$, is a triple $(H_C, \Theta, \varpi)$ (cf. Def.s 2.2.9-11) such that

1. $\Theta(c) = c$ for every $c \in Cons(C)$;

2. $\Theta(f)(t_1, ..., t_n) = f(t_1, ..., t_n)$ for all $t_1, ..., t_n \in H_C$, if $f \in Fun(C)$.

$H\mathcal{I}_C$ provides a *fixed* interpretation, as every constant symbol is interpreted as itself (i.e. $H\mathcal{I}_C$ maps every constant to itself), and every function symbol is interpreted as a term builder over $H_C$, or, in other words, as the function that applies it ($H\mathcal{I}_C$ maps every function symbol $f \in Fun(C)$ with arity $> 0$ to the $n$-ary function that maps every $n$-tuple $(t_1, ..., t_n)$ of terms $t_1, ..., t_n \in H_C$ to the term $f(t_1, ..., t_n)$). Moreover, because clauses are interpreted as closed formulae, $\varpi$ is irrelevant in $H\mathcal{I}_C$.

All this entails that we end up with a purely syntactical interpretation, being meant by this that the symbols in a set of clauses are interpreted independently of any domain.

## 4.3. Herbrand semantics

**4.3.4. (Def.)** A H-interpretation $HI_C$ is a subset $H'(C)$ of $H(C)$ such that the truth value $\mathtt{t}$ is assigned to all elements of $HI_C$ and the truth value $\mathtt{f}$ is assigned to all atoms in $H(C) - HI_C$. The subset $H'(C)$ is in fact a *Herbrand model (H-model)* $HM_C$ of $C$, because for an interpretation $HI_C$ and some $P \in Pred(C)$ we have

$$\Theta(P)(t_1, ..., t_n) = val_{HI}(P(t_1, ..., t_n))$$

so that for $t_i \in H(C)$ we have

$$HM_C = \{P(t_1, ..., t_n) \mid \Theta(P)(t_1, ..., t_n) = \mathtt{t}\}.$$

**Example 4.3.2.** Let $C = \{P(x) \vee Q(x), R(f(y))\}$. Then,

$$H_C = \{a, f(a), f(f(a)), ...\}$$

and

$$H(C) = \{P(a), Q(a), R(a), P(f(a)), Q(f(a)), R(f(a)), ...\}$$

The following are H-interpretations for $C$:

$HI_{C_1} = \{P(a), Q(a), R(a), P(f(a)), Q(f(a)), R(f(a)), ...\}$
$HI_{C_2} = \{\neg P(a), \neg Q(a), \neg R(a), \neg P(f(a)), \neg Q(f(a)), \neg R(f(a)), ...\}$
$HI_{C_3} = \{P(a), Q(a), \neg R(a), P(f(a)), Q(f(a)), \neg R(f(a)), ...\}$

It is easy to see that $C$ is satisfied by $HI_{C_1}$, but falsified by $HI_{C_2}$ and $HI_{C_3}$. Thus, only $HI_{C_1}$ is a H-model $HM_C$ of $C$.

We can now provide the following central result:

**Theorem 4.3.3.** *A set $C$ of clauses is unsatisfiable iff $C$ is false under all H-interpretations.*

*Proof:* Obvious from the above. **QED**

Given the above, this theorem simply states the fact that a set $C$ of clauses is unsatisfiable iff it has no H-model $HM_C$.

Given $n$ elements in $H(C)$, there will generally be $2^n$ H-interpretations. This, however, is where Herbrand's results might be problematic, as there are infinitely many H-interpretations for $n = \infty$. It so happens that the first requirement of an algorithm is that the number of steps that constitute it be finite. In order to organize all H-interpretations in a systematic way, we can apply the notion of a semantic tree.

## 4. Logical decisions

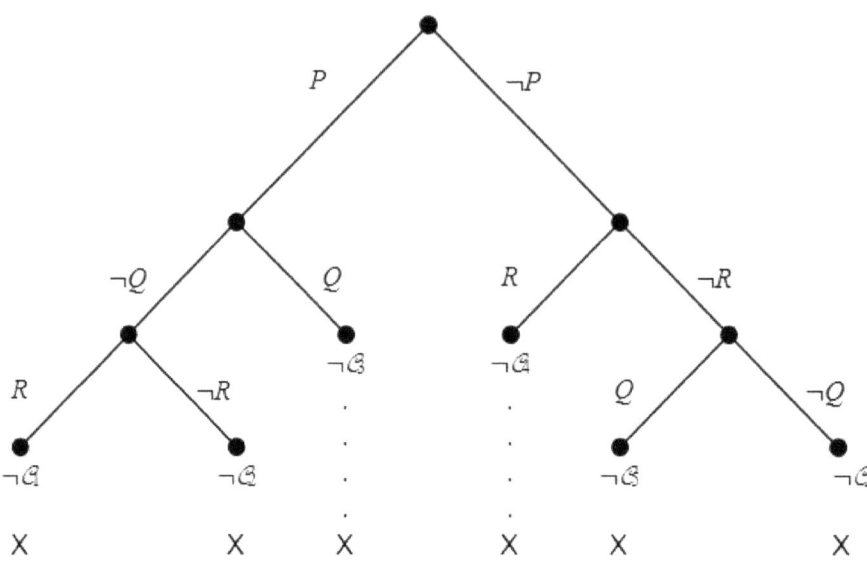

Figure 4.3.1.: Closed semantic tree of $C = \{C_1, C_2, C_3, C_4, C_5\}$ in Example 4.3.3.

**4.3.5. (Def.)** A *semantic tree* $\mathcal{T}$ for a set $C$ of clauses, denoted by $\mathcal{T}_C$, is a (downward growing) *labeled binary tree* in which each link is attached with a finite set of (negations of) atoms from $H(C)$ in such a way that:

1. For each node $N$ there are only finitely many immediate links $L_1, ..., L_n$ from $N$. For $\mathcal{E}_i$ the conjunction of all the literals in the set attached to $L_i, i = 1, ..., n$, $\mathcal{E}_1 \vee \mathcal{E}_2 \vee ... \vee \mathcal{E}_n$ is a valid formula.

2. For each node $N$, let $I(N)$ be the union of all the sets attached to the links of the branch of $\mathcal{T}_C$ down to and including $N$. Then $I(N)$ does not contain any complementary pair.

**4.3.6. (Def.)** Let $H(C) = \{A_1, A_2, ..., A_k, ...\}$. A semantic tree $\mathcal{T}_C$ is *complete* iff, for every tip node (leaf) $N$, $I(N)$ contains either $A_i$ or $\neg A_i$ for $i = 1, 2, ...$ $N$ is a *failure node* if $I(N)$ falsifies some ground instance of some $\mathcal{C}$ in $C$, but $I(N')$ does not falsify any ground instance of some $\mathcal{C}$ in $C$ for every ancestor node $N'$ of $N$. A branch of a semantic tree $\mathcal{T}_C$ *is closed* iff it terminates at a failure node; otherwise, it is said to be *open*. A semantic tree $\mathcal{T}_C$ is closed iff each of its branches is closed.

If $I(N)$, which is in fact a partial interpretation for $C$ (i.e. $I(N)$ can be seen as an assignment of truth values to ground atoms of $H(C)$),

## 4.3. Herbrand semantics

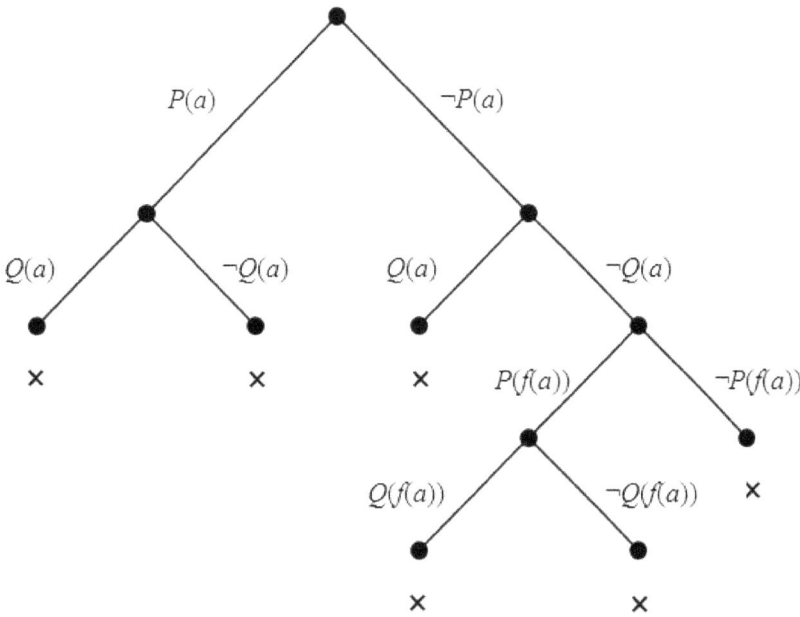

Figure 4.3.2.: A closed semantic tree.

falsifies $C$, then we can stop expanding nodes from $N$. This means that if $C$ is unsatisfiable, its semantic tree $\mathcal{T}_C$ cannot fail to be finite.

**Example 4.3.3.** Let $\mathcal{C}_1 = Q \vee \neg R$, $\mathcal{C}_2 = Q \vee R$, $\mathcal{C}_3 = \neg P \vee \neg Q$, $\mathcal{C}_4 = P \vee \neg R$, and $\mathcal{C}_5 = P \vee \neg Q \vee R$. The atom set of $C = \{\mathcal{C}_1, \mathcal{C}_2, \mathcal{C}_3, \mathcal{C}_4, \mathcal{C}_5\}$ is $A(C) = \{P, Q, R\}$. The above conditions 4.3.5.1-2 are satisfied for $C$. Figure 4.3.1 shows the closed semantic tree of $C$. A closed branch is marked with ×.

**Example 4.3.4.** Let $C = \{\neg P(x) \vee Q(x), P(f(a)), \neg Q(z)\}$. Then:
$H_C = \{a, f(a), f(f((a))), ...\}$
$H(C) = \{P(a), Q(a), P(f(a)), Q(f(a)), ...\}$
$C$ is unsatisfiable. Figure 4.3.2 shows the semantic tree for $C$. A failure node is marked with ×.

**Theorem 4.3.4.** *(Herbrand, 1930 - version 1)* A set $C$ of clauses is unsatisfiable iff corresponding to every complete semantic tree of $C$, there is a finite closed semantic tree.

*Proof:* The proof follows immediately from the above:

## 4. Logical decisions

($\Rightarrow$) Assume $C$ is unsatisfiable and $\mathcal{T}_C$ a complete semantic tree for $C$. For every branch there is the set of labels $I(N)$, which is an interpretation, because the tree is complete. Hence, $I(N)$ falsifies some ground instance $\mathcal{C}'$ of a clause $\mathcal{C} \in C$. Since there are only finitely many literals in $\mathcal{C}'$, there must be a failure node $N$ in a finite distance from the root. Since every branch terminates at a failure node, there is a corresponding closed semantic tree $\mathcal{T}'_C$ that is finite.

($\Leftarrow$) Assume that for every complete semantic tree $\mathcal{T}_C$ there is a corresponding finite closed tree $\mathcal{T}'_C$. Then, every branch terminates at a failure node and hence every interpretation $I(N)$ falsifies $C$. Hence, $C$ is unsatisfiable. **QED**

The reason why we chose last to introduce version 1 of Herbrand's theorem is that it is more directly connected to *resolution*, while version 2 has more to do with *refutation*. Nevertheless, the two versions make it clear that resolution is a refutation procedure, and a proof of version 1 can be greatly simplified if we already can apply the important definition of semantic tree:

*Proof:* (Herbrand's theorem, version 2). ($\Rightarrow$) Assume $C$ is unsatisfiable and $\mathcal{T}_C$ is a complete semantic tree for $C$. Then, by Herbrand's theorem, version 1, there is a finite closed semantic tree $\mathcal{T}'_C$ of $C$. Let $C'$ be the set of ground instances of clauses that are falsified at all failure nodes of $\mathcal{T}'_C$: $C'$ is finite and is falsified by every interpretation. It follows that $C'$ is unsatisfiable.

($\Leftarrow$) The proof is by contraposition. **QED**

**4.3.7.** Version 2 of Herbrand's theorem can be turned into a proof procedure by successively generating the sets $C'_0, C'_1, \ldots$ in which $C'_i$ is the set of all the ground instances of clauses of $C$, and by testing them for unsatisfiability by means available to propositional calculus: the theorem tells us that for some finite $N$ there is a set $C'_N$ that is unsatisfiable if $C$ is unsatisfiable. Gilmore (1960) was the first to implement this idea with the *multiplication method*: as each $C'_i$ is generated, it is multiplied out into a DNF; any conjunction in the DNF containing a complementary pair is then removed, and if some $C'_i$ is found to be empty, a proof for its unsatisfiability has been found. However, as said above, this method is highly inefficient, as for, say, a set of ten two-literal ground clauses, there are $2^{10}$ conjunctions.

## 4.4. Proving validity and satisfiability

Given Herbrand's theorem, we can check for validity and/or satisfiability by resorting to either semantical or syntactical techniques. We provide a few, brief, examples of validity- and satisfiability-proving systems and techniques, including clause logic ones. As our interest falls mainly on unsatisfiability-proving systems (Sections 4.5-6), the brevity of this elaboration is justified; nevertheless, we provide an elaboration long and complete enough, in order to facilitate a sufficient grasp of this important component of theorem proving. Understandably, no theorems or proofs will be provided in this Section; readers interested in a more comprehensive approach are referred to, for instance, Troelstra & Schwichtenberg (2000) for the material in Sections 4.4.2-4, and to Harrison (2009) for the topic in Section 4.4.5.

### 4.4.1. Truth tables

The truth-table method is ideal for validity checking (as well as for satisfiability checking, for that matter), but it is so only in two cases: for propositional logics and for a small number of subformulae. This leaves out the FO logics, and propositional formulae with a high number of subformulae. This problem is emphasized in the many-valued propositional logics (cf. Prop. 3.2.5).

See Section 2.1 for truth tables in CPL. See Part II for abundant examples of truth tables in many-valued propositional logics.

### 4.4.2. Axiom systems

Axiom systems can be applied to a larger plethora of cases as compared to truth tables.

**Example 4.4.1.** The following axiom schemata $\mathscr{L}$1-3, together with rule MP, constitute a proof system for classical propositional logic known as the Frege-Łukasiewicz system:

$$
\begin{array}{ll}
(\mathscr{L}1) & A \to (B \to A) \\
(\mathscr{L}2) & (A \to (B \to C)) \to ((A \to B) \to (A \to C)) \\
(\mathscr{L}3) & (\neg A \to \neg B) \to (B \to A)
\end{array}
$$

We denote this proof system by $\mathscr{L}$.

## 4. Logical decisions

Because the proof system $\mathscr{L}$ is composed mostly of axiom schemata, it is called an *axiom system*.[11] When checking for validity in an axiom system, every line of a proof is a theorem.

**Example 4.4.2.** Figure 4.4.1 shows a proof of the theorem $A \to A$ in $\mathscr{L}$. On the right, the axiom schemata and/or rule used in each step of the derivation are indicated. For instance, "MP (1,2)" denotes the application of the rule MP on steps 1 and 2.

| 1. | $(A \to ((A \to A) \to A)) \to ((A \to (A \to A)) \to (A \to A))$ | $\mathscr{L}2$ |
|---|---|---|
| 2. | $A \to ((A \to A) \to A)$ | $\mathscr{L}1$ |
| 3. | $(A \to (A \to A)) \to (A \to A)$ | MP (1, 2) |
| 4. | $A \to (A \to A)$ | $\mathscr{L}1$ |
| 5. | $A \to A$ | MP (3, 4) |

Figure 4.4.1.: Proof of $\vdash A \to A$ in $\mathscr{L}$.

It should be obvious from the example above that, while the axioms of such systems allow for a characterization of a logic as a whole, they do not allow for a characterization of the behavior of the individual connectives of the logic at hand. This shortcoming is overcome in systems of rules of inference, such as natural deduction and sequent calculi.

Further relevant axiom systems are those of Hilbert & Ackermann (1928) and Kleene (1952).

**Example 4.4.3.** If we add the following axiom schemata $\mathcal{L}$4-5 and the following rule of inference GEN (*generalization rule*) to the proof system $\mathcal{L}$ in Exercise 3.10, we obtain the FO axiom system $\mathcal{L}^*$:

($\mathcal{L}$4)  $\forall x A(x) \to A(t)$  (if $t$ is free for $x$ in $A(x)$)
($\mathcal{L}$5)  $\forall x (A \to B) \to (A \to \forall x B)$
  (if $A$ contains no free occurrence of $x$)

(GEN)  $A/\forall x A$

**Example 4.4.4.** We prove in $\mathcal{L}^*$ that $\forall x B$ follows from $A, \forall x A \to B$ in the FO proof system $\mathcal{L}^*$. Note how the premises are listed first in the proof (see Fig. 4.4.2).

---

[11] The label "axiom system" is, however, frequently applied to proof systems in general, even when these have no, or only very few, axioms. For this reason, the expressions "Hilbert system" or "Frege system" are often preferred when referring to axiom systems proper.

$$\begin{array}{lll}
1. & A & \text{Premise} \\
2. & \forall x A \to B & \text{Premise} \\
3. & \forall x A & \text{GEN (1)} \\
4. & B & \text{MP (2, 3)} \\
5. & \forall x B & \text{GEN (4)}
\end{array}$$

Figure 4.4.2.: $A, \forall x A \to B \vdash \forall x B$: Proof in the axiom system $\mathcal{L}^*$.

### 4.4.3. Natural deduction

Natural deduction,[12] denoted here by $\mathcal{NK}$, was invented independently by Gentzen (1934-5) and Jaśkowski (1934) as a proof system for classical logic over $\mathsf{L}^{(*)}$. $\mathcal{NK} = (RI, \emptyset)$, i.e. $\mathcal{NK}$ has a (small) set of inference rules and no axioms. The inference rules of natural deduction are rules for the *introduction* (denoted by $I$) and the *elimination* ($E$) of connectives and quantifiers. In what follows, $[\![\phi]\!]$ denotes that the *assumption* $\phi$ is *discharged* (see Example 4.4.5 for meanings), and $\bot$ denotes absurdity. The rules for the quantifiers implicitly apply substitutions such as $\phi_x^a$ ($\phi_t^x$), denoting the substitution in $\phi$ of the variable $x$ for the constant $a$ (the term $t$ free for the variable $x$, respectively).[13]

**4.4.1.** The following are the inference (or formation) rules of $\mathcal{NK}$:

1.
$$(\wedge I) \quad \frac{A \quad B}{A \wedge B}$$

2.
$$(\wedge E) \quad \frac{A \wedge B}{A} ; \frac{A \wedge B}{B}$$

---

[12]So called because it is believed to mirror *natural* ways of reasoning in mathematical proof.

[13]More strictly, $a$ is a *parameter*, an element $a \in Pm$ such that $Pm \cap Cons = \emptyset$ but $T = Pm \cup Cons$ (Prawitz, 1965; Smullyan, 1968). Then, $\phi_a^x$ does not allow for a direct valuation; $\forall x \phi(x) \to \phi(a)$ can be said to be valid, but only in the sense that the sentence $\forall x \phi(x) \to \phi(c)$ is true for every element $c \in \mathscr{D}$ assigned to $a$. This said, we may consider a parameter to be just a free variable, or a constant symbol not originally in $\mathsf{L}^*$.

## 4. Logical decisions

3.  $(\vee I)$  $\dfrac{A}{A \vee B} ; \dfrac{B}{A \vee B}$

4.  $(\vee E)$  $\dfrac{A \vee B \quad \begin{array}{c}[A]\\ C\end{array} \quad \begin{array}{c}[B]\\ C\end{array}}{C}$

5.  $(\to I)$  $\dfrac{\begin{array}{c}[A]\\ B\end{array}}{A \to B}$

6.  $(\to E)$  $\dfrac{A \quad A \to B}{B}$

7.  $(\neg I)$  $\dfrac{\begin{array}{c}[A]\\ \bot\end{array}}{\neg A}$

8.  $(\neg E)$  $\dfrac{A \quad \neg A}{\bot}$

9.  $(\forall I)$  $\dfrac{A(a)}{\forall x A(x)}$

    *Restriction*: $a$ must not occur in any assumption on which $A(a)$ depends.

10. $(\forall E)$  $\dfrac{\forall x A(x)}{A(t)}$

11. $(\exists I)$  $\dfrac{A(t)}{\exists x A(x)}$

12. $(\exists E)$  $\dfrac{\exists x A(x) \quad \begin{array}{c}[A(a)]\\ B\end{array}}{B}$

    *Restriction*: $a$ must not occur in either $\exists x A(x)$ or $B$, or in any assumption on which the upper occurrence of $B$ depends other than $A(a)$.

| | | |
|---|---|---|
| 1. | $(A \to B) \land (A \to C)$ | Assumption |
| 2. | $A$ | Assumption |
| 3. | $A \to B$ | $\land E1$ (1) |
| 4. | $A \to C$ | $\land E2$ (1) |
| 5. | $B$ | $\to E$ (3, 2) |
| 6. | $C$ | $\to E$ (4, 2) |
| 7. | $B \land C$ | $\land I$ (5, 6) |
| 8. | $A \to (B \land C)$ | $\to I$ (2, 7) |
| 9. | $((A \to B) \land (A \to C)) \to (A \to (B \land C))$ | $\to I$ (1, 8) |

Figure 4.4.3.: $\mathcal{NK}$ proof of $\vdash (A \to B) \land (A \to C) \to (A \to (B \land C))$.

13. (Considered by Prawitz (1965) instead of rules 7 and 8)

$$(\bot) \qquad \frac{\begin{array}{c}[\![\neg A]\!]\\ \bot\end{array}}{A}$$

*Restrictions*: $A$ should be different from $\bot$ and it should not have the form $B \to \bot$.

**Example 4.4.5.** Figure 4.4.3 shows a $\mathcal{NK}$ proof of the theorem $\vdash ((A \to B) \land (A \to C)) \to (A \to (B \land C))$. This is the proof of a conditional statement (i.e. the conditional is the main connective), so the best strategy is to apply $\to I$. We begin by *assuming* the antecedent (step 1). This antecedent has itself two conditional propositions in which $A$ is an antecedent, and we actually want to prove first that from this antecedent $(B \land C)$ can be proved. We thus *assume* $A$ (step 2) and enter a new level of the proof (a *subproof*), which is graphically represented by an indentation. Steps 3-7 represent the application of the rules of inference of $\mathcal{NK}$ with respect to the former steps to obtain a proof that $A \to (B \land C)$ (step 8). Note in 8 that the indentation started in 2 is closed; this represents graphically the fact that the assumption in 2 was *discharged*. Step 9 concludes the proof of the main conditional, and resumes the indentation in 1, i.e. it discharges the first assumption in 1. Note thus that a (sub)formula is proved iff it is proved under certain assumptions, unless it is a theorem.

## 4. Logical decisions

### 4.4.4. The sequent calculus $\mathcal{LK}$

Sequent calculi are rather more interesting for the analysis of proofs than for theorem proving, for which natural deduction or the tableaux calculus are more adequate. In any case, and also because they are related to the analytic tableaux calculus to be approached in depth below, we provide a more developed exposition of this proof calculus.

The sequent calculus for classical logic $\mathcal{LK}$ (abbreviating the German expression **k**lassische Prädikaten**l**ogik, "classical predicate logic" in English) was created by Gentzen (1934-5). Clearly, sequents are the central object of $\mathcal{LK}$, and we begin our short exposition of this calculus by providing its definition.

**4.4.2. (Def.)** A *sequent* $s$ is an expression of the form

$$(s) \quad A_1, ..., A_n \Rightarrow B_1, ..., B_k$$

where $A_i$, $0 \leq i \leq n$, and $B_j$, $0 \leq j \leq k$, are formulae, and $\Rightarrow$ is a new symbol denoting that the disjunction of the $B_j$ *follows from* the conjunction of the $A_i$, i.e.

$$(s) \quad \left(\bigwedge_{i=0}^{n} A_i\right) \Rightarrow \left(\bigvee_{j=0}^{k} B_j\right)$$

which, in turn, is equivalent to

$$(s') \quad \Rightarrow \left(\bigwedge_{i=0}^{n} A_i\right) \rightarrow \left(\bigvee_{j=0}^{k} B_j\right)$$

it being the case that a sequent of the form "$\Rightarrow A$" denotes a theorem. This is the same as saying that a sequent is actually a structure asserting that whenever all the $A_i$ are true at least one of the $B_j$ will also be true.[14] We call $\bigwedge_{i=0}^{n} A_i$ the *antecedent* and $\bigvee_{j=0}^{k} B_j$ the *succedent* (or *consequent*).

**4.4.3. (Def.)** A *context* (or *side formula*), denoted by $\Gamma, \Lambda, \Sigma, \Pi$, is a finite, possibly empty sequence of formulae. In the conclusion of each rule, the formula not in the context is the *principal* (or *main*) formula.

---

[14] Cf. Theorems 3.2.6 and 3.3.2. Cf. also the rule $\rightarrow$ R below. This means that one can unproblematically replace the symbol $\Rightarrow$ by the symbol for the consequence relation $\vdash$.

## 4.4. Proving validity and satisfiability

**4.4.4. (Def.)** A *sequent rule* **s** is of the general form

$$(\mathbf{s}) \quad \frac{s_n}{s_{n-1}}$$

and a *sequent axiom* $\mathbf{a}_s$ has the form

$$(\mathbf{a}_s) \quad \frac{}{s_n}.$$

Note the "upward direction" of the sequent rules and axioms.

**4.4.5. (Def.)** A *proof* in a sequent calculus is a (branching) sequence

$$s_0, s_{11}, ..., s_{ij}..., s_{nm}$$

where $s_0$ is the proved sequent / formula, and each of the sequents $s_{ij}$ in the proof is inferred (or derived) from earlier sequents in the sequence by means of rules of inference (or derivation). This sequence is an (upwards growing) *ordered finite tree*, with the *root* $s_0$, the sequents $s_{ij}$, $0 < i < n$ are the *i*-th *node* in the *j*-th *branch*, $0 < j \leq m$, and $s_{nj}$ is the *leaf* of the *j*-th branch.[15]

Although not necessary, one can provide the rules of inference, thus labeling the tree with the rules of inference and axioms (see Fig.s 4.4.4-5). Importantly, this proof structure without the rule CUT (see below) guarantees the following property of the sequent calculus:

**4.4.6. (Prop.)** *Subformula property* – All formulae occurring in a proof of a sequent $s_0$ are subformulae of the formulae in $s_0$.

*Proof:* Trivial. **QED**

In effect, the possibility of applying rules along each *j*-th branch stops when we have obtained atomic subformulae, the above $s_{nj}$ sequents. The semantical explanation for this proof structure is as follows for material implication (for example): if all branches terminate in sequents of the form $\Gamma', L \vdash L, \Lambda'$, then there is *no* interpretation $\mathcal{I}$ for which $val_\mathcal{I}(\Gamma) = \mathbf{t}$ and $val_\mathcal{I}(\Lambda) = \mathbf{f}$.

**4.4.7. (Def.)** A *sequent calculus* is thus a pair $(RI_s, AX_s)$ of rules of inference and axioms for sequents.

$\mathcal{LK}$ has several rules of inference (see below) and a single axiom:

**4.4.8.** *Axiom of identity:*

$$(\text{Ax}) \quad \frac{}{A \vdash A}$$

The inference rules are of two kinds, *structural* and *logical* (or op-

---
[15] There are variations to this structure; see, e.g., Troelstra & Schwichtenberg (2000).

## 4. Logical decisions

*erational*), the latter being rules for the use of the logical operators (connectives, as well as quantifiers). Each rule, structural or logical, has a right and left version, denoted by R and L, respectively.

**4.4.9.** The following are the structural rules of $\mathcal{LK}$:

1. *Weakening* (W):

$$\text{(WL)} \quad \frac{\Gamma \vdash \Lambda}{\Gamma, A \vdash \Lambda} \qquad \text{(WR)} \quad \frac{\Gamma \vdash \Lambda}{\Gamma \vdash A, \Lambda}$$

2. *Contraction* (C):

$$\text{(CL)} \quad \frac{\Gamma, A, A \vdash \Lambda}{\Gamma, A \vdash \Lambda} \qquad \text{(CR)} \quad \frac{\Gamma \vdash A, A, \Lambda}{\Gamma \vdash A, \Lambda}$$

3. *Permutation* (P):

$$\text{(PL)} \quad \frac{\Gamma, A, B, \Sigma \vdash \Lambda}{\Gamma, B, A, \Sigma \vdash \Lambda} \qquad \text{(PR)} \quad \frac{\Gamma \vdash \Lambda, A, B, \Pi}{\Gamma \vdash \Lambda, B, A, \Pi}$$

**4.4.10.** The following are the operational rules for the connectives of $\mathcal{LK}$:

1. $\wedge$:

$$(\wedge L_1) \quad \frac{\Gamma, A \vdash \Lambda}{\Gamma, A \wedge B \vdash \Lambda} \qquad (\wedge L_2) \quad \frac{\Gamma, B \vdash \Lambda}{\Gamma, A \wedge B \vdash \Lambda}$$

$$(\wedge R) \quad \frac{\Gamma \vdash A, \Lambda \quad \Sigma \vdash B, \Pi}{\Gamma, \Sigma \vdash A \wedge B, \Lambda, \Pi}$$

2. $\vee$:

$$(\vee L) \quad \frac{\Gamma, A \vdash \Lambda \quad \Sigma, B \vdash \Pi}{\Gamma, \Sigma, A \vee B \vdash \Lambda, \Pi}$$

$$(\vee R_1) \quad \frac{\Gamma \vdash A, \Lambda}{\Gamma \vdash A \vee B, \Lambda} \qquad (\vee R_2) \quad \frac{\Gamma \vdash B, \Lambda}{\Gamma \vdash A \vee B, \Lambda}$$

3. $\rightarrow$:

$$(\rightarrow L) \quad \frac{\Gamma \vdash A, \Lambda \quad \Sigma, B \vdash \Pi}{\Gamma, \Sigma, A \rightarrow B \vdash \Lambda, \Pi}$$

$$(\rightarrow R) \quad \frac{\Gamma, A \vdash B, \Lambda}{\Gamma \vdash A \rightarrow B, \Lambda}$$

4. $\neg$:

$$(\neg L) \quad \frac{\Gamma \vdash A, \Lambda}{\Gamma, \neg A \vdash \Lambda}$$

## 4.4. Proving validity and satisfiability

$$(\neg R) \quad \frac{\Gamma, A \vdash \Lambda}{\Gamma \vdash \neg A, \Lambda}$$

Note that, contrarily to natural deduction calculi, in which there are rules for the introduction and the elimination of the connectives, in the sequent calculi there are only rules for the introduction of the connectives in the antecedent or in the succedent of a sequent; the eliminations in the latter calculi take the form of introductions in the antecedent.

The sequent calculus $\mathcal{LK}$ can be extended to classical FOL, sufficing to that end to add two more operational rules for the quantifiers.

**4.4.11.** Let $t$ be a term free for $x$ and let $a$ be a constant (more strictly: a parameter). The following are the logical rules of $\mathcal{LK}$ for the quantifiers:

1. ($\forall$):

$$(\forall L) \quad \frac{\Gamma, A(t) \vdash \Lambda}{\Gamma, \forall x A(x) \vdash \Lambda}$$

$$(\forall R) \quad \frac{\Gamma \vdash A(a), \Lambda}{\Gamma \vdash \forall x A(x), \Lambda}$$

2. ($\exists$):

$$(\exists L) \quad \frac{\Gamma, A(a) \vdash \Lambda}{\Gamma, \exists x A(x) \vdash \Lambda}$$

$$(\exists R) \quad \frac{\Gamma \vdash A(t), \Lambda}{\Gamma \vdash \exists x A(x), \Lambda}$$

Restrictions apply to the rules $\forall$R and $\exists$L: $a$ must not occur within $\Gamma$ and $\Lambda$, *or* it must not appear anywhere in the respective lower sequents.

**Example 4.4.6.** Figure 4.4.4 shows a proof in $\mathcal{LK}$ of the FOL theorem $\vdash \forall x (A(x) \to B) \to \exists x (A(x) \to B)$. Figure 4.4.5 shows a proof in $\mathcal{LK}$ of axiom $\mathscr{L}2$. We make $t = a$.

103

4. *Logical decisions*

$$\dfrac{\dfrac{\dfrac{\dfrac{\dfrac{\dfrac{\overline{A(a) \vdash A(a), B}^{\text{Ax}} \quad \overline{A(a), B \vdash B}^{\text{Ax}}}{\vdash A(a), A(a) \to B\,(\to R) \qquad B \vdash A(a) \to B\,(\to R)}}{\vdash A(a), \exists x A(x) \to B\,(\exists R) \qquad B \vdash \exists x A(x) \to B\,(\exists R)}}{A(a) \to B \vdash \exists x A(x) \to B}(\to L)}{\forall x A(x) \to B \vdash \exists x A(x) \to B}(\forall L)}{\vdash \forall x (A(x) \to B) \to \exists x (A(x) \to B)}(\to R)$$

Figure 4.4.4.: Proof in $\mathcal{LK}$ of a FO theorem.

Finally, an additional structural rule of $\mathcal{LK}$ is the following:

**4.4.12.** *Cut*:

$$\text{(CUT)} \quad \dfrac{\Gamma \vdash \Lambda, A \quad A, \Sigma \vdash \Pi}{\Gamma, \Sigma \vdash \Lambda, \Pi}$$

CUT has motivated much work–if not furore–in modern logic, in particular in the field of automated deduction. Despite this, CUT plays no important role in this text, and we leave it at that. For its "special" character and status in the sequent calculus above, see, e.g., Fitting (1999).

### 4.4.5. The DPLL procedure

The Davis-Putnam-Logemann-Loveland procedure (Davis, Logemann, & Loveland, 1962), abbreviated as DPLL, is the basis for many classical propositional satisfiability solvers, or classes of algorithms for large subsets of SAT instances. It is actually a refinement of the resolution-based procedure first implemented by Davis and Putnam for propositional logic formulae (Davis & Putnam, 1960), and it can be straightforwardly extended to the CNF-SAT problem in many-valued logics, reason why we provide its essentials here.

**4.4.13.** *The CNF-SAT* – Given a formula $\phi$ in CNF, we want to know whether $\phi$ is satisfiable.

The DPLL procedure consists of a set of rules and a strategy to apply them.

**4.4.14.** Let $\phi$ be a formula in CNF and let $C_\phi$ be the set of clauses of $\phi$. The *DPLL1-8 rules* to apply to $C_\phi$ are as follows:

1. *UNSAT:* If $C_\phi$ contains the empty clause, then $\phi$ is unsatisfiable.

## 4.4. Proving validity and satisfiability

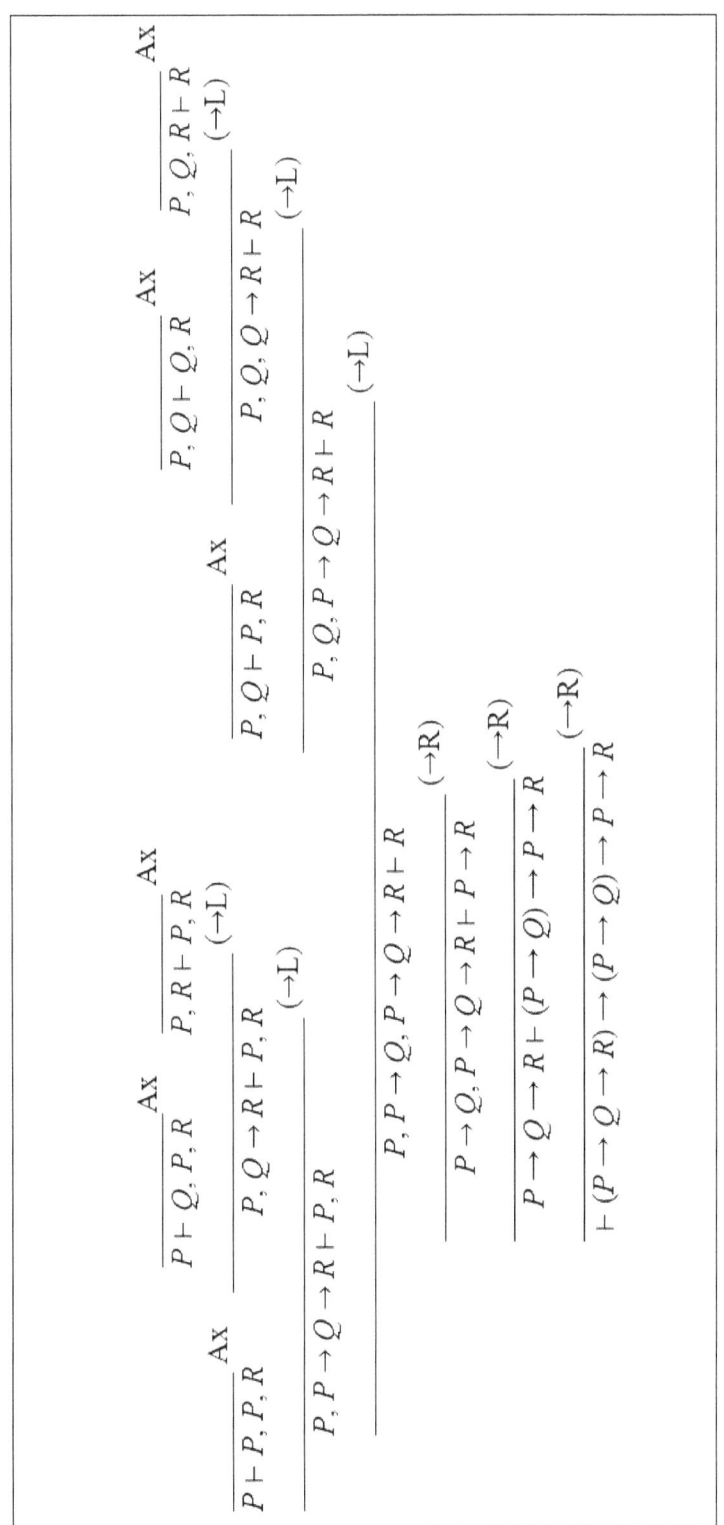

Figure 4.4.5.: Proof in $\mathcal{LK}$ of axiom $\mathcal{L}2$ of the axiom system $\mathcal{L}$.

4. Logical decisions

2. *SAT:* If $C_\phi$ is empty, then $\phi$ is satisfiable.

3. *MULT:* If a literal occurs more than once in a clause, then all but one of its occurrences can be deleted.

4. *SUBS:* If a clause $\mathcal{C}_1$ is a superset of a clause $\mathcal{C}_2$, $\mathcal{C}_1, \mathcal{C}_2 \subseteq C_\phi$, then $\mathcal{C}_1$ can be deleted.

5. *UNIT:* If $C_\phi$ contains $\|L\|$, then an element $\neg L \in \mathcal{C}_i \subset C_\phi$ can be deleted.

6. *TAUT:* If a clause $\mathcal{C}$ contains a literal and its complement, then it can be deleted.

7. *PURE:* If a clause $\mathcal{C}$ contains a literal $L$ and $\neg L$ does not occur in $\phi$, then $\mathcal{C}$ can be deleted.

8. *SPLIT:* If $C_\phi$ is semantically equivalent to a formula of the form

$$\{(\mathcal{C}_1 \vee L), ..., (\mathcal{C}_k \vee L), (\mathcal{C}_{k+1} \vee \neg L), ..., (\mathcal{C}_m \vee \neg L), \mathcal{C}_{m+1}, ..., \mathcal{C}_n\}$$

where neither $L$ nor $\neg L$ occur in $\mathcal{C}_i$, $1 < i < n$, then replace $C_\phi$ by

$$\{\mathcal{C}_1, ..., \mathcal{C}_k, \mathcal{C}_{m+1}, ..., \mathcal{C}_n\} \vee \{\mathcal{C}_{k+1}, ..., \mathcal{C}_m, \mathcal{C}_{m+1}, ..., \mathcal{C}_n\}.$$

**Example 4.4.7.** We provide examples of the DPLL rules above:
(DPLL1) $C = \{\Box, \mathcal{C}_1, ..., \mathcal{C}_n\} \equiv \Box$; UNSAT.
(DPLL2) $C = \{L_1, L_2, \neg L_1\} \equiv \{\ \}$; SAT.
(DPLL3) $\|L, L, L_1, ..., L_m\| \equiv \|L, L_1, ..., L_m\|$.
(DPLL4) $C = \{\|L_1, ..., L_m\|, \|L_1, ..., L_m, ..., L_k\|, \mathcal{C}_1, ..., \mathcal{C}_n\} \equiv \{\|L_1, ..., L_m\|, \mathcal{C}_1, ..., \mathcal{C}_n\}$.
(DPLL5) $C = \{\|L_1, ..., L_m, \neg L\|, \|L\|\} \equiv \{\|L_1, ..., L_m\|, \|L\|\}$.
(DPLL6) $C = \{\|L_1, ..., L_m, L, \neg L\|, \mathcal{C}_1, ..., \mathcal{C}_n\} \equiv \{\mathcal{C}_1, ..., \mathcal{C}_n\}$.
(DPLL7) $C = \{\|P, \neg Q\|, \|\neg R\|\}$ is *not* equivalent to $C' = \{\|P, \neg Q\|\}$, but $\{\mathcal{C}_1, ..., \mathcal{C}_m, \|L, L_1, ..., L_n\|\}$ is unsatisfiable iff $\{\mathcal{C}_1, ..., \mathcal{C}_m\}$ is unsatisfiable, where $\neg L$ occurs neither in $\mathcal{C}_i$, $1 < i < m$, nor in $L_j$, $1 < j < n$.
(DPLL8) $C = \{\|P, R\|, \|\neg R\|, \|Q\|\}$ and $C' = \{\|P\|, \|Q\|\} \vee \{\Box, \|Q\|\}$ are *not* equivalent, but the rule preserves unsatisfiability.

It is obvious that while rules DPLL3-6 are equivalence-preserving, rules DPLL7-8 are unsatisfiability-preserving, i.e. for a formula $\psi$ obtained from a formula $\phi$ by application of any of these two rules, we have it that $\phi$ is unsatisfiable iff $\psi$ is unsatisfiable.

## 4.4. Proving validity and satisfiability

---
**Algorithm 4.2** DPLL proof procedure

**Input:** A formula $A \in \mathsf{L}^*$ in CNF

**Output:** A proof of (un)satisfiability of $A$

---

1. Input formula $A$ in CNF.

2. Apply rules DPLL3 to DPLL8, until either SAT (DPLL2) or UN-SAT (DPLL1) can be applied.

3. If SAT is applicable, then terminate with "$A$ is satisfiable"; else, UNSAT is applicable and terminate with "$A$ is unsatisfiable".

---

**4.4.15. (Prop.)** The DPLL algorithm constitutes an adequate method of satisfiability testing.

*Proof:* Left as an exercise.

The DPLL algorithm has the good property that it always terminates, and it is sound and complete. The strategy we mentioned above applies to step 3 of the algorithm: rules DPLL3 to DPLL7 are to be applied eagerly, but wise decisions should be made when applying DPLL8 with regard to the literal to be eliminated.

**Example 4.4.8.** Let $C = (L \vee M) \wedge (O \vee \neg M \vee \neg N) \wedge (L \vee N) \wedge \neg O$. We apply the DPLL-algorithm to $C$ (see Fig. 4.4.6).

| | RULES |
|---|---|
| $\{\|L, M\|, \|O, \neg M, \neg N\|, \|\neg L, N\|, \|\neg O\|\}$ | Initialization |
| $\{\|L, M\|, \|O, \neg M, \neg N\|, \|\neg L, N\|, \|\neg O\|\}$ | $\|\neg O\|$, DPLL5 |
| $\{\|L, M\|, \|\neg M, \neg N\|, \|\neg L, N\|, \|\neg O\|\}$ | $\neg O$, DPLL 7 |
| $\{\|L, M\|, \|\neg M, \neg N\|, \|\neg L, N\|\}$ | $L$, DPLL8 |
| $\{\|M, \neg M, \neg N\|, \|N, \neg M, \neg N\|\}$ | $\|M, \neg M, \neg N\|$, DPLL6 |
| $\{\|N, \neg M, \neg N\|\}$ | $\|N, \neg M, \neg N\|$, DPLL6 |
| $\{\}$ | $\{\}$, DPLL2 |
| "Satisfiable" | |

Figure 4.4.6.: A DPLL proof procedure.

## 4.5. Refutation I: Analytic tableaux

This proof system was firstly conceived as *semantic tableaux* by Beth (1955) and Hintikka (1955), whose concerns were mostly semantical; it was later greatly simplified by Smullyan (1968) into the variant known as *analytic tableaux*. We concentrate on the latter.

Analytic tableaux is a remarkably efficient proof procedure for classical logic, and some non-classical logics as well, based on labeled binary trees. Briefly, analytic tableaux are binary trees whose nodes are formulae that are (sub)goals in the proofs; the tree structure concretizes the logical dependence among the (sub)goals. A tree constitutes a proof iff it is a closed tableau, i.e. a tree whose every branch has a contradictory pair of literals.

This is thus a proof of unsatisfiability of a set of formulae. Recall the definitions above (4.1.1) of (semi-)decidability. In fact, given a finite set of formulae $\Gamma$, the analytic tableaux proof is guaranteed to *terminate* in either a closed or an open tableau, being thus indeed a decision procedure–albeit only for propositional logic. If we allow for $\Gamma$ being infinite, then the tableau construction is guaranteed to terminate only if $\Gamma$ is unsatisfiable, running forever if $\Gamma$ is satisfiable. Thus, in the latter case we speak of semi-decidability.

### 4.5.1. Analytic tableaux as a propositional calculus

We now expand on and formalize the above in terms of a propositional calculus for classical logic.

**4.5.1. (Def.)** Let the propositional language $\mathsf{L}$ be given. A *tableau* (plural: tableaux) is a finite binary tree $\mathcal{T}$ whose nodes are formulae from $F_\mathsf{L}$.[16] A *tableaux proof system* (or *calculus*) over $F_\mathsf{L}$ is a set $RT = \{\mathsf{I}, \mathsf{A}, \mathsf{B}, \mathsf{X}\}$ of *initialization rule(s)* I, *expansion rules* A and B each having a set $\Gamma = F'_\mathsf{L}$ of formulae from $F_\mathsf{L}$ and optionally a tableau for $\Gamma$ as premises and another tableau for $\Gamma$ as a conclusion, and *closure rule(s)* X. For each $\Gamma$, the transitive closure of the rules in $RT$ defines a set of tableaux constructed with $RT$ for $\Gamma$.

**4.5.2. (Def.)** In a tableau $\mathcal{T}_\Gamma$ for a set of formulae $\Gamma = F'_\mathsf{L}$, a branch $\mathcal{B}$ of $\mathcal{T}_\Gamma$ is *closed*, denoted by X, iff $\mathcal{B} \cup \Gamma$ contains either a pair $(L, \neg L) \in F_\mathsf{L}$ or $\bot$. Otherwise, $\mathcal{B}$ is *open*. A tableau for $\Gamma$ is closed, denoted by $\mathcal{T}_\Gamma^\mathsf{X}$, iff all its branches are closed.

**4.5.3. (Def.)** A *tableau proof* of a set $\Gamma = F'_\mathsf{L}$ is a closed tableau $\mathcal{T}_\Gamma^\mathsf{X}$.

**4.5.4. (Prop.)** Algorithm 4.2 constitutes a refutation proof procedure.

---

[16] More properly put, a tableau is *implemented* by a tree (cf. Fitting, 1999).

## 4.5. Refutation I: Analytic tableaux

**Algorithm 4.3** Analytic tableaux proof procedure

**Input:** A formula $A$ (or an argument $\alpha = \chi_1, .., \chi_n/\psi$)

**Output:** A closed binary tree $\mathcal{T}_{\neg A}^{\times}$ ($\mathcal{T}_{\Gamma}^{\times}$) that constitutes a proof of validity of $A$ (or $\alpha$)

1. Place $\neg A$ (or the set $\Gamma = \{\chi_i\}_1^n \cup \{\neg \psi\} = \alpha_{\neg}$) in the *root* of a binary tree.

2. Apply the *expansion rules* to the formulae on the tree, thus adding new formulae and splitting branches.

3. *Close* contradictory branches (i.e. branches containing both $L$ and $\neg L$).

4. *Terminate* successfully iff all branches are closed; unsuccessfully, otherwise.

*Proof:* Left as an exercise. (Hint: $\not\models \Gamma$ is a refutation of $\Gamma$.)

We formalize Algorithm 4.2:

**4.5.5. (Def.)** Given a set $\Gamma = F_L'$ of formulae from $F_L$, a *tableau for* $\Gamma$ is defined as a tableau constructed according to the following rules:

1. *Initialization rule:* The tree $\mathcal{T}_\Gamma$ consisting of a single node **t** is a tableau for $\Gamma$.

2. *Expansion rule:* Let $\mathcal{T}_\Gamma$ be a tableau for $\Gamma$, and $\mathcal{B}$ a branch of $\mathcal{T}_\Gamma$. Let further $\phi$ be a formula in $\mathcal{B} \cup \Gamma$. Obtain the tree $\mathcal{T}_\Gamma'$ by expanding $\mathcal{B}$ with $n$ new subtrees whose nodes are the formulae in the expansion of the rule instance. Then, $\mathcal{T}_\Gamma'$ is a tableau for $\Gamma$.[17]

3. *Closure rule:* Given a tableau $\mathcal{T}_\Gamma$ and a branch $\mathcal{B}$ thereof, if we have $(L, \neg L) \in \mathcal{B} \cup \Gamma$, then $\mathcal{B}$ is a closed branch of $\mathcal{T}_\Gamma$. If all branches of $\mathcal{T}_\Gamma$ are closed, then $\mathcal{T}_\Gamma$ is a closed tableau for $\Gamma$.

The order of application of the expansion rules is non-deterministic. Moreover, one can repeat formulae, though this may increase the com-

---

[17] In the tableaux literature, the term "extension," originally used in Smullyan (1968), is also to be frequently found as a synonym for "expansion." Our choosing the latter is accounted for by our, distinct, use of the term "extension" in this text.

plexity of the tableau. We say that a tableau implementation is *fair* if we apply every rule that is applicable in the case at hand. The notions of unicity of decomposition and immediate subformula (cf. Prop. 2.1.4 and Def. 2.1.5) allow us to act in such a way as to have a final set of (negated) atoms by proceeding to a step-by-step decomposition of complex formulae. Underlying the construction of the tree is the rewriting of all the formulae into equivalent (negations of) conjunctions and disjunctions in the usual ways, being a tree thus in fact a disjunction of conjunctions.[18] Now, recall that for a literal $L$ we have $(L \wedge \neg L) = \bot$, and that in turn $(\bot \vee \bot)$ is also a contradiction; this explains why a disjunction of finitely many $\bot$-containing branches constitutes a closed tree, i.e. a contradiction of the negated formula.

**4.5.6.** The expansion rules and the closure rule, expressed in the language of set theory and accounted for in terms of the classical logical consequence relation, are as follows:

1.
$$(\wedge_{RE}) \quad \frac{\Gamma \cup \{\phi \wedge \psi\}}{\Gamma \cup \{\phi, \psi\}}$$

given that $\wedge$ is classical in terms of $\Vdash$ iff we have $\Gamma, \phi \wedge \psi \Vdash \chi$ iff $\Gamma, \phi, \psi \Vdash \chi$.

2.
$$(\vee_{RE}) \quad \frac{\Gamma \cup \{\phi \vee \psi\}}{\Gamma \cup \{\phi\} \mid \Gamma \cup \{\psi\}}$$

given that $\vee$ is classical in terms of $\Vdash$ iff we have $\Gamma, \phi \vee \psi \Vdash \chi$ iff $\Gamma, \phi \Vdash \chi$ and $\Gamma, \psi \Vdash \chi$.

3.
$$(\mathsf{X}) \quad \frac{\Gamma \cup \{L, \neg L\}}{\mathsf{X}}$$

because classically we have $\Gamma, \phi, \neg \phi \Vdash \chi$, i.e. *ex contradictione quodlibet*, or "explosion."[19]

**4.5.7.** All the expansion rules can be conveniently reduced to two, in what is known as the $\alpha\beta$-*classification* (Fig. 4.5.1).[20]

---

[18] Indeed, a tableau can be a conjunction of disjunctions, but this typically increases the size of the tableau, and it is thus a less efficient proof procedure.

[19] See Augusto (2017) for the classicality conditions for the connectives of L.

[20] By convention, doubly negated formulae are treated as formulae of type $\alpha$, rather than as an elimination rule. One can further treat $\neg\top$ and $\neg\bot$ as type-$\alpha$ formulae, resulting in the nodes labeled with $\bot$ and $\top$, respectively.

**4.5.8.** There are thus in effect only two expansion rules, A (for $\alpha$) and B (for $\beta$):

$$(A) \quad \frac{\alpha}{\begin{array}{c}\alpha_1\\\alpha_2\end{array}}$$

$$(B) \quad \frac{\beta}{\beta_1 \mid \beta_2}$$

Informally, these two rules mean that if $\alpha$ is a conjunction of $\alpha_1$ and $\alpha_2$, then both these two subformulae are logical consequences of $\alpha$ (rule A) and thus are nodes in the one and same branch of the tree (the analytic tableau); if $\beta$ is a disjunction of $\beta_1$ and $\beta_2$, then these subformulae originate two different branches as a logical consequence of $\beta$ (rule B). As said above, a tree is a disjunction of conjunctions; this is the result of giving priority to $\alpha$-type formulae over $\beta$-type formulae in terms of decomposing (Smullyan, 1968).

| $\alpha$ | $\alpha_1$ | $\alpha_2$ | $\beta$ | $\beta_1$ | $\beta_2$ |
|---|---|---|---|---|---|
| $A \wedge B$ | $A$ | $B$ | $\neg(A \wedge B)$ | $\neg A$ | $\neg B$ |
| $\neg(A \vee B)$ | $\neg A$ | $\neg B$ | $A \vee B$ | $A$ | $B$ |
| $\neg(A \rightarrow B)$ | $A$ | $\neg B$ | $A \rightarrow B$ | $\neg A$ | $B$ |
| $\neg\neg A$ | $A$ | $A$ | | | |

Figure 4.5.1.: Analytic tableaux expansion rules: $\alpha\beta$-classification.

**Theorem 4.5.1.** *For any propositional tableau, after a finite number of steps no more expansion rules will be applicable.*

*Proof:* (Sketch) The theorem holds assuming that we analyze each formula at most once. Let us begin by assuming that we have a formula with $n = 0$ connectives. Then, this is a propositional atom, and no expansion rules apply. Let us now assume that the theorem holds for any formula with at most $n$ connectives. Then we can prove it for a formula $\phi$ with $n + 1$ connectives. This can be done in two ways, depending on the type of the formula:

1. $\phi$ is a formula of type $\alpha$. We apply rule A, and we mark $\phi$ as analyzed once. Clearly, $\alpha_1$ and $\alpha_2$ contain each fewer connectives with relation to $\phi$; we apply the inductive hypothesis, and say that a tableau can be built such that each formula is analyzed at most once. After a finite number of steps, no more expansion rules can be applied. We combine the two proofs, and the theorem is proved for a type-$\alpha$ formula.

2. $\phi$ is a formula of type $\beta$. We apply rule B, and we mark $\phi$ as analyzed once. Clearly, $\beta_1$ and $\beta_2$ contain each fewer connectives with relation to $\phi$; we apply the inductive hypothesis and say that two tableaux can be built for $\beta_1$ and $\beta_2$ such that each formula is analyzed at most once. After a finite number of steps, no more expansion rules can be applied. We combine the two proofs, and the theorem is proved for a type-$\beta$ formula. **QED**

Recall the subformula property of a sequent calculus proof (cf. Prop. 4.4.6). It is evident that this property is at play in the proof above. In effect, we say that an expansion rule is *analytic* whenever its every application yields the subformula property. Formalizing this:

**4.5.9. (Def.)** We say that an expansion rule is *analytic* if every formula that occurs as a conclusion of the rule ($\alpha_i, \beta_i$, for $i = 1, 2$) is a subformula of the formula that occurs as the premise of the rule ($\alpha, \beta$).

This shows the interesting relation between the sequent calculus and analytic tableaux: a downward growing tree in the latter is an upward growing tree in the former. This has to do with the *invertibility* of the rules of the sequent calculi (e.g., D'Agostino, 1999).

**Example 4.5.1.** Figure 4.5.2 shows an analytic tableaux proof of $\vdash ((A \to B) \land ((A \land B) \to C)) \to (A \to C)$. Note the indication of the numbered steps on the left and the application of the corresponding expansion rules on the right.

Although analytic tableaux is a proof system, it actually is a hybrid system in the sense that a tableau proof is a *countermodel* $\overline{\mathcal{M}}$ with respect to some semantics. In effect, a tableaux proof procedure is so with respect to the negation of the formula one wishes to prove. Informally, a countermodel corresponds to a tree whose branches are partial descriptions of the model, where the $\alpha\beta$-classification is to be understood as follows:

**4.5.10. (Prop.)** Under any interpretation $\mathcal{I}$, the following facts clearly hold:

1. ($\alpha_T$)   $\alpha$ is true iff $\alpha_1, \alpha_2$ are both true;

2. ($\beta_T$)   $\beta$ is true iff at least one of $\beta_1, \beta_2$ is true.

*Proof:* Left as an exercise.

4.5. Refutation I: Analytic tableaux

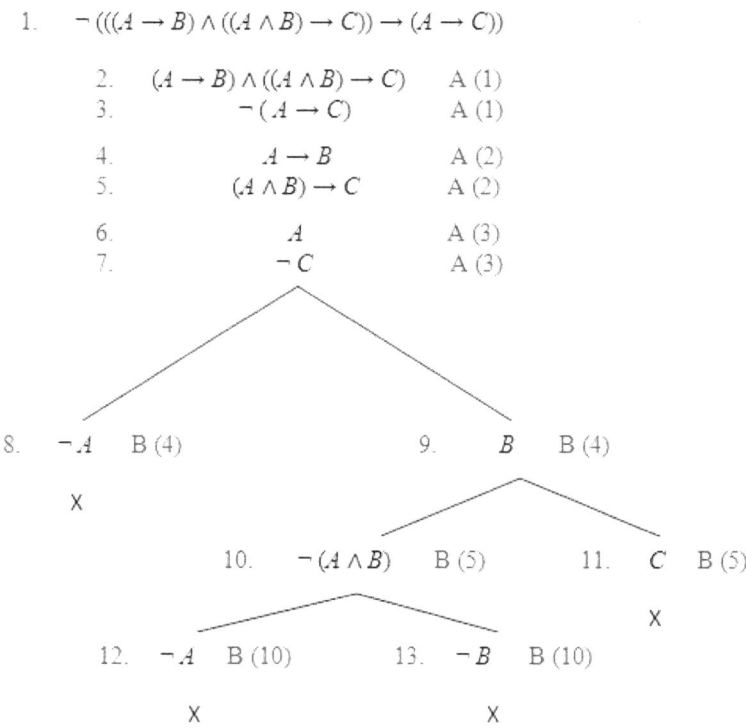

Figure 4.5.2.: ⊢ $((A \to B) \land ((A \land B) \to C)) \to (A \to C)$: A propositional tableau proof.

## 4. Logical decisions

As already known, a branch in a tree is said to close if both $L$ and $\neg L$ are in it (i.e. if they are nodes of the same branch), and the tree itself is said to close if all its branches close. We formalize this:

**4.5.11. (Prop.)** A tableau $\mathcal{T}_\Gamma$ for a set of formulae $\Gamma \subseteq F_L$ is satisfiable iff there is a model $\mathcal{M}_\Gamma$ of $\Gamma$ such that for every valuation $val$ of the formulae of $\Gamma$ there is a branch $\mathcal{B}$ of $\mathcal{T}_\Gamma$ with $\models_{val_\mathcal{M}} \mathcal{B}$. We then say that $\mathcal{M}$ is a model of $\mathcal{T}_\Gamma$, and we write $\models_\mathcal{M} \mathcal{T}_\Gamma$.
*Proof:* Left as an exercise.

**4.5.12. (Prop.)** It is evident that no model or valuation can satisfy a closed branch, so a closed tree $\mathcal{T}_\Gamma^\times$ is a proof of the unsatisfiability of $\Gamma$.
*Proof:* Left as an exercise.

In terms of satisfiability, we have the following facts with respect to the expansion rules for formulae of types $\alpha$ and $\beta$:

**4.5.13. (Prop.)** Let $X \subseteq F_L$ be a set of formulae. Then the following facts hold:

1. ($\alpha_{sat}$) If $X$ is satisfiable and $\alpha \in X$, then $\{X, \alpha_1, \alpha_2\}$ is satisfiable.

2. ($\beta_{sat}$) If $X$ is satisfiable and $\beta \in X$, then at least one of $\{X, \beta_1\}$, $\{X, \beta_2\}$ is satisfiable.

*Proof:* Left as an exercise.

Propositions 4.5.10 and 4.5.13 are fundamental in that they state that the decomposition of formulae in the propositional tableaux calculus preserves truth and satisfiability.

We now require a few further definitions in order to address the soundness and completeness of the propositional analytic tableaux calculus.

**4.5.14. (Def.)** Let $\mathcal{T}$ be a tableau. We say that a branch $\mathcal{B} \in \mathcal{T}$ is *complete* if for every $\alpha$ occurring in $\mathcal{B}$, both $\alpha_1$ and $\alpha_2$ occur in $\mathcal{B}$, and for every $\beta$ occurring in $\mathcal{B}$, at least one of $\beta_1, \beta_2$ occurs in $\mathcal{B}$. We call a tableau $\mathcal{T}$ *complete(d)* if every branch of $\mathcal{T}$ is either closed or complete.

**Theorem 4.5.2.** *(Soundness of propositional analytic tableaux)* Every formula provable by the analytic tableaux calculus is a tautology.

*Proof:* (Sketch) Let $\Gamma$ be a set of formulae, possibly a singleton, and let $\mathcal{I}$ be an interpretation. Let $\models_\mathcal{I} \mathcal{T}_\Gamma^{(0)}$, $\mathcal{T}_\Gamma^{(0)}$ is the completed tableau for $\Gamma$. Then, for any subtree $\mathcal{T}_\Gamma^i, 0 < i \leq n$, we have $\models_\mathcal{I} \mathcal{T}_\Gamma^i$ by Propositions

## 4.5. Refutation I: Analytic tableaux

4.5.10 and 4.5.13, i.e. any interpretation satisfying all the formulae in $\Gamma$ must also satisfy all the formulae contained in at least one complete branch of any completed tableau for $\Gamma$. Thus, for any completed tableau $\mathcal{T}$, and applying induction and the analyticity property of the tableaux calculus (Def. 4.5.9), if the root is true under an interpretation $\mathcal{I}$, then $\mathcal{T}$ must be true under $\mathcal{I}$. It is evident (by Proposition 4.5.12) that a closed tableau cannot be true under any interpretation, and therefore its root cannot be true under any interpretation. Hence, every formula provable by the analytic tableaux method must be a tautology. **QED**

**Corollary 4.5.3.** *(Consistency of propositional analytic tableaux)* The tableau method for classical propositional logic is consistent.

*Proof:* From the proof of the soundness theorem, we conclude that no formula and its negation are both provable in the analytic tableaux calculus. **QED**

We turn now to the completeness theorem of analytic tableaux. In order to prove this theorem, a few more fundamental notions are required.

**4.5.15. (Def.)** Let $X \subseteq F_L$. We say that $X$ is a *downward saturated set* iff

1. if $\alpha \in X$, then $\alpha_1 \in X$ and $\alpha_2 \in X$;

2. if $\beta \in X$, then $\beta_1 \in X$ or $\beta_2 \in X$.

Note that this entails that a tableau $\mathcal{T}_X$ for a downward saturated set $X$ is complete in the sense that every branch $\mathcal{B}$ of $\mathcal{T}_X$ is complete.

**4.5.16. (Def.)** *Hintikka set* – A downward saturated set that does not contain an (atomic) formula and its negation is called an *(atomic) Hintikka set*.

**Lemma 4.5.4.** *(Hintikka's lemma)* Every downward saturated set $X$ (whether finite or infinite) is satisfiable.

*Proof:* Let $X$ be a Hintikka set. We need to show that some Herbrand interpretation $H\mathcal{I}_X$ of $X$ is a model of $X$.[21] Let $H_X$ be the set of ground terms in $X$. Because $H(X)$ by definition (of $X$ as a Hintikka set) does not contain an atomic formula and its negation, it immediately defines a Herbrand interpretation $H\mathcal{I}_X$. Assume that $H\mathcal{I}_\phi = \mathbf{t}$ for all formulae

---
[21] Recall the material from Section 4.3.

## 4. Logical decisions

$\phi \in X$ with complexity less than $n$, where by complexity of a formula $\chi$ it is meant the number $n$ of occurrences of formulae in $\chi$. Now consider any complex formula $\psi \in X$ with complexity $n$. It is evident that the complexity of any tableau subformula of $\psi$ is less than $n$. Clearly, $\psi$ must be a $\alpha$- or a $\beta$-type formula, and we have two cases.

Case 1: If $\psi$ is a $\alpha$-type formula, then by definition of downward saturation every $\alpha_i$ is in $X$. By induction, and by $\alpha_\top$ (Prop. 4.5.10.1), we have $H\mathcal{I}_{\alpha_i} = \mathfrak{t}$, $H\mathcal{I}_{\phi,\psi} = \mathfrak{t}$, and of course $H\mathcal{I}_X = \mathfrak{t}$.

Case 2: If $\psi$ is a $\beta$-type formula, then by definition of downward saturation some $\beta_i$ is in $X$. By induction, and by $\beta_\top$ (Prop. 4.5.10.2), we have $H\mathcal{I}_{\beta_i} = \mathfrak{t}$, $H\mathcal{I}_{\phi,\psi} = \mathfrak{t}$, and of course $H\mathcal{I}_X = \mathfrak{t}$. **QED**

The proof of Hintikka's lemma, together with $\alpha_{sat}$ and $\beta_{sat}$ (Prop.s 4.5.13.1-2), actually proves the following important theorem:

**Theorem 4.5.5.** *(Smullyan, 1968) Any complete open branch of any tableau is (simultaneously) satisfiable.*

In turn, this theorem implies the completeness of the propositional analytic tableaux calculus.

**Theorem 4.5.6.** *(Completeness of propositional analytic tableaux) (a) If $\psi$ is a tautology, then every complete tableau starting with $\neg\psi$ must close. (b) Every tautology is provable in the tableaux calculus.*

The derivation of (a) from Theorem 4.5.5 runs as follows: Assume that $\mathcal{T}_\psi$ is a complete tableau starting with $\neg\psi$. By Theorem 4.5.5, if $\mathcal{T}_\psi$ is open, then $\neg\psi$ is satisfiable. Therefore, $\psi$ cannot be a tautology (recall Prop. 3.2.14.1). Hence, if $\psi$ is a tautology, then $\mathcal{T}_\psi$ must be closed.

The proofs of Theorems 4.5.5-6 are left as exercises.

### 4.5.2. Analytic tableaux as a predicate calculus

As seen above (Section 4.1.2), Lemma 3.4.6 sanctions the generalization of refutation as a proof procedure to classical FOL. The introduction of analytic tableaux rules for the quantifiers motivates some problems, in particular because the set of constants is infinite, a problem that actually accounts for the non-terminating character of tableaux proofs in FOL. In effect, Theorem 4.5.1 does *not* hold for FO tableaux. In any case, these problems can be overcome in several ways, so that the tableaux calculus is sound and complete for FOL.

## 4.5. Refutation I: Analytic tableaux

**4.5.17. (Def.)** Let $\Upsilon$ be a signature for the language $\mathsf{L}^*$. A *tableau (over $\Upsilon$)* is a finite tree whose nodes are formulae from $F_{\mathsf{L}^*}$. A *tableaux proof system (or calculus)* over $F_{\mathsf{L}^*}$ is a set $RT = \{\mathrm{I}, \mathrm{A}, \mathrm{B}, \mathrm{C}, \mathrm{D}, \mathrm{X}\}$ where I, A, B, X are essentially as in Definition 4.5.1 and C, D are additional *rules of expansion for FO formulae*.

Definitions 4.5.1-3 are thus generalizable to classical FOL; the same holds for formulae of types $\alpha$ and $\beta$ (with the proviso that now by "formula" it is understood "closed formula of quantification theory"; cf. Smullyan, 1968). This sets the formal scenario for approaching tableaux for FOL.

**4.5.18.** Just as in the case of propositional formulae, there are only two expansion rules when quantifiers are involved: this is called the $\gamma\delta$-*classification* (Fig. 4.5.3). It is evident that Definition 4.5.9 generalizes to FOL.

**4.5.19.** Formulae with the universal quantifier are classified as type $\gamma$, and those with the existential quantifier are of type $\delta$. The corresponding analytic tableaux rules for the quantified formulae are:

$$(\mathrm{C}) \quad \frac{\gamma}{\gamma(a)}$$

where $a$ is a ground term or a Skolem constant (or a parameter), and

$$(\mathrm{D}) \quad \frac{\delta}{\delta(a)}$$

where $a$ is new to the branch. By $\gamma(a)$ or $\delta(a)$ it is meant, for a formula $\phi$, $\phi_a^x$ or $\neg\phi_a^x$, i.e. the formula $\phi$ with $x$ substituted by $a$.

| $\gamma$ | $\gamma(a)$ |
|---|---|
| $\forall x A(x)$ | $A(a)$ |
| $\neg \exists x A(x)$ | $\neg A(a)$ |

| $\delta$ | $\delta(a)$ |
|---|---|
| $\exists x A(x)$ | $A(a)$ |
| $\neg \forall x A(x)$ | $\neg A(a)$ |

Figure 4.5.3.: Analytic tableaux expansion rules: $\gamma\delta$-classification.

Just as in the case of the $\alpha\beta$-classification, there is a semantical account for the expansion rules for formulae of types $\gamma$ and $\delta$ that assures us that the decomposition of these latter types of formulae preserves truth and satisfiability.

**4.5.20. (Prop.)** Under any interpretation $\mathcal{I}$ for a domain $\mathscr{D}$, the following facts hold:

1. ($\gamma_T$)   $\gamma$ is true iff $\gamma(a)$ is true for every $a \in \mathscr{D}$;

## 4. Logical decisions

2. ($\delta_T$)  $\delta$ is true iff $\delta(a)$ is true for at least one $a \in \mathcal{D}$.

*Proof:* Left as an exercise.

In terms of satisfiability, we have the following facts:

**4.5.21. (Prop.)** Let $X \in F_{L^*}$ be a set of formulae. Then,

1. ($\gamma_{sat}$)  If $X$ is satisfiable and $\gamma \in X$, then for every constant (or parameter) $a$ the set $\{X, \gamma(a)\}$ is satisfiable.

2. ($\delta_{sat}$)  If $X$ is satisfiable and $\delta \in X$, and if $a$ is a constant (or parameter) that does not occur in any element of $X$, then $\{X, \delta(a)\}$ is satisfiable.

*Proof:* Left as an exercise.

However, the above classifications are all too general. Importantly, we need to discuss separately tableaux for FOL with and without unification, as the expansion rules for the quantifiers are different for each.

### 4.5.2.1. FOL tableaux without unification

To begin with, some useful heuristics: whenever possible, apply propositional rules before quantifier rules; in the latter case, apply the tableau rules on $\delta$-formulae before $\gamma$-formulae. "Beyond this, you are on your own" (Fitting, 1996), a remark that works as a reminder that analytic tableaux is *not* a decision procedure for FOL, for the simple reason that no such procedure exists (see above). In particular, misapplication of the tableau rule for $\gamma$-formulae may cause the tableau not to close.

**4.5.22.** The analytic tableaux rules for the quantifiers without unification are:

$$(C) \quad \frac{\forall x \gamma(x)}{\gamma(t)}$$

where $t$ is any ground term, and

$$(D) \quad \frac{\exists x \delta(x)}{\delta(c)}$$

where the constant symbol $c$ is new to the branch. Note that in $\exists x \delta(x)$, the variable $x$ does not occur within the scope of a universal quantifier.

## 4.5. Refutation I: Analytic tableaux

Therefore, the skolemization of $\exists x \delta(x)$ generates solely a constant, i.e. a 0-ary function (cf. Def. 2.4.10.1).

**Example 4.5.2.** To prove by means of analytic tableaux the validity of the formula $\forall x \, (F(x) \wedge G(x)) \leftrightarrow \forall x \, (F(x)) \wedge \forall x \, (G(x))$ we apply rules C and D, i.e. the expansion rules for the universal quantifier without unification. See Fig. 4.5.4 for the respective tableau proof.

**4.5.23. (Def.)** Given a tree $\mathcal{T}_\Gamma$, for $\Gamma \subseteq F_{L^*}$, and a branch $\mathcal{B}$ thereof, let $(L, \neg L) \in \mathcal{B} \cup \Gamma$, $L$ is a literal. Then $\mathcal{B}$ is a closed branch of $\mathcal{T}_\Gamma$. If all branches of $\mathcal{T}_\Gamma$ are closed, then $\mathcal{T}_\Gamma^\times$ is a closed tableau for $\Gamma$.

The proof of soundness of the predicate analytic tableaux calculus without unification is as for the propositional case, but appealing now also to Propositions 4.5.20-1. In order to prove the completeness of tableaux for FOL we need to extend our definition of a downward saturated set (Def. 4.5.15) to sets containing $\gamma$- and $\delta$-type formulae:

**4.5.24. (Def.)** Let $X \subseteq F_{L^*}$. We say that $X$ is a downward saturated set iff conditions 1 and 2 of Def. 4.5.15 hold, and additionally:

1. if $\gamma \in X$, then $\gamma(t) \in X$ for all $\gamma(t)$ with $t \in H_X$;
2. if $\delta \in X$, then $\delta(c) \in X$ for at least one $c \in Cons \subseteq \Upsilon_X$.

The definition of Hintikka set (Def. 4.5.16) generalizes immediately to the sets above, and so does Lemma 4.5.4. Thus, the proof of completeness of the predicate analytic tableaux calculus without unification runs as for the propositional case, with the additional assumptions that $H\mathcal{I}_{\gamma(t)} = \mathsf{t}$ for any $t \in H_X$ and $H\mathcal{I}_{\delta(c)} = \mathsf{t}$ for some constant $c \in Cons \subseteq \Upsilon_X$. In the former case, let $\phi \in X$, $\phi$ is a formula of type $\gamma = \forall x \phi'$, where $\phi'$ is an *immediate tableau subformula of* $\phi$ (in this case, $\phi' = \phi[x/t]$); by the property of being a downward saturated set and by the induction assumption, we have $H\mathcal{I}_{\gamma(t)} = \mathsf{t}$ for any term $t \in H_X$. Because $H_X$ is the universe of $H\mathcal{I}_X$, and $H\mathcal{I}_X$ maps any term $t$ to itself (cf. Def. 4.3.3), for all variable assignments $\varpi$ to $H_X$ we have $H\mathcal{I}_\phi^\varpi = H\mathcal{I}_{\phi'[x/\varpi(x)]} = \mathsf{t}$. In the latter case, $\phi \in X$ is a $\delta$-formula, by downward saturation and the induction hypothesis we have it that $H\mathcal{I}_{\delta(c)} = \mathsf{t}$ for some constant $c \in T$ and therefore $H\mathcal{I}_\phi = \mathsf{t}$.

### 4.5.2.2. FOL tableaux with unification

Note that when applying rule C without unification we have the problem of choosing the ground term $t$. In effect, any possible ground term

## 4. Logical decisions

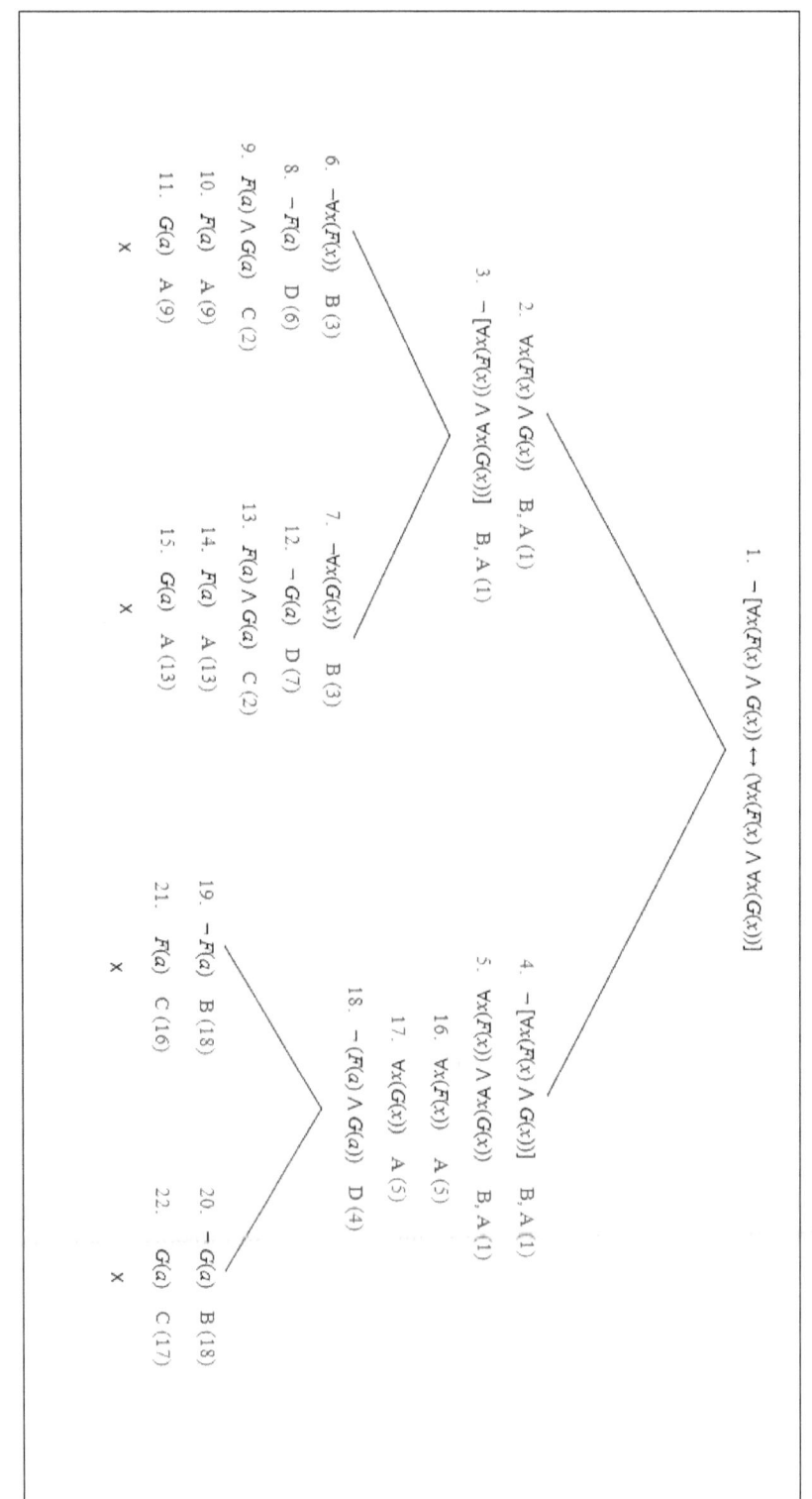

Figure 4.5.4.: An FO tableau proof.

## 4.5. Refutation I: Analytic tableaux

can be chosen, or even every one, but this would certainly not be relevant for the tableau proof. The problem is particularly acute in terms of automation, as this is basically impossible without guidance on what term(s) to choose. The strategy then is to delay the choice of the ground term until $\gamma(t)$ allows the closure of *at least* one of the branches of the tree, though evidently a substitution $\sigma$ that will *simultaneously* close *all* branches is what is required. To do this, we simply use a variable instead of a ground term, so that the rule now is:

**4.5.25.**
$$(C') \quad \frac{\forall x \gamma(x)}{\gamma(x')}$$

where $x'$ is a variable not occurring anywhere else in the tableau.

Just as for rule D, we apply skolemization again, but now Skolem terms must not be constants, because the application of unification may create free variables that are implicitly universally quantified. This means that a formula $\exists x \delta(x)$ can now be within the scope of one or more universal quantifiers. According to Def. 2.4.10.2, we now have:

**4.5.26.**
$$(D') \quad \frac{\exists x \delta(x)}{\delta(f(x_1, ..., x_n))}$$

where $f$ is a new function symbol and $x_1, ..., x_n$ are the $n$ free variables of $\delta$.

The skolemization procedure can be made technically more rigorous by making it explicit that $f$ is a function assigning to each $\delta \in \check{\Upsilon}_{F^*}$ a symbol $\ast \in Fun_{Sko}$, the set of Skolem functions (cf. Def. 2.4.11), $Fun_{Sko} \cap \Upsilon_{F^*} = \emptyset$, $\check{\Upsilon}_{F^*} = \Upsilon_{F^*} \cup Fun_{Sko}$, such that $\ast > f$ for all $f \in Fun_{Sko}$ occurring in $\delta$ for $>$ an arbitrary but fixed ordering on $Fun_{Sko}$, and for all $\delta, \delta' \in \Upsilon_{F^*}$ the symbols $\ast$ and $\ast'$ are identical iff $\delta$ and $\delta'$ are identical up to variable renaming, so that we have the rule

$$(D') \quad \frac{\exists x \delta(x)}{\delta(\ast(x_1, ..., x_n))}$$

where $x_1, ..., x_n$ are the free variables in $\delta$. This $\delta$-rule is formulated in Beckert, Hähnle, & Schmitt (1993).

**4.5.27. (Def.)** The above entail a new rule for FO tableaux, known as *substitution rule for FO tableaux*:

$$(\mathcal{T}Sub) \quad \text{Modify } \mathcal{T} \text{ to } \mathcal{T}_\sigma$$

where $\sigma$ is free for all formulae in $\mathcal{T}$.

$\mathcal{T}Sub$ is specific to FO tableaux with unification. As a matter of fact, it is a "destructive" rule, whereas all previous rules are conservative in

## 4. Logical decisions

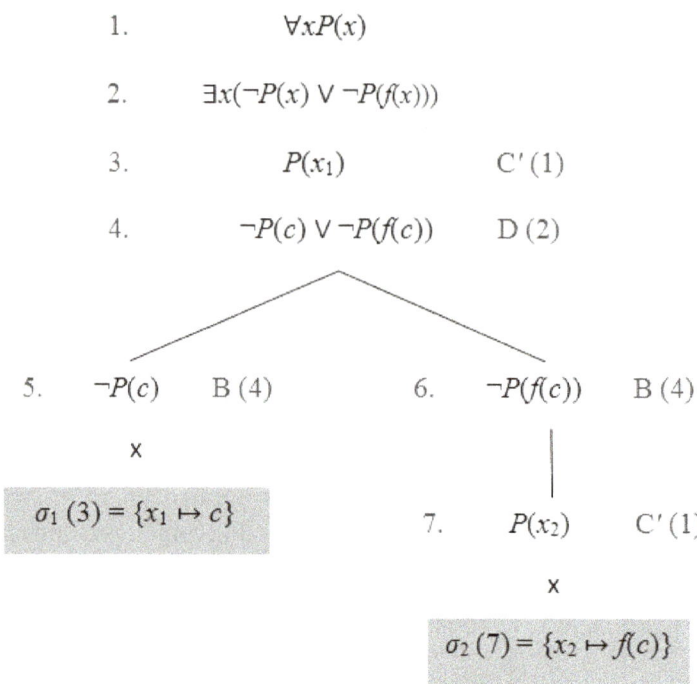

Figure 4.5.5.: An FO tableau with unification.

the sense that the initial tableau is not modified but merely expanded.[22]

In the case of free-variable tableaux, a reformulation of satisfiability is required (Hähnle & Schmitt, 1994):

**4.5.28.** (Def.) $\forall$-*satisfiability* – A collection $\mathscr{C}$ of sets of FO formulae is said to be $\forall$-satisfiable if there is an interpretation $\mathcal{I}$ such that, for every variable assignment $\varpi$, $\mathcal{I}$ is a $\{\varpi\}$-model for some element of $\mathscr{C}$.

Then, for closed formulae, for some element of $\mathscr{C}$ we have it that $\forall$-satisfiability coincides with satisfiability. Propositions 4.5.13 and 4.5.21 can easily be reformulated for $\forall$-satisfiability.

We finally rephrase the closure rule above (Def. 4.5.23) for our FO language $\mathsf{L}^*$ with formulae with free variables as follows:

**4.5.29.** (Def.) Given a tableau $\mathcal{T}_\Gamma$ for $\Gamma \subseteq F_{\mathsf{L}^*}$ and a branch $\mathcal{B}$ thereof, let $L, \neg L \in (\mathcal{B} \cup \Gamma)$, $L$ is a literal. Let further $L$ and $\neg L$ be unifiable by means of the MGU $\sigma$. Then $\mathcal{B}$ is a closed branch of $\mathcal{T}_{\sigma\Gamma}$, the tree for $\sigma\Gamma$ constructed by applying $\sigma$ to all formulae of $\Gamma$. If all branches of $\mathcal{T}_{\sigma\Gamma}$ are closed, then $\mathcal{T}^\times_{\sigma\Gamma}$ is a closed tableau for $\sigma\Gamma$.

**Example 4.5.3.** Figure 4.5.5 shows a proof by means of a tableau

---
[22] See Letz (1999) for an elaboration.

with unification that the set $A = \{\forall x\, (P(x)), \exists x\, (\neg P(x) \vee \neg P(f(x)))\}$ is unsatisfiable. Rule D' was not applied, as $\exists x \delta(x)$ does not fall within the scope of a universal quantifier.

## 4.6. Refutation II: Resolution

We introduce this Section by giving the general algorithm for binary resolution; the specifications for the specific sub-calculi will be given in the corresponding sub-sections. We follow closely the "classic" Chang & Lee (1973), to which we refer the reader for a comprehensive textbook on resolution.

### 4.6.1. The resolution principle for propositional logic

Davis and Putnam (1960) addressed the computational inefficiency in 4.3.7 by means of the four rules mentioned above in 4.2.2. One of them, the one-literal rule, must be briefly discussed, as the resolution principle is an extension of this rule.

**4.6.1.** The *one-literal rule* states that if there is a unit ground clause $\mathcal{C} = \|L\|$ in a set of clauses $C$, we can obtain $C'$ from $C$ by deleting the ground clauses in $C$ that contain $L$. There are then two possible cases:

1. If $C'$ is empty, then $C$ is satisfiable.

2. If $C'$ is not empty, we obtain a set $C''$ from $C'$ by deleting $\neg L$ from $C'$. $C''$ is unsatisfiable iff $C$ is.

**Example 4.6.1.** Let $C = \{\neg P \vee Q \vee \neg R, \neg P \vee \neg Q, P, R, U\}$. We have the set $C' = \{\neg P \vee Q \vee \neg R, \neg P \vee \neg Q, R, U\}$ by applying the one-literal rule for $P$, and then $C'' = \{Q \vee \neg R, \neg Q, R, U\}$ by 4.6.1.2. We have $C \equiv_{sat} C''$. We repeat the rule for $R$ in $C''$ and obtain the set $C''' = \{Q, \neg Q, U\}$. We have $C \equiv_{sat} C'''$. As $C'''$ contains the empty clause ($Q \wedge \neg Q = \square$; cf. 3.5.4), $C$ is unsatisfiable because $C'''$ is unsatisfiable.

Recall the definition of contradiction (Def. 3.2.11.2). Recall also that the empty clause is always false (Prop. 2.4.3), thus equating with a contradiction (cf. $\square_{df}$). Given this, we have from Theorem 3.2.8 that a set of clauses is unsatisfiable iff we can derive from it the empty clause.

**4.6.2.** By extending the one-literal rule to *any* pair of clauses, we obtain the *resolution principle*: For any two clauses $\mathcal{C}_1$ and $\mathcal{C}_2$ and two complementary literals $L_1 \in \mathcal{C}_1$ and $L_2 \in \mathcal{C}_2$, delete $L_1$ and $L_2$ from

## 4. Logical decisions

---
**Algorithm 4.4** Binary resolution proof procedure
---

**Input:** A set of clauses $C$

**Output:** An inverted binary tree $\mathcal{T}_C$ for $C$ that constitutes a proof of validity of $C$

---

1. Input the (factorized) clauses $\mathcal{C}_1, ..., \mathcal{C}_n \in C$, for finite $n \geq 2$, as the *leaves* of an inverted binary tree.

2. Resolve a pair of clauses $\mathcal{C}_i, \mathcal{C}_j$ for $1 \leq i \leq n, 1 \leq j \leq n$, and $i \neq j$, for some $l, m$ such that $L_l \in \mathcal{C}_i$ and $L_m \in \mathcal{C}_j$ are complementary literals.

3. The derived clause $\mathcal{C}_{n+1}$–called resolvent–is a *node* in the tree and is added to the search space of Step 2, so that now we have $\mathcal{C}_1, ..., \mathcal{C}_{n+1}$.

4. Repeat Steps 2 and 3 until (i) the empty clause is derived or (ii) there are no more pairs of clauses to be resolved, in which cases the *root* of the tree is respectively (i) $\square$ or (ii) some non-empty resolvent $\mathcal{C}_{n+r}$ for some $r$.

    a) If (i), then $C$ is unsatisfiable.

    b) Otherwise, $C$ is satisfiable.

$C_1$ and $C_2$, respectively, and construct the disjunction of the remaining clauses $C_1' \vee C_2'$.

**4.6.3. (Def.)** The constructed clause, $C_1' \vee C_2'$, is called a *resolvent* of $C_1$ and $C_2$.

**Theorem 4.6.1.** *A resolvent* $C = C_1' \vee C_2'$ *of two clauses* $C_1 = C_1' \vee L$ *and* $C_2 = C_2' \vee \neg L$ *is a logical consequence of* $C_1 \wedge C_2$, *i.e.*

$$\frac{C_1' \vee L \quad C_2' \vee \neg L}{C_1' \vee C_2'}.$$

*Proof:* Let $C_1 = L \vee C_1'$, $C_2 = \neg L \vee C_2'$, and $C = C_1' \vee C_2'$, $C_1'$ and $C_2'$ are disjunctions of literals. Supposing that $C_1$ and $C_2$ are both true in an interpretation $\mathcal{I}$, their resolvent $C$ must also be true in $\mathcal{I}$. Obviously, either $L$ or $\neg L$ is false in $\mathcal{I}$. Assume $L$ is false in $\mathcal{I}$; then $C_1$ must not be a unit clause, otherwise it would be false in $\mathcal{I}$. Hence, $C_1'$ must be true in $\mathcal{I}$, and the resolvent $C_1' \vee C_2'$ is true in $\mathcal{I}$. Assume $\neg L$ is false in $\mathcal{I}$ and proceed in the same way. Hence, $C_1' \vee C_2'$ is true in $\mathcal{I}$. **QED**

**4.6.4. (Def.)** A *resolution deduction* of $C$ from a set of clauses $C$, denoted by $C \vdash_{res} C$, is a finite sequence $C_1, C_2, ..., C_k$ of clauses such that each $C_i$ is either a clause in $C$ or a resolvent of clauses preceding $C_i$, and $C_k = C$. We call the deduction of the empty set $\square$ from $C$ a *refutation*, or *proof* of $C$.[23]

**4.6.5.** We can represent a resolution deduction by means of a *deduction*, or *refutation*, *tree*.

**Example 4.6.2.** Let $C = \{P \vee Q, \neg P \vee Q, P \vee \neg Q, \neg P \vee \neg Q\}$. Figure 4.6.1 shows the refutation tree for $C$.

### 4.6.2. The resolution principle for FOL

We now consider the resolution principle for FOL. In order to do so we require the definitions of substitution and unification (cf. Section 2.6).

**4.6.6. (Def.)** A *factor* of a clause $C$ is a clause $C\sigma$, where $\sigma$ is a MGU of some $C' \subseteq C$. If $C\sigma$ is a unit clause, then it is called a *unit factor* of $C$.

**4.6.7. (Def.)** Let $C_1$ and $C_2$ be two clauses (called *parent clauses*) with no variables in common. Let $L_1$ and $L_2$ be two complementary literals in $C_1$ and $C_2$, respectively. Then the clause

---
[23] Compare with Def.s 3.3.4-5.

## 4. Logical decisions

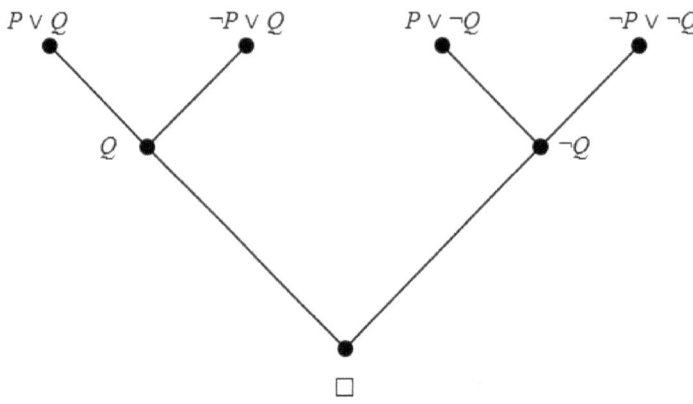

Figure 4.6.1.: A refutation tree.

$$(C_1 - L_1)\sigma \cup (C_2 - L_2)\sigma$$

where $\sigma$ is a MGU of $L_1$ and $L_2$, is called a *binary resolvent* of $C_1$ and $C_2$, and the literals $L_1$ and $L_2$ are the *literals resolved upon*.

**4.6.8. (Def.)** A *resolvent* of (parent) clauses $C_1$ and $C_2$ is one of the following binary resolvents:

1. a binary resolvent of $C_1$ and $C_2$.

2. a binary resolvent of $C_1$ and a factor of $C_2$.

3. a binary resolvent of a factor of $C_1$ and $C_2$.

4. a binary resolvent of a factor of $C_1$ and a factor of $C_2$.

**Theorem 4.6.2.** *(Binary resolution).* A resolvent $C = C'_1 \vee C'_2$ of two (parent) clauses $C_1 = C'_1 \vee L_1$ and $C_2 = C'_2 \vee L_2$ of first-order logic is a logical consequence of $C_1$ and $C_2$, if there is a substitution $\sigma$ such that $\sigma$ unifies the pair of complementary literals $L$ and $\neg L$, i.e.

$$\frac{C'_1 \vee L \quad C'_2 \vee \neg L}{(C'_1 \vee C'_2)\sigma}.$$

*Proof:* The proof is left as an exercise.

## 4.6. Refutation II: Resolution

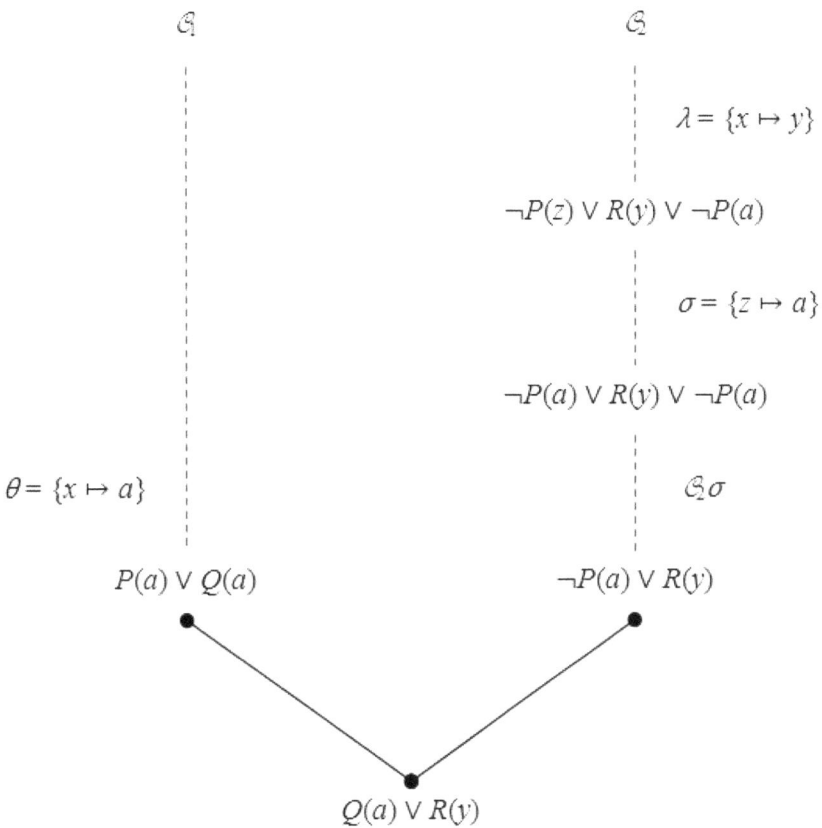

Figure 4.6.2.: A refutation-failure tree.

**Example 4.6.3.** Let $C_1 = P(x) \vee Q(x)$ and $C_2 = \neg P(z) \vee R(x) \vee \neg P(a)$. In order to apply binary resolution to this pair of clauses we first must rename the variable $x$ in $C_2$; renaming $x$ as $y$ will do. We next factor $\neg P(z)$ and $\neg P(a)$ in $C_2$, namely by applying the substitution set $\sigma = \{z \mapsto a\}$. We now have $C_1 = P(x) \vee Q(x)$ and $C_2 = \neg P(a) \vee R(y)$. Given the MGU $\theta = \{x \mapsto a\}$, we then have the binary resolvent $Q(a) \vee R(y)$. Obviously, the set $C = (C_1, C_2)$ is satisfiable. Figure 4.6.2 shows the refutation failure in a deduction tree.

### 4.6.3. Completeness of the resolution principle

Two rules of inference describe in an essential way the resolution calculus: binary resolution (Theorem 4.6.2) and (positive) factoring (Def. 4.6.6). In order to prove the completeness of the resolution principle we

4. Logical decisions

require a further theorem known as the lifting lemma:

**Theorem 4.6.3.** *(Lifting lemma) Let $C_1', C_2'$ be instances of $C_1$ and $C_2$, respectively. If $C'$ is a resolvent of $C_1'$ and $C_2'$, then there exists a resolvent $C$ of $C_1$ and $C_2$ such that $C'$ is an instance of $C$.*

*Proof:* (Sketch) Let

$$C' = \left(C_1'\gamma - L_1'\gamma\right) \cup \left(C_2'\gamma - L_2'\gamma\right)$$

where $\gamma$ is a MGU of $L_1'$ and $\neg L_2'$. We need to show that $C'$ is an instance of $C$,

$$C = ((C_1\lambda)\sigma - L_1\sigma) \cup ((C_2\lambda)\sigma - L_2\sigma)$$

where $\sigma$ is a MGU of $L_1$ and $\neg L_2$ and $\lambda_i$ is a MGU for $\{L_i^1, ..., L_i^{r_i}\}$ with $L_i = L_i^1 \lambda_i$, $i = 1, 2$, if $r_i > 1$; if $r_i = 1$, then let $\lambda_i = \epsilon$ and $L_i = L_i^1 \lambda_i$. In order to do so it suffices to show that there is a substitution $\theta$ such that $C_1' = C_1\theta$, $C_2' = C_2\theta$ and $(\lambda \circ \sigma) \leq_s (\theta \circ \gamma)$. **QED**

**Theorem 4.6.4.** *(Completeness of the resolution principle) A set $C$ of clauses is unsatisfiable iff there is a deduction of $\square$ from $C$.*

*Proof:* We provide a sketch of the proof.

($\Rightarrow$) Given that a contradiction (the empty clause) can be deduced from any unsatisfiable set of clauses, the search for one proceeds by saturating the clause set, i.e. by systematically and exhaustively applying the inference rules until the empty clause is derived. In terms of a semantic tree, this means that the tree for a set $C$ consisting only of the root node is generated, after, by Theorem 4.6.3, a process of obtaining subsequently smaller trees $\mathcal{T}_C', \mathcal{T}_C'', ...$ for $C \cup \{\mathcal{C}\}$, where $\mathcal{C}$ is a resolvent of clauses $C_1$ and $C_2$. The root node is generated only when $\square$ is derived. Therefore, there is a deduction of $\square$ from $C$.

($\Leftarrow$) Suppose there is a deduction of $\square$ from $C$. Let $R_1, ..., R_k$ be the resultants in the deduction. Assume $C$ is satisfiable. Then there is a model $\mathcal{M}$ of $C$. By Theorem 4.6.1, $\mathcal{M}$ satisfies $R_1, ..., R_k$. But this is impossible, because one of these resolvents is $\square$. Therefore, $C$ must be unsatisfiable. **QED**

**4.6.9. (Prop.)** Let now $\Psi$ be a complete resolution prover. Applying $\Psi$ to a set of clauses $C$ will produce *one* of the following results:

1. Derivation of $\square$, i.e. $\square \in \Psi(C)$. Obviously, $C$ is unsatisfiable.

2. $\Psi(C)$ is finite (i.e. given $C$, $\Psi$ is verified to terminate) and does not contain a refutation of $C$. Given that $\Psi$ is complete, we know that $C$ is satisfiable.

3. $\Psi(C)$ is infinite and does not contain a refutation of $C$. Given $C$, it is verified that $\Psi$ does not terminate. Given that $\Psi$ is complete, $C$ must be satisfiable.

*Proof:* Left as an exercise.

### 4.6.4. Resolution refinements

Although binary resolution is indeed a very efficient proof procedure, it often produces redundant resolvents. In order to minimize this problem we apply resolution refinements.

**4.6.10. (Def.)** Let $Res_{rf}$ be a mapping from the set $\mathscr{C}$ of all finite sets of clauses to the set $Res$ of all resolution deductions (i.e. Res-deductions). Let further $\varrho$ denote a ground resolution. We say that $Res_{rf}$ is a *resolution refinement* iff for every set of clauses $C \in \mathscr{C}$,

1. we have $Res_{rf}(C) \subseteq Res(C)$;
2. the set $\{\varrho | \varrho \in Res_{rf}(C)\}$ is decidable;
3. there is an algorithm $\xi$ that constructs $Res_{rf}(C)$;
4. if $C_1 \subseteq C_2$, then we have $Res_{rf}(C_1) \subseteq Res_{rf}(C_2)$.

There are actually many resolution refinements, but we discuss here only those that are relevant for resolution in many-valued logics (Chapter 8). See Chang & Lee (1973) and Leitsch (1997) for many other important resolution refinements.

**4.6.11. (Def.)** A resolution refinement $Res_{rf}$ is said to be *complete* if, given an arbitrary unsatisfiable set of clauses $C$, $Res_{rf}(C)$ contains a refutation of $C$.

Recall Proposition 4.6.9.

**4.6.12. (Prop.)** Let $Res_{rf}$ be a complete resolution refinement. Given a set of clauses $C$ as input, there are three possible outputs:

1. $Res_{rf}(C)$ terminates with $\square$, i.e. $\square \in Res_{rf}(C)$. Obviously, $C$ is unsatisfiable.

2. $Res_{rf}(C)$ terminates but does not produce $\square$. Given that $Res_{rf}$ is complete, we know that $C$ is satisfiable.

4. Logical decisions

3. $Res_{rf}(C)$ does not terminate. $Res_{rf}(C)$ is infinite and $\Box \notin Res_{rf}(C)$. Then $C$ must be satisfiable, but $Res_{rf}(C)$ does not allow us to detect this property.

*Proof:* Left as an exercise.

### 4.6.4.1. A-ordering

An important refinement in the resolution calculus is *A-ordering* that, as the very expression indicates, has to do with an order imposed on atoms.[24]

**4.6.13. (Def.)** Let $F_0$ be the set of all atomic formulae and let $\sigma$ be some substitution. An *A-ordering* (i.e. *atom ordering*) $<_A$ is a binary relation on $F_0$ such that

1. $<_A$ is

   a) irreflexive
   b) transitive

2. for every $A, B \in F_0$, $A < B$ entails $\sigma A <_A \sigma B$.

Given condition 4.6.13.1(a), it is obvious that $A$ and $B$ are *not* unifiable, i.e. $\sigma A \neq \sigma B$. The most important point with respect to $<_A$ is the specification of the ordering; this falls typically on the complexity of the terms and of the atom variables.

**4.6.14. (Def.)** We denote by $\vartheta(\cdot)$ and define *term depth*

1. of a term as
$$\vartheta(t) = 0$$
for $t \in (Vi \cup Cons)$; for $f \in Fun$, $t_1, ..., t_n$ are terms, we have
$$\vartheta(f(t_1, ..., t_n)) = 1 + max\{\vartheta(t_i) | i = 1, ..., n\}.$$

---
[24] We follow here Leitsch (1997) very closely, including examples, as we see no need to improve on his discussion of this refinement.

2. of a literal $L$ as

$$\vartheta(L) = max\{\vartheta(t) \,|\, t \in arg(L)\}.$$

3. of a clause $\mathcal{C}$ as

$$\vartheta(\mathcal{C}) = max\{\vartheta(L) \,|\, L \in \mathcal{C}\}.$$

4. of a set of clauses $C$ as

$$\vartheta(C) = max\{\vartheta(\mathcal{C}) \,|\, \mathcal{C} \in C\}.$$

**4.6.15. (Def.)** We denote by $\vartheta_{max}(x, E)$ and define *maximal depth of a variable $x$ in an expression $E$* as,

1. for $E$ a term $t$,

$$\vartheta_{max}(x,t) = \begin{cases} 0 & \text{if } * \\ 1 + max\{\vartheta_{max}(x, t_i) \,|\, i = 1, ..., n\} & \text{if } ** \end{cases},$$

$$* = x \neq Vi(t) \text{ or } x = t$$
$$** = x \in Vi(t) \text{ and } t = f(t_1, ..., t_n), f \in Fun.$$

2. for $E$ an atom $P(t_1, ..., t_n)$,

$$\vartheta_{max}(x, P(t_1, ..., t_n)) = max\{(x, t_i) \,|\, i = 1, ..., n\}.$$

3. for $E$ a literal $L$,

$$\vartheta_{max}(x, L) = \vartheta_{max}(x, arg(L)).$$

4. for $E$ a clause $\mathcal{C} = L_1 \vee ... \vee L_n$,

$$\vartheta_{max}(x, L_1 \vee ... \vee L_n) = max\{\vartheta_{max}(x, L_i) \,|\, i = 1, ..., n\}.$$

**Example 4.6.4.** Let $A = P(x, f(f(y)))$, $B = Q(f(x))$, and $\mathcal{C} = \|A, \neg B\|$. Then, $\vartheta(A) = 2$, $\vartheta(B) = 1$, $\vartheta_{max}(x, \mathcal{C}) = 1$, and $\vartheta_{max}(y, \mathcal{C}) = 2$.

**4.6.16.** Given these definitions, we can now specify an ordering $<_\alpha$ such that, for every atoms $A$ and $B$ we have $A <_\alpha B$ iff

4. Logical decisions

1. $\vartheta(A) < \vartheta(B)$, and

2. for every $x \in Vi(A)$ we have $\vartheta_{max}(x, A) < \vartheta_{max}(x, B)$ (entailing that $Vi(A) \subseteq Vi(B)$).

**4.6.17.** The above conditions (4.6.16.1-2) guarantee the irreflexivity and the transitivity of $<_\alpha$. If both conditions are satisfied, then for all substitutions $\sigma$ and for all $y \in Vi(\sigma A)$ we have

1. $\vartheta(\sigma A) < \vartheta(\sigma B)$ and

2. $\vartheta_{max}(y, \sigma A) < \vartheta_{max}(y, \sigma B)$.

**Example 4.6.5.** $P(x, x) <_\alpha Q(f(x), y)$ because $\vartheta(P(x, x)) = 0$, $\vartheta(Q(f(x), y)) = 1$, $\vartheta_{max}(P(x, x)) = 0$, and $\vartheta_{max}(Q(f(x), y)) = 1$; also, for any substitution $\sigma = \{x \mapsto t\}$ for any term $t$, $\vartheta(\sigma[P(x, x)]) < \vartheta(\sigma[Q(f(x), y)])$ and $\vartheta_{max}(\sigma[P(x, x)]) < \vartheta_{max}(\sigma[Q(f(x), y)])$.

**Example 4.6.6.** $P(x, f(a)) \not<_\alpha Q(x, f(x))$ because condition 4.6.16.1 is violated. $P(x, a) \not<_\alpha P(f(a), x)$, since condition 4.6.16.2 is violated.

**4.6.18. (Def.)** We say that a clause $C$ is *condensed* if there is no factor of $C$ that is a proper subclass of $C$, and we say that $C'$ is *the condensation of* $C$ if $C'$ is a condensation of $C$ such that $C' \subseteq C$.

**4.6.19.** Condensations are unique up to renaming. The *condensation rule* stipulates that in a proof procedure a clause be immediately replaced by its condensation, if it exists.

**Example 4.6.7.** The clause $\|P(x, y), P(y, x)\|$ is condensed. On the contrary, $\|P(x, y), P(x, a)\|$ is not condensed; its condensation is $\|P(x, a)\|$.

**4.6.20. (Def.)** Let $C$ be a set of condensed clauses and $<_A$ an A-ordering. Let $C$ be a resolvent of clauses $C_1, C_2 \in C$. Then, (the condensation of) $C$ is a $<_A$-*resolvent* of $C_1$ and $C_2$ if there is no literal $L$ in $C$ such that $B <_A L$ for $B$ the resolved atom in the resolution of $C_1$ and $C_2$.

We denote the fact that $C$ is a $<_A$-resolvent of a set of clauses $C$ by $C \in \rho <_A (C)$.

**Example 4.6.8.** Let $C_1 = \|Q(f(x_1), x_1), \neg R(f(x_1))\|$ and $C_2 = \|R(f(x_1)), \neg Q(x_1, x_2)\|$. Clearly, $C_1$ and $C_2$ are condensed. Let now $C = \{C_1, C_2\}$; we want to obtain a resolvent $C \in \rho <_A (C)$.

We begin by renaming the variables, obtaining, for instance, $C'_1 = \|Q(f(x),x), \neg R(f(x))\|$ and $C'_2 = \|R(f(y)), \neg Q(y,z)\|$. There are now two possible ways to obtain a resolvent $C \in \rho <_\alpha (C)$: given $\sigma = \{y \mapsto x\}$, $\sigma = mgu\left(C'_1, C'_2\right)$, we obtain $C^* = Q(f(x), x) \vee \neg Q(x, z)$; with $\theta = \{y \mapsto f(x), z \mapsto x\}$, $\theta = mgu\left(C'_1, C'_2\right)$, we obtain $C_* = \neg R(f(x)) \vee R(f(f(x)))$. However, only for $C^*$ do we have $C^* \in \rho <_\alpha (C)$, as $R(f(x)) \not<_\alpha L$ for $L = Q(f(x), x)$ or $L = \neg Q(x, z)$; on the other hand, $C_* \notin \rho <_\alpha (C)$ because $Q(f(x), x) <_\alpha R(f(f(x)))$.

This last example shows that we have to use a-posteriori criteria and that, among these, we have to choose the strongest one. An a-priori criterion for $<_\alpha$ would not have barred the resolution of $C_*$, even because a priori there is no $<_\alpha$-ordering relation between the atoms of $C_1$ and $C_2$.

Recall from above the definition of a proof or derivation (Def. 3.3.5). The following should be read with this definition in mind.

**4.6.21. (Def.)** Let $C$ be a set of condensed clauses and $<_A$ an A-ordering. A *Res $<_A$-deduction* of a condensed clause $C$ is a sequence $C_1, ..., C_n$ such that

1. $C_n = C$,

2. and for every $i = 1, ..., n$, either $C_i \in C$ or $C_i \in \rho <_A(\{C_j, C_k\})$ for any $j, k < i$.

**Theorem 4.6.5.** *(Completeness of Res $<_A$-deduction)* Let $C$ be a finite set of condensed clauses and $<_A$ an A-ordering. If $C$ is unsatisfiable, then there is a Res $<_A$-refutation of $C$.

*Proof:* ($\Rightarrow$) We build a closed semantic tree (cf. Def. 4.3.5): Given the set of atoms $At(C) = \{A_1, ..., A_n\}$ and an ordering such that $A_i < A_j$ for $i < j \leq n$, we can build a semantic tree with the labels $A_1$ and $\neg A_1$ in the two edges starting immediately at the root node and with the labels $A_n$ and $\neg A_n$ in the edges of failure nodes, it being the case that every branch has a failure node. Recall Theorem 4.3.4. By this theorem, every failure node falsifies a clause of $C$. A semantic tree $\mathcal{T}_C$ is closed if all its branches end in a failure node. If we take a closed semantic tree for $C$, by reverting its building process up to its collapse in the empty clause (its root), i.e. $\square \in Res <_A$, we show that there cannot fail to exist a *Res $<_A$-deduction* from $C$ (cf. Theorem 4.6.4).

4. Logical decisions

($\Leftarrow$) The proof runs as for Theorem 4.6.4, namely by Definition 4.6.20 and by the fact that $\square \in Res <_A$ for any $C_i = \square$. **QED**

### 4.6.4.2. Hyper-resolution and semantic resolution

*Hyper-resolution* is a kind of *macro-resolution*, the contraction of a sequence of resolution steps into a single inference.

**4.6.22. (Def.)** Let $C$ be any non-positive clause and let $\mathcal{D}_1, ..., \mathcal{D}_n$ be positive clauses, $\mathcal{C}, \mathcal{D}_1, ..., \mathcal{D}_n \in C$. Then, $\Xi = (\mathcal{C}; \mathcal{D}_1, ..., \mathcal{D}_n)$ is a *clash sequence* in which $\mathcal{C}$ is the *nucleus* and $\mathcal{D}_1, ..., \mathcal{D}_n$ are the *satellites*. Let $C_0 = \mathcal{C}$ and $C_{i+1} \in Res(\{C_i, \mathcal{D}_{i+1}\})$ for $i = 1, ..., n-1$. If $C_n$ is defined and positive, then we say that $C_n$ is a *hyper-resolvent* of $\Xi$. We denote by $R_H(C)$ the set of all the hyper-resolvents of a set of clauses $C$ and we say that the corresponding operator $R_H^*$ is a *hyper-resolution operator*.

**Example 4.6.9.** Let $C = \{C_1, C_2, C_3, C_4\}$, $C_1 = P(a, b)$, $C_2 = P(b, a)$, $C_3 = \neg P(x, y) \vee \neg P(y, z) \vee P(x, z)$, and $C_4 = \neg P(a, a)$. In the following resolution refutation (Fig. 4.6.3), one of the resolving clauses is always positive. In effect, it is a resolution refutation of $\Xi = (C_3; C_1, C_2)$, in which $C_1$ and $C_2$ are the satellites (or electrons).

We say that $C_6 = P(a, a)$ is a hyper-resolvent of $\Xi = (C_3; C_1, C_2)$ insofar as we can say that $C_5 = \neg P(x, b) \vee P(x, a)$ is an intermediate result with respect to $C_6$. Note that in $C_5$ the negative literal belongs to the nucleus and the positive literal is in fact the satellite $(C_1) \lambda$ (or $(C_2) \lambda$) for $\lambda = \{b \mapsto a\}$. The *macro-resolvents* of $C$ are thus $C_6$ and $C_7$. We have then $R_H^*(C) = C \cup \{P(a, a), \square\}$, it being the case that all the produced clauses have at most one atom.

**4.6.23.** This is an example of *positive hyper-resolution*, because all the electrons and all the hyper-resolvents are positive. In the case of *negative hyper-resolution*, the resolvents and the electrons are negative. In either case, it is all about imposing an interpretation when resolving a set of clauses, reason why hyper-resolution is in fact a kind of *semantic resolution*.

**4.6.24. (Def.)** Given a set of clauses $C$ and the set $Pred(C)$ of predicates of $C$, let $\mathcal{I} = (\mathcal{D}, \Theta, \varpi)$ be an interpretation such that for every $m$-place predicate $P \in Pred(C)$ we have $\Theta(P)(val_\mathcal{I}(a_1), ..., val_\mathcal{I}(a_m)) = $ f for every $a_1, ..., a_m \in \mathcal{D}$. Then all the positive clauses are false in $\mathcal{I}$ and the remaining clauses are true in $\mathcal{I}$.

**4.6.25. (Def.)** Let $C$ be a set of clauses and $\mathcal{I}$ an interpretation for $C$. Let $C_1$ and $C_2$ be clauses in the signature of $C$ such that at least one

## 4.6. Refutation II: Resolution

| | | |
|---|---|---|
| $C_1$ | $P(a,b)$ | |
| $C_2$ | $P(b,a)$ | |
| $C_3$ | $\neg P(x,y) \vee \neg P(y,z) \vee P(x,z)$ | |
| $C_4$ | $\neg P(a,a)$ | |
| $C_5$ | $\neg P(x,b) \vee P(x,a)$ | Resolvent of $(C_2, C_3)\,\sigma$, $\sigma = \{y \mapsto b, z \mapsto a\}$ |
| $C_6$ | $P(a,a)$ | Resolvent of $(C_1, C_5)\,\theta$, $\theta = \{x \mapsto a\}$ |
| $C_7$ | $\square$ | Resolvent of $C_4$ and $C_6$ |

Figure 4.6.3.: Hyper-resolution of $\Xi = (C_3; C_1, C_2)$.

of $\{C_1, C_2\}$ is false in $\mathcal{I}$. Then, we say that all the resolvents of $C_1$ and $C_2$ are $\mathcal{I}$-*semantic resolvents*. Consider now the set $C \cup D$ where $C$ is the set of true clauses in $\mathcal{I}$ and $D$ the set of false clauses in $\mathcal{I}$. Let

$$\ddagger_{\mathcal{I}}(D) = \{E | E \text{ is a } \mathcal{I} - \text{semantic resolvent of } D\}.$$

We define the *semantic resolution operator* $\ddagger_{\mathcal{I}}^*$ as

$$\ddagger_{\mathcal{I}}^*(D) = D \cup \ddagger_{\mathcal{I}}(D).$$

**4.6.26. (Def.)** Given a set of clauses $C$, let $\mathcal{I}$ be an interpretation and $\mathscr{P}$ an ordering of $P, Q, \ldots \in Pred(C)$. A finite set of clauses of $C$ constitutes a *semantic clash* (or a *PI-clash*) $\Xi_{\mathscr{P}\mathcal{I}} = (N; E_1, \ldots, E_n)$ in which $N$ denotes the nucleus and $E_1, \ldots, E_n$ the electrons or satellites of $\Xi(C)$ iff

1. $E_1, \ldots, E_n$ are false in $\mathcal{I}$;

2. Let $C_1 = N$. Por every $i = 1, \ldots, n$ there is a resolvent $C_{i+1}$ of $C_i$ and $E_i$;

3. The resolved literal of $E_i$ contains the highest predicate symbol in $E_i$, $i = 1, \ldots, n$;

4. $C_{n+1}$ is false in $\mathcal{I}$.[25]

$C_{n+1}$ is called the *PI-resolvent* of $\Xi_{\mathscr{P}\mathcal{I}}(C) = (N; E_1, \ldots, E_n)$.

---
[25]Recall that the empty clause is always false in any interpretation.

4. Logical decisions

**Example 4.6.10.** Let $E_1 = P \vee R$, $E_2 = Q \vee R$, and $N = \neg P \vee \neg Q \vee R$. Given the interpretation $\mathcal{I} = \{\neg P, \neg Q, \neg R\}$ and the ordering $P > Q > R$, $(N; E_1, E_2)$ is a PI-clash and the resolvent of this clash is $R$. Note that $R$ is false in $\mathcal{I}$.

Besides the restriction concerning the signs of the clash sequence, hyper-resolution imposes other restrictions on the space of the selection of the clauses to resolve, hindering in this way the production of many redundant formulae and their addition to the search space.

We prove solely the completeness of semantic resolution, of which hyper-resolution is a kind. Because by applying any interpretation $\mathcal{I}$ and any ordering $\mathscr{P}$ we can always obtain a PI-deduction of the empty clause from an unsatisfiable set of ground clauses, we prove the completeness of semantic resolution via a proof of PI-resolution.

**Theorem 4.6.6.** *(Completeness of PI-resolution) If $\mathscr{P}$ is an ordering of predicate symbols in an unsatisfiable finite set $C$ of clauses and if $\mathcal{I}$ is an interpretation on $C$, then there is a PI-deduction of $\square$ from $C$.*

*Proof:* The proof is by induction on the number of atoms of $C$. Let $C$ be an unsatisfiable set of ground clauses. Let $At(C) = \{P\}$. Then, $C = \{P, \neg P\} = C'$. Clearly, the resolvent from $C'$ is $\square$, regardless of the interpretation $\mathcal{I}$ (i.e. $\mathcal{I} = \{P\}$ or $\mathcal{I} = \{\neg P\}$). Therefore, either $P$ or $\neg P$ is false in $\mathcal{I}$ and $\square$ is equally false in $\mathcal{I}$, and we have it that $\square$ is a PI-resolvent.

We thus showed that the theorem holds for $n = 1$. Assume now that the theorem holds for $|At(C)| = i$, $1 \leq i \leq n$. In order to complete the induction, we shall consider that $|At(C)| = n+1$. We start by searching for a unit clause $C = L$ that is false in $\mathcal{I}$.

(i) $C$ contains a unit clause $\|L\|$ that is false in $\mathcal{I}$. Then, by deleting in $C$ the clauses containing $L$ and by deleting $\neg L$ from the remaining clauses of $C$, we obtain $C'$. Clearly, $C'$ is unsatisfiable (cf. one-literal rule: 4.6.1). $C'$ contains $n$ or fewer than $n$ atoms, so by the induction hypothesis there is a PI-deduction of $\square$ from $C'$. Let us denote this PI-deduction by $D'$. We can obtain from $C$ a PI-deduction of $\square$ from $D'$: it suffices to replace every clash sequence $\left(N'; E'_1, ..., E'_q\right)$, in which $N', E'_1, ..., E'_q$ are clauses connected to the initial nodes of $D'$ and $N'$ was obtained from $N$ by deleting $\neg L$, by the PI-clash sequence $\left(N, L; E'_1, ...E'_q\right)$. If $E'_i$ was obtained from $E_i$ by deleting $\neg L$ in it, we add the PI-clash $(L; E_i)$ above the node of $E'_i$. In this way, we obtain a PI-deduction of $\square$ from $C$.

(ii) $C$ does not contain a unit clause $\|L\|$ that is false in $\mathcal{I}$. Then, we can obtain a PI-deduction $D'$ of $\square$ from $C'$, in which $C'$ is obtained by applying the one-literal rule to a literal $L$ that is the symbol of the lowest predicate in some set $\{P, \neg P\} \subseteq At(C)$ and is false in $\mathcal{I}$. Clearly, $C'$ is unsatisfiable. $C'$ contains $n$ or fewer than $n$ atoms, so by the induction hypothesis there is a PI-deduction $D'$ of $\square$ from $C'$. Replace now again literal $L$ in the clauses from which it was firstly removed and denote by $D_1$ the deduction obtained from $D'$ by means of this operation: $D_1$ remains a PI-deduction, given that $L$ contains the lowest predicate symbol and is false in $\mathcal{I}$. It is evident that either $D_1 = \square$ or $D_1 = L$. In the first case, the proof is finished. In the second case, by (i) we obtain a PI-deduction $D_2 = \square$ from $C \cup \{L\}$ in which $L$ is a unit clause and false in $\mathcal{I}$. By the combination of $D_1$ and $D_2$ we obtain a PI-deduction of $\square$ from $C$. **QED**

Herbrand's theorem (version 2; Theorem 4.3.2) and the lifting lemma (Theorem 4.6.3) assure us that this result (the completeness of PI-resolution) holds for any unsatisfiable set of clauses, i.e. a set of non-ground clauses (cf. Chang & Lee, 1973, p. 107).

### 4.6.5. Implementation of resolution in Prover9-Mace4

The fundamental aspects of resolution proof procedures were given above. As said, the group at the Argonne National Laboratory developed a refutation-based prover for first-order logic, Otter, which worked by implementing resolution proof procedures. Its present-day successor is known as Prover9-Mace 4,[26] and it implements *binary resolution* and *unit deletion*, a first-order generalization of the Davis-Putnam one-literal rule (it also implements proof procedures for equational logic, i.e. paramodulation). Importantly, Prover9-Mace4 is highly efficient for resolution theorem proving, as besides making the necessary transformations into clause logic, it also carries out a unification algorithm, being thus completely autonomous as far as resolution proof procedures are concerned.

We show how Prover9-Mace4 implements resolution in theorem proving in FOL by beginning with a basic geometric property (Example 4.6.11) and by ending with a notable case of FOL complexity that succumbs to resolution (Example 4.6.13).

---

[26] Prover9-Mace4 is a free, and therefore open source, software available at:
https://www.cs.unm.edu/~mccune/prover9/download/

## 4. Logical decisions

---
Assumptions
```
% Definition of a trapezoid
all x all y all u all v (T(x,y,u,v)->(P(x,y,u,v)).
% Alternate interior angles of parallel lines are equal
all x all y all u all v (P(x,y,u,v)->E(x,y,v,u,v,y)).
% Trapezoid in consideration
T(a,b,c,d).
```

Goals
```
% Alternate interior angles formed by a diagonal of a
trapezoid are equal
E(a,b,d,c,d,b).
```
---

Figure 4.6.4.: Input in Prover9-Mace4.

**Example 4.6.11.** We want to prove that alternate interior angles formed by a diagonal of a trapezoid are equal. Let the alternate angles be $abd$ and $cdb$. Let $T(x, y, u, v)$ mean that $xyuv$ is a trapezoid with upper-left vertex $x$, upper-right vertex $y$, lower-right vertex $u$, and lower-left vertex $v$; we represent the property that the line segment $xy$ is parallel to the line segment $uv$ as $P(x, y, u, v)$. We represent the equality of the angles $xyz$ and $uvw$ as $E(x, y, z, u, v, w)$. We then have the following axioms:

$$(T_1) \quad \forall x \forall y \forall u \forall v (T(x,y,u,v) \to P(x,y,u,v))$$
$$(T_2) \quad \forall x \forall y \forall u \forall v (P(x,y,u,v) \to E(x,y,v,u,v,y))$$
$$(T_3) \quad T(a,b,c,d)$$

And we want to prove that $E(a, b, d, c, d, b)$ is true, and

$$\vdash (T_1 \land T_2 \land T_3) \to E(a,b,d,c,d,b).$$

is a valid formula. We insert the axioms and the conclusion to be proved in Prover9-Mace4 as shown in Fig. 4.6.4 in the window Formulas (of course, one can omit the comments, the lines starting with %).

Because Prover9-Mace4 negates the goal, we can just leave the goals box empty and enter the conclusion to be proved in the negated form after all the axioms have been entered, i.e. as

$$T_1 \land T_2 \land T_3 \land \neg E(a,b,d,c,d,b).$$

An analysis of the output of Prover9-Mace4 in Figure 4.6.5, which is easily readable, shows the implementation of the resolution principle; in

## 4.6. Refutation II: Resolution

```
% -------- Comments from original proof --------
% Proof 1 at 0.01 (+ 0.08) seconds.
% Length of proof is 10.
% Level of proof is 4.
% Maximum clause weight is 0.
% Given clauses 0.

1 (all x all y all z all u (T(x,y,z,u) -> P(x,y,z,u))) #
  label(non_clause). [assumption].
2 (all x all y all z all u (P(x,y,z,u) -> E(x,y,u,z,u,y)))
  # label(non_clause). [assumption].
3 E(a,b,d,c,d,b) # label(non_clause) # label(goal).
  [goal].
4 T(a,b,c,d). [assumption].
5 -T(x,y,z,u) | P(x,y,z,u). [clausify(1)].
6 P(a,b,c,d). [resolve(4,a,5,a)].
7 -P(x,y,z,u) | E(x,y,u,z,u,y). [clausify(2)].
8 E(a,b,d,c,d,b). [resolve(6,a,7,a)].
9 -E(a,b,d,c,d,b). [deny(3)].
10 $F. [resolve(8,a,9,a)].
```

Figure 4.6.5.: Output by Prover9-Mace4.

effect, the prover deduces the empty clause (denoted by $) from the set of propositions (assumptions and conclusion) that constitute the theory and which it clausifies.

**Example 4.6.12.** Let the following theory be given:

$(P_1)$ $\quad \forall x \left( F(x) \to \exists y \left( G(y) \wedge H(x,y) \right) \wedge \exists y \left( G(y) \wedge \neg H(x,y) \right) \right)$
$(P_2)$ $\quad \exists x \left( J(x) \wedge \forall y \left( G(y) \to H(x,y) \right) \right)$
$(P_3)$ $\quad \exists x \left( J(x) \wedge \neg F(x) \right)$

We want to know whether the conclusion is valid. The repetition of atoms in $P_1$ makes this a somehow complex theory to prove "by hand," reason why Prover9-Mace4 comes in handy. In merely 14 steps, and by applying binary resolution alone, we have a proof of validity (Fig. 4.6.6).

**Example 4.6.13.** We now increase the degree of complexity of a theory. In 1978, L. Schubert proposed a challenge to automated theorem proving that would become known in the literature as *Schubert's steamroller*. This is indeed a complex theory, namely in combinatorial

## 4. Logical decisions

```
% -------- Comments from original proof --------
% Proof 1 at 0.00 (+ 0.06) seconds.
% Length of proof is 14.
% Level of proof is 5.
% Maximum clause weight is 0.
% Given clauses 0.

1 (all x (F(x) -> (exists y (G(y) & H(x,y))) & (exists y
(G(y) &
-H(x,y))))) # label(non_clause). [assumption].
2 (exists x (J(x) & (all y (G(y) -> H(x,y))))) # la-
bel(non_clause).
[assumption].
3 (exists x (J(x) & -F(x))) # label(non_clause) # la-
bel(goal).
[goal].
4 -J(x) | F(x).  [deny(3)].
7 -F(x) | G(f2(x)).  [clausify(1)].
8 -F(x) | -H(x,f2(x)).  [clausify(1)].
10 J(c1).  [clausify(2)].
12 -J(x) | G(f2(x)).  [resolve(4,b,7,a)].
13 -J(x) | -H(x,f2(x)).  [resolve(4,b,8,a)].
15 -G(x) | H(c1,x).  [clausify(2)].
16 G(f2(c1)).  [resolve(12,a,10,a)].
17 -H(c1,f2(c1)).  [resolve(13,a,10,a)].
19 H(c1,f2(c1)).  [resolve(16,a,15,a)].
20 $F.  [resolve(19,a,17,a)].
```

Figure 4.6.6.: Output by Prover9-Mace4.

## 4.6. Refutation II: Resolution

> *Wolves, foxes, birds, caterpillars, and snails are animals, and there are some of each of them. Also there are some grains, and grains are plants. Every animal either likes to eat all plants or all animals much smaller than itself that like to eat some plants. Caterpillars and snails are much smaller than birds, which are much smaller than foxes, which are in turn much smaller than wolves. Wolves do not like to eat foxes or grains, while birds like to eat caterpillars but not snails. Caterpillars and snails like to eat some plants. Prove there is an animal that likes to eat a grain-eating animal.*

Figure 4.6.7.: Schubert's steamroller in natural language.

terms. Despite this complexity, the challenge was solved by a resolution calculus in Walther (1985). We provide the text in English (Fig. 4.6.7) and our translation into FOL (Fig. 4.6.8). Figure 4.6.9 shows the output of Prover9-Mace4.

This proof is particularly interesting as, despite the complexity of the theory, it is massively carried out by means of resolution, with only a few applications of *unit deletion*, and of *unit resulting resolution*, which, just like hyper-resolution, allows for the resolution in a single step of more than one nucleus clause with other satellite clauses.

## 4. Logical decisions

(P$_1$)   $\forall x \left( W\left( x\right) \rightarrow A\left( x\right) \right) \land \exists x\left( W\left( x\right) \right)$
(P$_2$)   $\forall x \left( F\left( x\right) \rightarrow A\left( x\right) \right) \land \exists x\left( F\left( x\right) \right)$
(P$_3$)   $\forall x \left( B\left( x\right) \rightarrow A\left( x\right) \right) \land \exists x\left( B\left( x\right) \right)$
(P$_4$)   $\forall x \left( C\left( x\right) \rightarrow A\left( x\right) \right) \land \exists x\left( C\left( x\right) \right)$
(P$_5$)   $\forall x \left( S\left( x\right) \rightarrow A\left( x\right) \right) \land \exists x\left( S\left( x\right) \right)$
(P$_6$)   $\exists x \left( G\left( x\right) \right) \land \forall x \left( G\left( x\right) \rightarrow P\left( x\right) \right)$
(P$_7$)   $\forall x \left( A\left( x\right) \rightarrow \left( \forall y \left( P\left( y\right) \rightarrow Eats\left( x,y\right) \right) \lor \forall y \left( \left( A\left( y\right) \land Smaller\left( y,x\right) \land \exists z \left( P\left( z\right) \land Eats\left( y,z\right) \right) \right) \rightarrow Eats\left( x,y\right) \right) \right) \right)$
(P$_8$)   $\forall x \forall y \left( C\left( x\right) \land B\left( y\right) \rightarrow Smaller\left( x,y\right) \right)$
(P$_9$)   $\forall x \forall y \left( S\left( x\right) \land B\left( y\right) \rightarrow Smaller\left( x,y\right) \right)$
(P$_{10}$)  $\forall x \forall y \left( B\left( x\right) \land F\left( y\right) \rightarrow Smaller\left( x,y\right) \right)$
(P$_{11}$)  $\forall x \forall y \left( F\left( x\right) \land W\left( y\right) \rightarrow Smaller\left( x,y\right) \right)$
(P$_{12}$)  $\forall x \forall y \left( W\left( x\right) \land \left( F\left( y\right) \lor G\left( y\right) \right) \rightarrow \neg Eats\left( x,y\right) \right)$
(P$_{13}$)  $\forall x \forall y \left( B\left( x\right) \land C\left( y\right) \rightarrow Eats\left( x,y\right) \right)$
(P$_{14}$)  $\forall x \forall y \left( B\left( x\right) \land S\left( y\right) \rightarrow \neg Eats\left( x,y\right) \right)$
(P$_{15}$)  $\forall x \left( C\left( x\right) \rightarrow \exists y \left( P\left( y\right) \land Eats\left( x,y\right) \right) \right)$
(P$_{16}$)  $\forall x \left( S\left( x\right) \rightarrow \exists y \left( P\left( y\right) \land Eats\left( x,y\right) \right) \right)$
(P$_{17}$)  $\underline{\exists x \exists y \left( A\left( x\right) \land A\left( y\right) \land \exists z \left( G\left( z\right) \land \overline{Eats}\left( y,z\right) \right) \right) \land Eats\left( x,y\right) )}$

Figure 4.6.8.: Schubert's steamroller in FOL.

## 4.6. Refutation II: Resolution

```
% -------- Comments from original proof --------
% Proof 1 at 0.01 (+ 0.06) seconds.
% Length of proof is 65.
% Level of proof is 9.
% Maximum clause weight is 13.
% Given clauses 31.

1 (all x (W(x) -> A(x))) & (exists x W(x)) # label(non_clause).  [assumption].
2 (all x (F(x) -> A(x))) & (exists x F(x)) # label(non_clause).  [assumption].
3 (all x (B(x) -> A(x))) & (exists x B(x)) # label(non_clause).  [assumption].
5 (all x (S(x) -> A(x))) & (exists x S(x)) # label(non_clause).  [assumption].
6 (exists x G(x)) & (all x (G(x) -> P(x))) # label(non_clause).  [assumption].
7 (all x (A(x) -> (all y (P(y) -> Eats(x,y))) | (all y (A(y) & Smaller(y,x) & (exists z (P(z) & Eats(y,z)))
-> Eats(x,y))))
# label(non_clause).  [assumption].
9 (all x all y (S(x) & B(y) -> Smaller(x,y))) # label(non_clause).  [assumption].
10 (all x all y (B(x) & F(y) -> Smaller(x,y))) # label(non_clause).  [assumption].
11 (all x all y (F(x) & W(y) -> Smaller(x,y))) # label(non_clause).  [assumption].
12 (all x all y (W(x) & (F(y) | G(y)) -> -Eats(x,y))) # label(non_clause).  [assumption].
14 (all x all y (B(x) & S(y) -> -Eats(x,y))) # label(non_clause).  [assumption].
16 (all x (S(x) -> (exists y (P(y) & Eats(x,y))))) # label(non_clause).  [assumption].
17 (exists x exists y (A(x) & A(y) & (exists z (G(z) & Eats(y,z))) & Eats(x,y))) # label(non_clause) # label(goal).  [goal].
18 W(c1).  [clausify(1)].
19 -W(x) | A(x).  [clausify(1)].
20 -F(x) | -W(y) | Smaller(x,y).  [clausify(11)].
21 -W(x) | -F(y) | -Eats(x,y).  [clausify(12)].
22 -W(x) | -G(y) | -Eats(x,y).  [clausify(12)].
23 F(c2).  [clausify(2)].
24 -F(x) | A(x).  [clausify(2)].
25 -B(x) | -F(y) | Smaller(x,y).  [clausify(10)].
26 -F(x) | Smaller(x,c1).  [resolve(20,b,18,a)].
27 -F(x) | -Eats(c1,x).  [resolve(21,a,18,a)].
28 B(c3).  [clausify(3)].
29 -B(x) | A(x).  [clausify(3)].
31 -S(x) | -B(y) | Smaller(x,y).  [clausify(9)].
33 -B(x) | -S(y) | -Eats(x,y).  [clausify(14)].
34 -B(x) | Smaller(x,c2).  [resolve(25,b,23,a)].
41 S(c5).  [clausify(5)].
42 -S(x) | A(x).  [clausify(5)].
43 -S(x) | P(f2(x)).  [clausify(16)].
44 -S(x) | Eats(x,f2(x)).  [clausify(16)].
45 -S(x) | Smaller(x,c3).  [resolve(31,b,28,a)].
46 -S(x) | -Eats(c3,x).  [resolve(33,a,28,a)].
47 -G(x) | P(x).  [clausify(6)].
48 G(c6).  [clausify(6)].
49 -A(x) | -A(y) | -G(z) | -Eats(y,z) | -Eats(x,y).  [deny(17)].
50 -G(x) | -Eats(c1,x).  [resolve(22,a,18,a)].
51 Smaller(c2,c1).  [resolve(26,a,23,a)].
52 -A(x) | -P(y) | Eats(x,y) | -A(z) | -Smaller(z,x) | -P(u) | -Eats(z,u) | Eats(x,z).  [clausify(7)].
53 Smaller(c3,c2).  [resolve(34,a,28,a)].
54 Smaller(c5,c3).  [resolve(45,a,41,a)].
55 A(c1).  [resolve(18,a,19,a)].
56 A(c2).  [resolve(23,a,24,a)].
57 -Eats(c1,c2).  [resolve(27,a,23,a)].
58 A(c3).  [resolve(28,a,29,a)].
63 A(c5).  [resolve(41,a,42,a)].
64 P(f2(c5)).  [resolve(43,a,41,a)].
65 Eats(c5,f2(c5)).  [resolve(44,a,41,a)].
66 -Eats(c3,c5).  [resolve(46,a,41,a)].
67 P(c6).  [resolve(47,a,48,a)].
68 -A(x) | -A(y) | -Eats(y,c6) | -Eats(x,y).  [resolve(49,c,48,a)].
69 -Eats(c1,c6).  [resolve(50,a,48,a)].
70 -A(c1) | -P(x) | Eats(c1,x) | -A(c2) | -P(y) | -Eats(c2,y) | Eats(c1,c2).  [resolve(51,a,52,e)].
71 -P(x) | Eats(c1,x) | -P(y) | -Eats(c2,y).[copy(70),unit_del(a,55),unit_del(d,56),unit_del(g,57)].
72 -A(c2) | -P(x) | Eats(c2,x) | -A(c3) | -P(y) | -Eats(c3,y) | Eats(c2,c3).  [resolve(53,a,52,e)].
73 -P(x) | Eats(c2,x) | -P(y) | -Eats(c3,y) | Eats(c2,c3).  [copy(72),unit_del(a,56),unit_del(d,58)].
74 -A(c3) | -P(x) | Eats(c3,x) | -A(c5) | -P(y) | -Eats(c5,y) | Eats(c3,c5).  [resolve(54,a,52,e)].
75 -P(x) | Eats(c3,x) | -P(y) | -Eats(c5,y).  [copy(74),unit_del(a,58),unit_del(d,63),unit_del(g,66)].
79 -P(x) | Eats(c2,x) | -Eats(c3,x) | Eats(c2,c3).  [factor(73,a,c)].
86 -Eats(c2,c6).  [ur(71,a,67,a,b,69,a,c,67,b)].
92 -P(x) | Eats(c3,x).  [resolve(75,d,65,a),unit_del(c,64)].
94 Eats(c3,c6).  [resolve(92,a,67,a)].
98 Eats(c2,c3).  [resolve(94,a,79,c),unit_del(a,67),unit_del(b,86)].
103 $F.  [ur(68,a,56,a,b,58,a,c,94,a),unit_del(a,98)].
```

Figure 4.6.9.: Proof of Schubert's steamroller by Prover9-Mace4.

## 4. Logical decisions

## Exercises

**Exercise 4.1.** Decide on the truth value in CPL of the following formulae by means of truth tables:

1. $\neg(((A \leftrightarrow B) \leftrightarrow (\neg C \wedge D)) \vee \neg E)$
2. $(\neg(P \vee Q) \wedge \neg(R \vee S)) \rightarrow (R \rightarrow \neg S)$
3. $((P \vee \neg(R \wedge S)) \rightarrow (Q \vee T)) \vee (\neg(P \wedge \neg S))$
4. $(A \rightarrow (B \wedge C)) \rightarrow (A \rightarrow C)$
5. $(A \leftrightarrow \neg(A \vee B)) \wedge \neg A$
6. $(P \vee Q) \rightarrow (P \vee \neg P)$
7. $((A \rightarrow B) \rightarrow A) \rightarrow B$

**Exercise 4.2.** Decide whether the following derivations are valid in CPL by means of the different validity proof systems in Section 4.4:

1. $A \leftrightarrow B, C \leftrightarrow D \vdash A \leftrightarrow D$
2. $\neg A \vee B \vdash C$
3. $\neg A \vdash A \rightarrow \neg A$
4. $A \wedge B \vdash B \leftrightarrow A$
5. $\neg \phi \rightarrow \phi \vdash \phi$
6. $\phi \wedge \psi \models \phi \rightarrow \psi$
7. $\neg P \vee Q \models \neg(P \wedge \neg Q)$
8. $P \rightarrow (Q \wedge R), (\neg Q \vee \neg R) \Vdash \neg P$

**Exercise 4.3.** Verify whether the following facts hold (or not) in the deductive system $\mathsf{L} = (\mathsf{L}, \mathscr{L})$:

1. $\vdash_\mathsf{L} A \rightarrow A$ (we omit the subscript L henceforth)
2. $\vdash \neg A \rightarrow (A \rightarrow B)$
3. $A \rightarrow B, \neg A \rightarrow B \vdash B$

4. $A \to C, C \to B \vdash \neg(\neg B \land A)$

5. $\vdash \neg(A \land C) \to (C \to \neg A)$

6. $\neg A \to \neg B \vdash B \to A$

7. $A \to C \vdash (B \land A) \to (B \land C)$

8. $\vdash (\neg A \to A) \to A$

9. $\vdash \neg A \to (\neg C \to \neg(A \lor C))$

10. $\vdash \neg\neg\neg A \to \neg A$

**Exercise 4.4.** Prove the following arguments in the axiom system $\mathscr{L}^*$ over the language L*:

1. All men are mortal. All philosophers are men. ∴ All philosophers are mortal.

2. Some cats have no tail. All cats are mammals. ∴ Some mammals have no tail.

3. No logic exercise is fun. Some thoughts are logic exercises. ∴ Some thoughts are not fun.

**Exercise 4.5.** Show that the following formulae are theorems in L* by means of the proof system $\mathcal{L}^*$:

1. $\forall x (A \to B) \to (\exists x A \to \exists x B)$

2. $\neg \forall x A \to \exists x (\neg A)$

3. $\forall x (A \to B) \to (\forall x A \to \forall x B)$

4. $\forall x A \to \forall x (A \lor B)$

5. $(A \to \forall x B) \leftrightarrow \forall x (A \to B)$

**Exercise 4.6.** Give proofs in $\mathcal{LK}$ of the following formulae:

1. $A \to (B \to (A \land B))$

2. $A \to (B \to A)$

## 4. Logical decisions

3. $(A \to B) \lor (B \to A)$

4. $\forall x A(x) \to A(t)$

5. $B(t) \to \exists x B(x)$

6. $\exists x A(x) \to \forall y A(y)$

**Exercise 4.7.** Check the validity of the following formulae and arguments by means of the axiom system $\mathcal{L}^*$ and of the calculi $\mathcal{NK}$ and $\mathcal{LK}$:

1. $\forall x F(x) \to \neg \exists x (\neg F(x))$

2. $\forall x (F(x) \to G(x)) \to (\forall x (F(x)) \to \forall x (G(x)))$

3. $\exists y \forall x (P(x,y) \leftrightarrow P(x,x)) \to \neg \forall x \exists y \forall z (P(z,y) \leftrightarrow \neg P(z,x))$

4. $\forall x ((P(x) \lor Q(x)) \to \neg R(x)) \vdash \forall x P(x)$

5. $\forall x ((R(x) \to \neg P(x)) \to (Q(x) \land S(x))) \vdash \forall x P(x)$

6. $\exists x \exists y (P(x,y)) \to \forall x \exists x (P(x,y))$

7. $\exists x (Q(x) \to \forall x Q(x))$

8. $\forall x \exists y \forall z \exists w (P(x,y) \lor \neg P(w,z))$

**Exercise 4.8.** Check the validity of the formulae and arguments in Exercises 4.1-7 by means of analytic tableaux.

**Exercise 4.9.** Formulate and prove the soundness and completeness theorems for FO tableaux with unification.

**Exercise 4.10.** Reformulate Propositions 4.5.13 and 4.5.21 for $\forall$-satisfiability. Prove the reformulated propositions.

**Exercise 4.11.** For each of the following sets of clauses $C$, find the Herbrand universe $H_C$ and the Herbrand base $H(C)$:

1. $C = \{P(x,y) \lor \neg Q(b), \neg P(a,x) \lor Q(b)\}$

2. $C = \{P(x, f(y)), P(z, g(z))\}$

3. $C = \{P(a, f(x,y)), P(b, f(x,y))\}$

**Exercise 4.12.** Find the resolvents of the following sets of clauses:

1. $\{\neg P(x) \vee Q(x, b), P(a) \vee Q(a, b)\}$
2. $\{\neg P(x) \vee Q(x, y), \neg Q(a, g(a))\}$
3. $\{\neg S(x, y, u) \vee \neg S(y, z.v) \vee \neg S(x, v, w) \vee S(u, z, w), S(h(x, y), x, y)\}$
4. $\{\neg R(v, z, v) \vee R(w, z, w), R(a, f(b, b), a)\}$

**Exercise 4.13.** Check the validity of the formulae and arguments in Exercises 4.1-7 by means of the resolution calculus.

**Exercise 4.14.** Let $C = \{P, \neg P \vee Q, R \vee \neg P, \neg P \vee \neg Q \vee \neg R\}$. Prove that the set $C$ is unsatisfiable by means of semantic resolution. The following cases should be considered:

1. $\mathcal{I} = \{\neg P, \neg Q, \neg R\}$; $R < Q < P$.
2. $\mathcal{I} = \{P, Q, R\}$; $R < P < Q$.
3. $\mathcal{I} = \{\neg P, \neg Q, R\}$; $P < Q < R$.

**Exercise 4.15.** Prove the (in)validity of the following theories by using the prover Prover9-Mace4. Then, analyze the output of the prover in detail.

1. $\Theta_1$:

    $Q \to R$
    $R \to (P \wedge Q)$
    $P \to (Q \vee R)$
    $\therefore P \leftrightarrow Q$

2. $\Theta_2$:

    $\exists x (P \to F(x))$
    $\exists x (F(x) \to P)$

## 4. Logical decisions

$$\therefore \exists x \, (P \leftrightarrow F(x))$$

3. $\Theta_3$:

$$\neg \exists x \, (S(x) \wedge Q(x))$$
$$\forall x \, (P(x) \rightarrow (Q(x) \vee R(x)))$$
$$\neg \exists x P(x) \rightarrow \exists x Q(x)$$
$$\forall x \, (Q(x) \vee R(x) \rightarrow S(x))$$
$$\therefore \exists x \, (P(x) \wedge R(x))$$

4. $\Theta_4$:

$$\exists x P(x) \leftrightarrow \exists x Q(x)$$
$$\forall x \forall y \, (P(x) \wedge Q(y) \rightarrow (R(x) \leftrightarrow S(y)))$$
$$\therefore \forall x \, (P(x) \rightarrow R(x)) \leftrightarrow \forall x \, (Q(x) \rightarrow S(x))$$

5. $\Theta_5$:

$$\forall z \exists w \forall x \exists y \, ((P(x,z) \rightarrow P(y,w)) \wedge P(y,z) \wedge (P(y,w) \rightarrow \exists u Q(u,w)))$$
$$\forall x \forall z \, (\neg P(x,z) \rightarrow \exists y \, (Q(y,z)))$$
$$\exists x \exists y \, (Q(x,y)) \rightarrow \forall x \, (R(x,x))$$
$$\therefore \forall x \exists y \, (R(x,y))$$

**Exercise 4.16.** Prove (Complete the proof of) the propositions and theorems in this Chapter that were left without a proof (with a sketchy proof, respectively).

# Part II.
# MANY-VALUED LOGICS

# 5. Many-valued logics

In this Chapter, we provide a comprehensive elaboration on both relevant aspects with respect to the many-valued logics, from *historical remarks* (Section 5.1) to *interpretation issues* and *structural features* (Sections 5.2-3), and many of the most important *many-valued logic systems* and respective *calculi* and *semantics* (Sections 5.4-6). We end this Chapter with the central topic of *quantification* in many-valued logics (Section 5.7).

## 5.1. Some historical notes

Interestingly enough, Aristotle, credited in western civilization for conceiving bivalent logic, was also the first to question the bivalent status of propositions about future events. In his *De interpretatione* (IX, 19a30), he wrote:

> A sea-fight must either take place tomorrow or not, but it is not necessary that it should take place tomorrow, neither is it necessary that it should not take place, yet it is necessary that it either should or should not take place tomorrow.

These propositions appear to be neither necessarily true nor necessarily false, which challenges the principle of bivalence and other pillars of classical logic, namely the principle of excluded middle (PEM; cf. Prop. 3.5.3, T1), stating that for any proposition, either it is true, or its negation is. In fact, in the above passage Aristotle seems to conceive a third, intermediate, truth value that can be expressed as *indeterminate* or *undetermined*.

Generally known as *problem of future contingents*, this was a much discussed topic in medieval logic, but theoretical progress can be said to have taken place only in the early 20th century, after G. Boole and G. Frege's fundamental formalization developments in mathematical logic. For instance, C. S. Peirce, one of "the founding fathers of many-valued logic,"[1] wrote some notes about his thoughts on a third truth value he called *limit*:

---

[1] Together with H. MacColl and N. Vasilev, according to Rescher (1969).

## 5. Many-valued logics

> Triadic logic is that logic which, though not rejecting entirely the Principle of Excluded Middle, nevertheless recognizes that every proposition, $S$ is $P$, is either true, or false, or else $S$ has a lower mode of being such that it can neither be determinately $P$, nor determinately not $P$, but is at the limit between $P$ and not $P$. (Peirce, 1909/1966, MS 339)

Despite this and a few other incipient conceptions (see Rescher, 1969), real progress in the establishment of many-valued logics was made only with the Polish logician J. Lukasiewicz around 1920. His motivations were philosophical in essence, concerning fatalism and determinism, but the mentioned formal developments also allowed more (meta-)mathematical preoccupations to be formulated in, and approached from, a many-valued logical perspective, as was the case with E. Post (1921) and, later on, S. Kleene (1938, 1952).

While both problems remain at the basis of the contemporary motivations of many-valued logics, these have grown in number and diversity: in effect, there seem not only to be *truth-value gaps*, as in the case of future contingents, but also *truth-value gluts*, or true contradictions (e.g., inconsistent laws; paradoxes of self-reference), aspects that raise important philosophical questions that bear on both legal and scientific matters (e.g., Priest, 2008). These are joined by further (meta-)mathematical preoccupations, many of which have increasingly to do with computational properties and (practical applications in) computer science in general (e.g., Bolc & Borowik, 2003; Epstein, 1993; Fitting & Orłowska, 2003).

Be it as it may, the many-valued logics are today essential elements of mathematical logic, and many are provided with adequate formalizations, though some interpretation problems remain open. We start our discussion of the most relevant many-valued logics with these.

### 5.2. Many-valuedness and interpretation

#### 5.2.1. Suszko's Thesis

Recall from Section 3.2.3 that a semantics for some algebra of formulae (a propositional language) $\mathfrak{L} = (F_\mathsf{L}, o_1, ..., o_m)$ can be provided by an interpretation structure $\mathfrak{A} = (A, f_1, ..., f_m)$, where $A$ is the range of semantical correlates of $\mathfrak{L}$. This is highly convenient when $|W| > 2$, because (via the homomorphism $Hom\,(\mathfrak{L},\mathfrak{A}))$ the semantical correlates of propositions may be set against the logical values proper, believed to be only two, to wit, `truth` and `falsity` (i.e. the range of $A$ can be reduced to the set $\{0,1\}$; see Fig. 5.2.1). For the same reason, this is highly problematic, as this appears to support the notion that logic proper is bivalent, casting doubt upon the metalogical status of many-valued systems. The problem is obviously not posed with $W_2$, as in this case the semantical correlate in $\mathfrak{A}_2$ of a formula in $\mathfrak{K}$ is necessarily either *true* or *false*.

Interpretation in many-valued logics is an open problem in contemporary logic that actually may impact on the automation of deduction. Therefore, some remarks are called for. To begin with, we note that Theorem 3.2.9 expresses the fact that every Tarskian logic has a many-valued semantics. On the other hand, Suszko, among others, has shown that any many-valued semantics can be reduced to a two-valued one, and ranted against the "multiplication of logical values" (Suszko, 1977).

**5.2.1. (Def.)** In effect, given a propositional language $\mathfrak{L}$ and a matrix $\mathfrak{M} = (\mathfrak{A}, D)$ for the same, $\phi \in F_\mathsf{L}$ is an arbitrary formula, we define the set of valuations $Val_\mathfrak{M} = \{val_h | h \in Hom\,(\mathfrak{L},\mathfrak{A})\}$ where

$$val_h(\phi) = \begin{cases} 1 & \text{if } h(\phi) \in D \\ 0 & \text{if } h(\phi) \notin D \end{cases}.$$

**5.2.2. (Prop.)** Consequently,

$$X \models_\mathfrak{M} \phi \text{ iff, for every } val_h \in Val_\mathfrak{M},$$
$$val_h(\phi) = 1 \text{ whenever } val_h(X) \subseteq \{1\}.$$

By repeating the definition of logical valuations with respect to any structural consequence operation $Cn$ by using its Lindenbaum bundle (cf. Prop. 3.2.24; Theorem 3.2.10), we obtain the important result that every structural logic $\mathsf{L} = (\mathsf{L}, Cn)$ is logically two-valued.

**5.2.3.** *Suszko's Thesis* – Suszko's stand according to which there are only two logical values, viz., `true` and `false`, is known more or less informally as Suszko's Thesis, and he took care to make this thesis nontrivial by distinguishing the two *logical* truth values from the *algebraic* truth values (cf. Def. 3.2.2) that have, for him, merely a referential role and can be in infinite number. In this view, many-valued logics are obviously also two-valued, but for $|W| > 2$ they are only *referentially* many-valued (vs. *logically* many-valued). It so happens that according to the *Fregean Axiom* (Prop. 3.2.1), two propositions in CPL with the same truth value describe the same object or fact, i.e. have the same

## 5. Many-valued logics

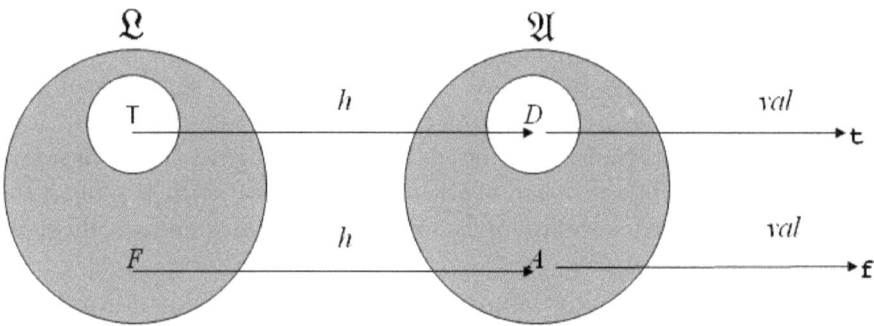

Figure 5.2.1.: The homomorphic interpretation $h \in Hom(\mathfrak{L}, \mathfrak{A})$ and the valuation $val_h : F \longrightarrow W_2$.

reference or denotation; as a consequence, from the viewpoint of classical logic, the logical truth value is the only attribute of a proposition that is of significance.[2] In this view, the logical values and the referential values coincide in CPL, but they are obviously distinct in the many-valued logics.

Suszko's concerns probably did not include the multiple applications of many-valued logics that are today essential (see Introduction), reason why he seemed so ready to dispense with them, but he highlighted the fact that a logical system is, in a perhaps fundamental way, a system of preservation of truth against the equally "noble" falsity. In a related line of argumentation, Rescher (1969) argued that while purely abstract many-valued logics may have important applications (e.g., proofs of independence of theorems in propositional logics), they may in fact be no more than mere calculi in the narrow sense of the term.

### 5.2.2. Non-trivial many-valuedness

It is obvious that logical two-valuedness of any given structural logic is a direct consequence of dividing the elements of $\mathfrak{A}$ into the two complementary sets of designated and undesignated values, $D$ and $A-D$, respectively, in turn a consequence of matrix construction (see Fig. 5.2.1). Nevertheless, it is precisely via the notion of logical matrix that we can

---

[2]This is expressed by means of the connective of material equivalence as

$$x \leftrightarrow y \in \{1\} \text{ iff } x = y$$

where the identity on the right side is the identity of the logical values (vs. identity of the propositions).

provide non-trivial criteria of many-valuedness.

In effect, logical many-valuedness just is the case when $|W| > 2$, but this is by itself not a non-trivial criterion of many-valuedness, as Suszko's Thesis claims. The most natural way of obtaining non-trivial many-valuedness is by means of logical matrices as they were defined above (Section 3.2.3). Once a propositional language–an algebra of formulae–is chosen (typically, $\mathfrak{L} = (F_L, \neg, \rightarrow, \wedge, \vee, \leftrightarrow)$ or a reduct thereof), a multiple-element algebra $\mathfrak{A} = (A, \neg, \rightarrow, \wedge, \vee, \leftrightarrow)$ similar to $\mathfrak{L}$ is defined together with the set of distinguished values $D \subsetneq A$, where $A$ is the range of semantical correlates of $\mathfrak{L}$, i.e. $\mathfrak{A}$ is an interpretation structure for $\mathfrak{L}$. We say that the resulting matrix $\mathfrak{M} = (\mathfrak{A}, D)$ defines a many-valued logic whenever the content of $\mathfrak{M}$, denoted by $E(\mathfrak{M})$, cannot be described by any two-element matrix.

In fact, two criteria are used to decide whether a matrix $\mathfrak{M}$ determines a many-valued logic:

**5.2.4. (Prop.)** A logical matrix $\mathfrak{M}$ determines a many-valued logic iff for *no* two-element matrix $\mathfrak{N}$ for $\mathfrak{L}$ it is the case that

1. $E(\mathfrak{M}) = E(\mathfrak{N})$;

2. $\models_{\mathfrak{M}} = \models_{\mathfrak{N}}$ (or, which is the same, $Cn_{\mathfrak{M}} = Cn_{\mathfrak{N}}$).

*Proof:* Left as an exercise.

Given the definitions above of $E(\mathfrak{M})$ (cf. Def. 3.2.20) and matrix consequence operation / relation (cf. Def. 3.2.21), both criteria above are in fact equivalent.

However, it is important to keep in mind that, as pointed out above, the notion of logical matrix can actually support Suszko's Thesis. With this in mind, there have been attempts to construct matrices that are not characterizable by complementary sets (e.g., Malinowski, 2012). As a matter of fact, be it as it may, it is still useful (if not actually necessary) to work in a classical bivalent framework under certain circumstances, namely with the automation of deduction of many-valued logics in view.

## 5.2.3. Classical generalizations to the many-valued logics

The same result of classical two-valuedness can be obtained by considering the truth-functionality principle at the root of the matrix description of CPL via its generalization by means of the *interpretation principle*

## 5. Many-valued logics

in Proposition 3.2.1 above. Given that most many-valued propositional logics are in fact conservative extensions of CPL matrices, we can determine the standard conditions that allow us to characterize them as bivalent. We follow Rosser & Turquette (1952).

**5.2.5. (Def.)** Given any matrix of the form

$$\mathfrak{M}_{n,k} = (\mathfrak{U}_n, D_k)$$

where $\mathfrak{U}_n = (G_n, f_1, f_2, ..., f_m)$, $G_n = \{1, 2, ..., n\}$ and $D_k = \{1, 2, ..., k\}$ for $1 \leq k < n$, $n \in \mathbb{N}$, and $n \geq 2$, $1 \leq k < n$ conveys a decreasing degree of truth with 1 always referring to truth and $n$ to falsity, we have the *general* standard conditions. The *specific* standard conditions describe the connectives that are seen as standard: Let $\neg, \rightarrow, \vee, \wedge, \leftrightarrow$ be functions of a given matrix $\mathfrak{M}_{n,k}$ and let $\{j_1, j_2, ..., j_n\}$ be a family of unary functions of $\mathfrak{U}_n$. If for any $x, y \in G_n$ and $i \in \{1, 2, ..., n\}$

| | | |
|---|---|---|
| $\neg x \in D_k$ | iff | $x \notin D_k$ |
| $x \rightarrow y \notin D_k$ | iff | $x \in D_k$ and $y \notin D_k$ |
| $x \vee y \in D_k$ | iff | $x \in D_k$ or $y \in D_k$ |
| $x \wedge y \in D_k$ | iff | $x \in D_k$ and $y \in D_k$ |
| $x \leftrightarrow y \in D_k$ | iff | either $x, y \in D_k$ or $x, y \notin D_k$ |
| $j_i(x) \in D_k$ | iff | $x = i$ |

where the $j_i$ are logical value identifiers, the $j_1, ..., j_k$ are assertions and the $j_{k+1}, ..., j_n$ are negations, then we say that the respective connectives of $\mathfrak{M}_{n,k}$ satisfy the standard conditions. Now let $O = \{\neg, \rightarrow, \vee, \wedge, \leftrightarrow\}$ and $J_n = \{j_1, j_2, ..., j_n\}$: the matrix $\mathfrak{M}_{n,k}$ with a set of definable functions $Y \subseteq O \cup J_n$ satisfying the corresponding standard conditions is called *Y-standard*, and *standard* if it is Y-standard for $Y = O \cup J_n$.

**5.2.6. (Prop.)** In turn, any O-standard matrix

$$\mathfrak{K}_{n,k} = (G_n, \neg, \rightarrow, \vee, \wedge, \leftrightarrow, D_k)$$

is epimorphic to the classical matrix $\mathfrak{M}_2$ via a mapping $h : G_n \longrightarrow W_2$ defined by

$$h(x) = \begin{cases} 1 & \text{when } x \in D_k \\ 0 & \text{when } x \notin D_k \end{cases}.$$

The result follows:

**5.2.7. (Prop.)**

$$E(\mathfrak{K}_{n,k}) = E(\mathfrak{M}_2) = \top(K).$$

## 5.2. Many-valuedness and interpretation

**5.2.8.** Rosser and Turquette showed that the content $E(\mathfrak{M}_{n,k})$ of any $(\{\to\} \cup J_n)$-standard matrix $\mathfrak{M}_{n,k}$ is axiomatizable with the following set of axioms and the rules MP and SUB:

($\mathcal{A}$1) $(B \to (A \to B))$
($\mathcal{A}$2) $(A \to (B \to C)) \to (B \to (A \to C))$
($\mathcal{A}$3) $(A \to B) \to ((B \to C) \to (A \to C))$
($\mathcal{A}$4) $(j_i(A) \to (j_i(A) \to B)) \to (j_i(A) \to C)$
($\mathcal{A}$5) $(j_n(A) \to B) \to ((j_{n-1}(A) \to B) \to (... \to (j_1(A) \to B) \to B)...)$
($\mathcal{A}$6) $j_i(A) \to A$ where $i = 1, 2, ..., k$
($\mathcal{A}$7) $j_{i(r)}(A_r) \to (j_{i(r-1)}(A_{r-1}) \to (... \to j_{i(1)}(A_1) \to j_f(\heartsuit(A_1, ..., A_r)))...)$
where $f = f(i(1), ..., i(r))$, $f$ is some function of $\mathfrak{M}_{n,k}$ and $\heartsuit$ is a propositional connective associated with $f$

The above axiom schemata and rules in fact provide an adequate axiomatization for Lukasiewicz's logics $Ł_n$ and Post's logics $P_n$. Regardless of these results, it is obvious that multiple truth values seriously challenge the important properties of tautologousness and contradictoriness and, by reflection, the important notion of entailment (semantic consequence), if the concepts of tautologousness and contradictoriness (Section 3.3) are readily generalizable (as indeed they are) to many-valued logical systems. It is easy to see that if in a many-valued logical system $S_n$ a truth-value assignment is always of some $v_i \notin W_2$, then the system has neither tautologies nor contradictions in the classical sense.

Nevertheless, we can "mitigate" this problem by proceeding as follows.

**5.2.9. (Def.)** Let $D$ be a set of designated values. Then, a formula that invariably takes (or can take)

1. some $v_i \in D$ for any and every given assignment of $v_0, v_1, ..., v_{n-1}$, $n > 2$, truth values to its propositional variables is said to be *quasi-tautologous*.

2. some $v_i \notin D$ for any and every given assignment of $v_0, v_1, ..., v_{n-1}$, $n > 2$, truth values to its propositional variables is said to be *quasi-contradictory*.

However, given that the set of non-designated values is not necessarily established as $A - D$ (that is, the *undesignated* truth values need not be the remaining truth values once the designated values have been selected; Rescher, 1969), we shall refer to the set of designated values

## 5. Many-valued logics

as $D^+$ and to the set of *anti-designated* values as $D^-$. We can now rephrase Definition 5.2.9:

**5.2.10. (Def.)** Let the sets $D^+$ and $D^-$ be the sets of designated and anti-designated values, respectively. Then, a formula that invariably takes (or can take)

1. some $v_i \in D^+$ for any and every given assignment of $v_0, v_1, ..., v_{n-1}$, $n > 2$, truth values to its propositional variables is said to be *quasi-tautologous*.

2. some $v_i \in D^-$ for any and every given assignment of $v_0, v_1, ..., v_{n-1}$, $n > 2$, truth values to its propositional variables is said to be *quasi-contradictory*.

We shall speak of designated set and use simply $D$ if distinction from the set of anti-designated values is not considered. We remark that for some $v_i \in W_n$, $n > 2$, it need not be the case that $v_i \in D^+$ iff $v_i \notin D^-$. This allows us to equate the sets of tautologies and contradictions of many-valued logical systems with those of CPL (cf. Rescher, 1969).

**Example 5.2.1.** The following connectives are such that the system they define ($B_3$; see below) can have only two-valued tautologies and only two-valued contradictions; moreover,

$$\top(B_3) = \top(K)$$

$$\bot(B_3) = \bot(K)$$

where $\top(S)$ and $\bot(S)$ designate the sets of tautologies and contradictions, respectively, of a logical system S. (Recall that K is the classical system.)

| A | ¬A | A\B | $\wedge$ | | | $\vee$ | | | $\to$ | | | $\leftrightarrow$ | | |
|---|---|---|---|---|---|---|---|---|---|---|---|---|---|---|
|   |    |    | t | i | f | t | i | f | t | i | f | t | i | f |
| +t | f | t | t | i | f | t | i | t | t | i | f | t | i | f |
| ±i | i | i | i | i | i | i | i | i | i | i | i | i | i | i |
| −f | t | f | f | i | f | t | i | f | t | i | t | f | i | t |

**5.2.11 (Def.)** *Quasi-truth-functionality (Rescher, 1969)* – A system S of propositional logic defined by truth tables that are not single-valued but in which multiple values may occur at least in some places is characterized as *quasi-truth-functional*.

## 5.2. Many-valuedness and interpretation

**Example 5.2.2.** (Rescher, 1969). The following connectives define a quasi-truth-functional system:

| A | ¬A | | A\B | ∧ t | i | f | ∨ t | i | f | → t | i | f | ↔ t | i | f |
|---|---|---|---|---|---|---|---|---|---|---|---|---|---|---|---|
| +t | f | | t | t | i | f | t | t | t | t | i | f | t | i | f |
| i | i | | i | i | i,f | f | t | i,t | i | t | i,t | i | i | i,t | i |
| −f | t | | f | f | f | f | t | i | f | t | t | t | f | i | t |

Clearly, all the above challenges the generalization of classical entailment to many-valued logics; in particular, the notion of validity needs reappraisal. From the definitions above of quasi-tautologousness and quasi-contradictoriness, the notion of *quasi-entailment* naturally follows, but we find the following definition more relevant, as it somehow joins the notions of entailment and quasi-entailment (Bergmann, 2008):

**5.2.12 (Def.)** *Degree-entailment* – A set of formulae $X$ degree-entails a formula $\phi$, denoted by $X \models_{dgr} \phi$, if the truth value of $\phi$ can never be less than the least value of all the $\chi \in X$.

**5.2.13. (Prop.)** An argument is *degree-valid* in a many-valued logical system $S_n$ if its set of premises degree-entails its conclusion.

*Proof:* Left as an exercise.

This can readily be generalized to infinitely many-valued logics:

**5.2.14. (Def.)** *n-degree-entailment* – A set of formulae $X$ n-degree-entails a formula $\phi$, denoted by $X \models_n \phi$, if $1 - n$, $n \subseteq [0, 1]$, is the maximum downward distance between $X$ and $\phi$.

The notion of *n-degree-validity* follows obviously. We remark that 1-degree-entailment (1-degree-validity) coincides with degree-entailment (degree-validity, respectively).

However, the choice of a designated set allows us to make a more general definition of validity (vs. a degree-validity definition):

**5.2.15. (Def.)** *Validity in many-valued logics* – Given a designated set $D \subset W_n, D \neq \emptyset$, we say that an inference in a many-valued system

## 5. Many-valued logics

$S_n$ is *quasi-valid*[3] if it preserves designated values, i.e. if we have:

$$X \models_D \phi \quad \text{iff for every interpretation } \mathcal{I},$$

whenever $val_\mathcal{I}(\chi) \in D$, for all $\chi \in X$, then $val_\mathcal{I}(\phi) \in D$.

### 5.3. Structural properties of many-valued logics

Above we approached *interpretational* relations between K and many-valued logics; we now introduce some important structural properties of many-valued logics, which put them in a *structural* relation with K.[4] We follow Rescher (1969) more or less closely.

**5.3.1. (Def.)** *Normality* – In a *normal* many-valued truth table, a classical truth-value assignment behaves exactly as it does in classical propositional logic: every formula that is `true` on that assignment in the many-valued logic is also `true` on that assignment in K, and every formula that is `false` on that assignment in the many-valued logic is also `false` on that assignment in K. A many-valued logic is said to be a *normal logic* if the truth tables for all its basic connectives are normal.

**5.3.2. (Def.)** *Uniformity* – A connective in a many-valued logic is said to be *uniform* if, whenever both entries (of a row/column) in the classically restricted truth table agree in the same classical truth value (`t` or `f`) for a complex formula, then the entire row/column also shows the same truth value. A many-valued logic whose truth tables for all its basic connectives are uniform is said to be a *uniform logic*.

We next introduce another important structural property of many-valued logics that will be clearer once we approach Kleene's 3-valued

---

[3]In the above exposition, the prefix *quasi* was introduced to call attention to the fact that the generalization of classical semantical notions to the many-valued logics is not without some issues. Nevertheless, the many-valued logics, whether with finite or infinite sets of truth values, are typically truth-functional logics, and, as seen above, they can be adequately characterized by a "two-valued" logical matrix. This characterization allows us to generalize notions such as validity and tautology to the many-valued logics whenever specifications are not made (e.g., *degree-validity*). Contrary to Suszko (see above), we do not think that "algebraic truth values" lack some sort of "logical" status; perhaps on the contrary, the "logical truth values" require an algebraic foundation–a discussion we are not taking up here (but see footnote to 2.1.9). This justifies the scarce–and henceforth bracketed–use of this prefix in this text.

[4]"Structural" refers here to *structure*; not to be confused with "structural" as in Definition 3.1.5.1.

logic:

**5.3.3. (Def.)** *K-regularity* – The truth table for a many-valued connective is *K-regular* if it is never the case that a classical truth value is assigned when one of the constituents of the formula is assigned the truth value i (undefined, or indeterminate), unless the classical truth value occurs uniformly in a i-row/-column. A many-valued logic is said to be a *K-regular logic* if the truth tables for all its basic connectives are so.

**5.3.4. (Def.)** *Decisiveness* – The truth table for a propositional connective of a logical system $S_n$, $n \geq 2$, is *decisive* if, for its set of formulae $F_S$ it is the case that $val : F_S \longrightarrow W_2$. A (many-valued) logic is said to be a *decisive logic* if its connectives are decisive.

**5.3.5. (Prop.)** K is decisive.
*Proof:* Left as an exercise.

## 5.4. The Łukasiewicz propositional logics

As said above, although Łukasiewicz did not exactly inaugurate the field of many-valued logics, he was the first to build thoroughly many-valued systems. We begin by introducing Łukasiewicz's 3-valued propositional logic Ł$_3$. This will allow us to introduce important generalizations concerning both finitely and infinitely many-valued logics alike.

### 5.4.1. Łukasiewicz's 3-valued propositional logic Ł$_3$

The work on many-valued logic of the Polish logician Jan Łukasiewicz dates from at least 1918, and it was in the period between 1920 and 1930 (Łukasiewicz, 1920; 1930) that the many-valued system known as Ł$_3$ was established.

**5.4.1.** Ł$_3$ has three truth values that we shall designate by f, i, and t as abbreviations for false, neither true nor false (i.e. intermediate or indeterminate), and true, respectively. (Łukasiewicz first conceived the set of truth values $\{0, 1/2, 1\}$, and we shall use it when necessary.)

The truth value i (originally: 1/2) was motivated by future-contingent propositions like "I shall be in Warsaw in a year," which appear to be *neither true nor false* (a truth-value gap) at the moment they are uttered, as it is possible, though not necessary, that the speaker will indeed be in Warsaw at the given time. In fact, if the proposition were true at the moment of speaking, then it would be necessary, which contradicts the assumption; on the other hand, if it were false, then it would be impossible that the speaker be in Warsaw at the given time, which likewise

## 5. Many-valued logics

contradicts the assumption. Thus Łukasiewicz concluded that a third, different, truth value representing "the possible" and joining "the true" and "the false" was accounted for (cf. Łukasiewicz, 1930; see transl.: 1967, p. 53).

**5.4.2.** With the future-contingency interpretation in mind, Łukasiewicz actually introduced a modal operator for *possibility* and *necessity*, which we denote by $\Diamond$ and $\Box$, respectively, for which he provided the following truth table:

| $A$ | $\Diamond A$ | $\Box A$ |
|---|---|---|
| t | t | t |
| i | t | f |
| f | f | f |

It is obvious that if we eliminate the truth value i from this table, the introduction of the modal operators becomes trivial, with $A \leftrightarrow \Diamond A$, $\Box A \leftrightarrow A$, and $\Box A \leftrightarrow \Diamond A$ all being tautologies.[5]

In order to construct Ł3 Łukasiewicz took the connectives $\neg$ and $\rightarrow$ as primitives, and, by extending their classical interpretation, put forward the truth tables for $\neg_{L3}$ and $\rightarrow_{L3}$ below.

**5.4.3.** The following are the truth tables for the primitive connectives of the logical system Ł3:

| $A$ | $\neg_{L3} A$ |
|---|---|
| t | f |
| i | i |
| f | t |

| $\rightarrow_{L3}$ | t | i | f |
|---|---|---|---|
| t | t | i | f |
| i | t | t | i |
| f | t | t | t |

**5.4.4.** The remaining three logical connectives of Ł3 were defined as follows:

$$A \vee_{L3} B \quad := \quad (A \rightarrow_{L3} B) \rightarrow_{L3} B$$
$$A \wedge_{L3} B \quad := \quad \neg_{L3}(\neg_{L3} A \vee_{L3} \neg_{L3} B)$$
$$A \leftrightarrow_{L3} B \quad := \quad (A \rightarrow_{L3} B) \wedge_{L3} (B \rightarrow_{L3} A)$$

---

[5]Malinowski (1993), p. 19ff, further discusses this topic.

## 5.4. The Łukasiewicz propositional logics

As a matter of fact, Ł3 is not functionally complete: the truth tables for $\to_{L3}$ and $\leftrightarrow_{L3}$ assign the truth value t to a formula $\phi = \chi \overset{\to}{\leftrightarrow} \psi$ when both subformulae $\chi$ and $\psi$ have the truth value i, which means that none of these connectives can be defined in terms of the set $\{\neg_{L3}, \wedge_{L3}, \vee_{L3}\}$. In particular, $\neg_{L3} A \to_{L3} B :\neq A \vee_{L3} B$.

**5.4.5.** The truth tables of the secondary connectives of Ł3 were built in the following way:

| $\vee_{L3}$ | t | i | f |
|---|---|---|---|
| t | t | t | t |
| i | t | i | i |
| f | t | i | f |

| $\wedge_{L3}$ | t | i | f |
|---|---|---|---|
| t | t | i | f |
| i | i | i | f |
| f | f | f | f |

| $\leftrightarrow_{L3}$ | t | i | f |
|---|---|---|---|
| t | t | i | f |
| i | i | t | i |
| f | f | i | t |

**5.4.6.** Wajsberg (1931) showed that the $(\neg, \to)$-fragment of Ł3 is axiomatizable in what we shall designate by $\mathscr{L}_3$ (i.e. $\mathscr{L}_3$ is sound and complete for Ł3), constituted by the axioms $\mathscr{L}_3$1-4 below and by the inference rules MP and SUB (we omit the subscripts in the connectives):

$(\mathscr{L}_3 1)$ $\quad A \to (B \to A)$
$(\mathscr{L}_3 2)$ $\quad (A \to B) \to ((B \to C) \to (A \to C))$
$(\mathscr{L}_3 3)$ $\quad (\neg A \to \neg B) \to (B \to A)$
$(\mathscr{L}_3 4)$ $\quad ((A \to \neg A) \to A) \to A$

Clearly, given the definitions above of the remaining connectives of Ł3, $\mathscr{L}_3$ extends to the whole of Ł3. Bergmann (2008) provides a complete description of $\mathscr{L}_3$, of which we shall state an important result (two further important results will be stated below):

**5.4.7. (Prop.)** Any derivation in K not involving axiom $\mathscr{L}2$ (cf. Example 4.4.1) is a derivation in $\mathscr{L}_3$, and any axiom that can be derived in K without using $\mathscr{L}2$ is a derived axiom in $\mathscr{L}_3$.
*Proof:* Left as an exercise.

**Example 5.4.1.** We show in Figure 5.4.1 the proof of $\vdash_{L3} \neg P \to (P \to Q)$ in $\mathscr{L}_3$. By, for example, $\phi/\neg P$, we indicate that $\phi$ in an axiom schema is replaced by the formula $\neg P$.

**Example 5.4.2.** The theorem $\vdash_{L3} \neg P \to (P \to Q)$ (see Example 5.4.1) is a *derived axiom* of $\mathscr{L}_3$. Further derived axioms in this axiom system are as follows:

163

## 5. Many-valued logics

1. $\neg P \to (\neg Q \to \neg P)$      $\mathscr{L}_3 1$, $A/\neg P$, $B/\neg Q$
2. $(\neg Q \to \neg P) \to (P \to Q)$      $\mathscr{L}_3 3$, $A/Q$, $B/P$
3. $(\neg P \to (\neg Q \to \neg P)) \to (((\neg Q \to \neg P) \to (P \to Q)) \to (\neg P \to (P \to Q)))$      $\mathscr{L}_3 2$, $A/\neg P$, $B/(\neg Q \to \neg P)$, $C/(P \to Q)$
4. $((\neg Q \to \neg P) \to (P \to Q)) \to (\neg P \to (P \to Q))$      MP (1, 3)
5. $\neg P \to (P \to Q)$      MP (2, 4)

Figure 5.4.1.: Proof of $\vdash_{\text{L3}} \neg P \to (P \to Q)$ in $\mathscr{L}_3$.

1. $\neg\neg P \to P$

2. $P \to \neg\neg P$

3. $P \to P$

4. $((P \to P) \to Q) \to Q$

5. $P \to ((P \to Q) \to Q)$

6. $(P \to (Q \to R)) \to (Q \to (P \to R))$

7. $\neg(P \to Q) \to P$

8. $(P \vee Q) \to (Q \vee P)$

9. $(P \to Q) \vee (Q \to P)$

10. $(P \to (P \to (P \to Q))) \to (P \to (P \to Q))$

**Example 5.4.3.** Some *derived rules* of $\mathscr{L}_3$ are:[6]

1. *Hypothetical syllogism* (HS): $A \to B, B \to C \vdash A \to C$

2. *Generalized hypothetical syllogism* (GHS):
$$(A_1 \to (A_2 \to ... \to (A_{n-1} \to A_n))), A_n \to B \vdash$$
$$(A_1 \to (A_2 \to ... \to (A_{n-1} \to B)))$$

3. *Contraposition* (ConP): $\neg A \to \neg B \vdash B \to A$

4. *Generalized contraposition* (GConP): Given a formula $A$ containing a subformula $D = B \to C$, replace one or more occurrences of $D$ in $A$ by $\neg C \to \neg B$ (and vice-versa) and infer $A'$ as a result.

5. *Double negation* (LDN): Given a formula $A$ and a subformula $B$ of $A$, replace one or more occurrences of $B$ in $A$ by $\neg\neg B$ (and vice-versa) and infer $A'$ as a result.

6. *Disjunctive consequence* (DCn): $A \to C, B \to C \vdash (A \vee B) \to C$

7. *Transposition* (TR): Given a formula $A$ with a subformula $E = B \to (C \to D)$, replace one or more occurrences of $E$ by $C \to (B \to D)$ and infer $A'$ as a result.

---
[6]See Bergmann (2008) for a more exhaustive list.

## 5. Many-valued logics

8. *Generalized MP* (GMP): Let $A = (A_1 \to (A_2 \to ... \to (A_{n-1} \to A_n)))$. From $A$ and $A_i \in A$, $1 \leq i \leq n-1$, delete $A_i$ (and associated connective and parentheses) and infer the resulting conditional.

9. *Substitution* (SBT): Given a formula $A$ containing $C$ as a subformula, from $A$, $C \to B$, and $B \to C$ infer any formula $A'$ as a result of replacing one or more occurrences of $C$ in $A$ with $B$.[7]

By augmenting the set of primitives of Ł3 with an additional connective **T**, Słupecki (1936) obtained an axiom system for the functionally complete 3-valued logic Ł3. In more detail, in order to realize that the three-element Łukasiewicz algebra $\mathscr{L}_3 = (\{0, 1/2, 1\}, \neg, \to, \vee, \wedge, \leftrightarrow)$ is not functionally complete it suffices to see that the constant function $\mathbf{T} : \mathbf{T}(x) = 1/2$ for any $x \in \{0, 1/2, 1\}$; however, adding **T** to the set of functions of Ł3 leads to a three-element functionally complete algebra, and consequently by adding the following two to the axioms of $\mathscr{L}_3$ an axiom system for the functionally complete 3-valued logic Ł3 is obtained:

$$(\mathscr{L}_3 5) \quad \mathbf{T}A \to \neg \mathbf{T}A$$
$$(\mathscr{L}_3 6) \quad \neg \mathbf{T}A \to \mathbf{T}A$$

An inspection of the truth tables for the connectives of Ł3 reveals that when the two classical truth values alone are being considered, the truth tables for each of the connectives are exactly those of K. In other words, whenever the connectives operate on formulae with classical truth values, they behave exactly as they do in classical logic.

**5.4.8. (Prop.)** Ł3 is normal.
*Proof:* The proof is by checking the truth tables of K and those of Ł3 for the assignments of the truth values **f** and **t** for the connectives of both logics. **QED**

**5.4.9. (Prop.)** Ł3 is uniform.
*Proof:* Checking the truth table of Ł3 for the connectives $\to$, $\wedge$, and $\vee$ shows that these connectives are uniform. Furthermore, the connective $\leftrightarrow$, which is not uniform, can be defined by means of $\to$ and $\wedge$. Hence, Ł3 is a uniform logic. **QED**

**5.4.10. (Prop.)** Ł3 is not K-regular.
*Proof:* The proof is by inspection of the truth tables of Ł3. **QED**

---
[7] Not to be confused with SUB, of which SBT is a particular case.

## 5.4.2. Tautologousness, contradictoriness, and entailment in Ł₃

Given the original motivations of Ł₃ (see above), it is justified that the law of the excluded middle (LEM; cf. Prop. 3.5.3, T1) does not hold in it; but neither does the principle of non-contradiction (T2), for that matter. Inversely, the formula $A \leftrightarrow \neg A$ ceases to be a contradiction in Ł₃. The following result is of import:

**5.4.11. (Prop.)** Every formula that is a tautology in Ł₃ is also a tautology in K, and every formula that is a contradiction in Ł₃ is also a contradiction in K.

*Proof:* The proof follows directly from the fact that Ł₃ is normal. **QED**

**5.4.12. (Prop.)** If $X \models_{Ł3} \phi$, then $X \models \phi$ (for unsubscripted $\models$ designating classical consequence).

*Proof:* As above. **QED**

However, it is obvious from the above that the inverse of Propositions 5.4.11-12 does not hold, i.e. it is not generally the case that (i) every formula that is a tautology (a contradiction) in K is also a tautology (a contradiction) in Ł₃, and (ii) if $X \models \phi$, then $X \models_{Ł3} \phi$.

This shows that:

**5.4.13. (Prop.)**
$$\top(Ł_3) \subsetneq \top(K)$$
and
$$\bot(Ł_3) \subsetneq \bot(K).$$

*Proof:* Left as an exercise.

## 5.4.3. N-valued generalizations of Ł₃

In the early 1930s, Łukasiewicz generalized Ł₃ to both finitely and infinitely many-valued systems. A convenient way to capture this generalization is by means of a logical matrix.

**5.4.14. (Def.)** A matrix of the form

$$\mathfrak{M}_n = (L_n, \neg, \rightarrow, \vee, \wedge, \leftrightarrow, \{1\})$$

for (i) $n \in \mathbb{N}, n \geq 2$ or (ii) $n = \aleph_0$ or $n = \aleph_1$ is called a *Łukasiewicz*

## 5. Many-valued logics

$n$-valued matrix provided that

$$L_n = \begin{cases} \{0, \frac{1}{n-1}, \frac{2}{n-1}, ..., \frac{n-2}{n-1}, 1\} & \text{if } n \in \mathbb{N}, n \geq 2 \\ \{s/w | 0 \leq s \leq w, s, w \in \mathbb{N}, w \neq 0\} & \text{if } n = \aleph_0 \\ [0, 1] & \text{if } n = \aleph_1 \end{cases}$$

for the unary function $\neg$ and the binary functions $\rightarrow, \vee, \wedge, \leftrightarrow$ ($\neg$ and $\rightarrow$ are primitives) defined on $L_n$ as follows:

1. $\neg x = 1 - x$

2. $x \rightarrow y = min(1, 1-x+y)$, or $x \rightarrow y = \begin{cases} 1 & \text{if } x \leq y \\ 1 - x + y & \text{if } x > y \end{cases}$,
   equivalently

3. $x \vee y = max(x, y)$

4. $x \wedge y = min(x, y)$

5. $x \leftrightarrow y = 1 - |x - y|$

**5.4.15. (Prop.)** If we identify truth and falsity with respectively 1 and 0 from the series

$$1 = \frac{n-1}{n-1}, \frac{n-2}{n-1}, ..., \frac{2}{n-1}, \frac{1}{n-1}, \frac{0}{n-1} = 0$$

and identify any of the intermediate values $\frac{n-2}{n-1}, ..., \frac{2}{n-1}, \frac{1}{n-1}$ with $\mathbf{i}_i$, $0 < i < 1$, we easily see that

1. $L_2 = W_2$

2. $L_3 = W_{L3}$

3. $E(\mathfrak{M}_n) \subseteq E(\mathfrak{M}_2)$, i.e. $\mathfrak{M}_2$ is a submatrix of any $\mathfrak{M}_n$.

*Proof:* Left as an exercise.

The contents of finite matrices can be related by the following theorem by Łukasiewicz & Tarski (1930):

**Theorem 5.4.1.** *(The Lindenbaum condition). For finite $n, m \in \mathbb{N}$, the following conditions are equivalent:*

## 5.4. The Łukasiewicz propositional logics

1. $E(\mathfrak{M}_n) \subseteq E(\mathfrak{M}_m)$.

2. $m-1$ *is a divisor of* $n-1$, *i.e.* $(m-1)\,|\,(n-1)$.

**5.4.16.** For $n > 3$, $n$ is finite, $L_n$ can be axiomatized by adding to $\mathscr{L}_{\aleph}1\text{-}4$ in Proposition 5.4.20 below (which for this purpose we can designate by $\mathscr{L}_n1\text{-}\mathscr{L}_n4$) the following two further axioms (Grigolia, 1977) for $1 < j < n-1$ and $j \nmid n-1$, i.e. $j$ does not divide $n-1$, and for the additional connectives $\phi + \psi := \neg\phi \to \psi$ and $\phi \cdot \psi := \neg(\phi \to \neg\psi)$, $k\phi$ is a substitution for $\underbrace{\phi + \phi + \ldots + \phi}_{k \text{ times}}$ and $\phi^k$ is a substitution for $\underbrace{\phi \cdot \phi \cdot \ldots \cdot \phi}_{k \text{ times}}$:

$$(\mathscr{L}_n 5) \quad nA \to (n-1)A$$
$$(\mathscr{L}_n 6) \quad (n-1)\left((\neg A)^j + (A \cdot (j-1)A)\right)$$

Now with respect to infinite $n$, two important results are proved in Malinowski (1993):

**5.4.17. (Prop.)**
$$E(\mathfrak{M}_{\aleph_0}) = E(\mathfrak{M}_{\aleph_1}).$$

*Proof:* Left as an exercise.

This allows us to designate $\mathfrak{M}_{\aleph_0}$ or $\mathfrak{M}_{\aleph_1}$ simply as $\mathfrak{M}_{\aleph}$.

**5.4.18. (Prop.)**
$$E(\mathfrak{M}_{\aleph_0}) = \cap\{E(\mathfrak{M}_n)\,|\,n \geq 2, n \in \mathbb{N}\}.$$

*Proof:* Left as an exercise.

By making $n - 1 = k(m - 1)$, Rescher (1969) generalizes the results of the Lindenbaum condition in the following way:

**5.4.19. (Prop.)** For every prime number $k$ we have the containment series:

$$E(\mathfrak{M}_{\aleph}) \subset E(\mathfrak{M}_{n(k+1)}) \subset \ldots \subset E(\mathfrak{M}_{2(k+1)}) \subset E(\mathfrak{M}_{1(k+1)}) \subset (E(\mathfrak{M}_2) = \top(K))$$

*Proof:* Left as an exercise.

It was conjectured by Łukasiewicz and Tarski (1930) and shown by Wajsberg (1931) that the infinitely many-valued propositional calculus $L_{\aleph}$ can be axiomatized.

## 5. Many-valued logics

**5.4.20.** With MP and SUB, the following axiom schemata constitute an axiom system for Ł$_\aleph$:[8]

$(\mathscr{L}_\aleph 1)$    $A \to (B \to A)$
$(\mathscr{L}_\aleph 2)$    $(A \to B) \to ((B \to C) \to (A \to C))$
$(\mathscr{L}_\aleph 3)$    $(\neg A \to \neg B) \to (B \to A)$
$(\mathscr{L}_\aleph 4)$    $((A \to B) \to B) \to ((B \to A) \to A)$
$(\mathscr{L}_\aleph 5)$    $((A \to B) \to (B \to A)) \to (B \to A)$

**Example 5.4.4.** We prove (Fig. 5.4.2) the theorem $\vdash_{Ł\aleph} P \to ((P \to Q) \to Q)$ in $\mathscr{L}_\aleph$.

---
1. $P \to ((Q \to P) \to P)$      $\mathscr{L}_\aleph 1$, $A/P$, $B/(Q \to P)$
2. $((Q \to P) \to P) \to ((P \to Q) \to Q)$      $\mathscr{L}_\aleph 4$, $A/Q$, $B/P$
3. $P \to ((P \to Q) \to Q)$      HS (1, 2)
---

Figure 5.4.2.: Proof of $\vdash_{Ł\aleph} P \to ((P \to Q) \to Q)$ in $\mathscr{L}_\aleph$.

**5.4.21. (Prop.)** Any derivation in $\mathscr{L}_3$ that does not use axiom $\mathscr{L}_3 4$ is a derivation in $\mathscr{L}_\aleph$.

*Proof:* Left as an exercise.

## 5.5. Finitely many-valued propositional logics

We now discuss the most important finitely many-valued propositional logics other than Łukasiewicz's. The discussion of each of the selected logics is shorter than that for Łukasiewicz's logics, and it is left as an exercise for the reader to determine the structural properties of most of these logics. However brief this discussion might be, we provide axiom systems, as well as other relevant information, for each of them.[9]

---
[8] Note that the three first axioms are shared by Ł$_\aleph$ and Ł$_3$ alike; Ł$_\aleph 5$ is dependent on the other axioms.
[9] Because Bochvar's and Kleene's connectives can be defined by means of Łukasiewicz's, inferences in their systems can be proven by means of $\mathscr{L}_3$.

## 5.5. Finitely many-valued propositional logics

### 5.5.1. Bochvar's 3-valued system

As in the case of Łukasiewicz, Bochvar's (1938) main motivation was rather philosophical, but whereas the former was concerned with truth-value gaps, the latter was more interested in tackling (apparent) truth-value gluts, cases in which it appears that a proposition is both true and false. We begin with Bochvar's *internal system*, which we shall designate by $B_3^I$ (see below the reasons for this label).

**5.5.1.** Given the truth table for conjunction (cf. Example 5.2.1),[10] and the definition of negation as in Ł3, conjunction and negation are the primitive connectives of $B_3^I$, its other connectives have the following definitions (we omit subscripts):

$$
\begin{aligned}
A \vee B &:= \neg(\neg A \wedge \neg B) \\
A \to B &:= \neg(A \wedge \neg B) \\
A \leftrightarrow B &:= (A \to B) \wedge (B \to A)
\end{aligned}
$$

**5.5.2.** A superficial inspection of the truth tables for the connectives of Bochvar's logic in Example 5.2.1 above shows that all the binary connectives of $B_3^I$, denoted by $\widetilde{\heartsuit}^2_{BI3}$, have the following pattern (besides being normal):

| $\widetilde{\heartsuit}^2_{BI3}$ |   | i |   |
|---|---|---|---|
|   |   | i |   |
| i | i | i | i |
|   |   | i |   |

This pattern concretizes the structural property defined above as K-regularity (Def. 5.3.3), a property that impacts directly on tautologousness and contradictoriness. In effect, any formula in which one of the two components is assigned the truth value i is for that reason alone assigned the truth value i, thus making those notions inoperant. Bochvar's third value, which we represent as i for reasons of uniformity, is thus so to say contagious. This is meta-theoretically explained by the fact that conjunction is a primitive connective and that Bochvar's object is the classical semantical paradoxes: generally,[11] a proposition such as "This sentence is false" cannot be attributed a single truth value (indeed, if true, this proposition is false; if false, true), being therefore *meaningless,*

---
[10] Ignore the signs +, ±, and −.
[11] More specifically, paradoxes such as Russell's and Grelling-Nelson's.

## 5. Many-valued logics

*paradoxical*, or, at best, *undecidable*. According to Bochvar, whenever a conjunct is valuated as i, this should make the whole conjunction meaningless or paradoxical.

Some important results follow.

**5.5.3. (Prop.)** $B_3^I$ has neither tautologies nor contradictions.
*Proof:* Left as an exercise.

**5.5.4. (Prop.)** If $X \models_{BI} \phi$, then $X \models \phi$.
*Proof:* Left as an exercise.

But not every entailment that holds in K holds in $B_3^I$. This shows that $B_3^I \subset K$.

Regarding Proposition 5.5.3, one could apply the concepts of quasi-tautology (quasi-contradiction) as a formula that never assumes the truth value f (respectively, t). However, Bochvar had in mind two levels of assertion in the same logical system, it being the case that besides the ordinary assertion of a formula $A$ as simply $A$ (the "internal" assertion), he conceived a second mode of assertion according to which a formula $A$ is "externally" asserted as being true or false. To this end, he set forward a special external assertion operator $\mathbf{a} : \mathbf{a}A$, and characterized these two modes of assertion by means of truth tables.

**5.5.5.** Bochvar's truth tables for *external* and *internal assertion* are as follows:

| $A$ | Internal assertion $A$ | External assertion $\mathbf{a}A$ |
|---|---|---|
| t | t | t |
| i | i | f |
| f | f | f |

**5.5.6. (Def.)** Given this, the *external connectives* (distinguished by means of *) are defined as follows, where $\stackrel{*}{\neg} A$ is read "$A$ is false", $A \stackrel{*}{\rightarrow} B$ as "if $A$ is true then $B$ is true", $A \stackrel{*}{\vee} B$ as "$A$ is true or $B$ is true", etc.:

$$\stackrel{*}{\neg} A := \neg \mathbf{a}A$$
$$A \stackrel{*}{\vee} B := \mathbf{a}A \vee \mathbf{a}B$$
$$A \stackrel{*}{\wedge} B := \mathbf{a}A \wedge \mathbf{a}B$$
$$A \stackrel{*}{\rightarrow} B := \mathbf{a}A \rightarrow \mathbf{a}B$$
$$A \stackrel{*}{\leftrightarrow} B := \mathbf{a}A \leftrightarrow \mathbf{a}B$$

## 5.5. Finitely many-valued propositional logics

**5.5.7.** A few easy calculations produce the following truth tables for what we shall call Bochvar's *external system*, designated by $B_3^E$:

| $A$ | $\neg A$ | $A \backslash B$ | $\stackrel{*}{\wedge}$ | | | $\stackrel{*}{\vee}$ | | | $\stackrel{*}{\rightarrow}$ | | | $\stackrel{*}{\leftrightarrow}$ | | |
|---|---|---|---|---|---|---|---|---|---|---|---|---|---|---|
| | | | t | i | f | t | i | f | t | i | f | t | i | f |
| t | f | t | t | f | f | t | t | t | t | f | f | t | f | f |
| i | t | i | f | f | f | t | f | f | t | t | t | f | t | t |
| f | t | f | f | f | f | t | f | f | t | t | t | f | t | t |

The following results can easily be proved:

**5.5.8. (Prop.)** $B_3^E$ connectives are both normal and uniform, but not K-regular.
*Proof:* Left as an exercise.

Hence, $B_3^E$ connectives yield a system with both tautologies as well as contradictions. As a matter of fact, we have:

**5.5.9. (Prop.)**
$$\top(B_3^E) = \top(K)$$
$$\bot(B_3^E) = \bot(K)$$

*Proof:* Left as an exercise.

**5.5.10. (Prop.)** $X \models_{BE} \phi$ iff $X \models \phi$.
*Proof:* Left as an exercise.

The important result follows: $B_3^E$ in a way extends K by augmenting $W_2$ ($W_2 \subset W_{BE}$), without for that changing the classical valuation function $val : F \longrightarrow W_2$, i.e.

$$val_{BE} : F_{BE} \longrightarrow W_2.$$

We thus have the obvious result:

**5.5.11. (Prop.)** $B_3^E$ connectives are decisive.
*Proof:* Left as an exercise.

**5.5.12.** It is thus the case that tautologousness, contradictoriness, and entailment all coincide for K and $B_3^E$. In matrix terms, this is so because the matrix[12]

$$\mathfrak{B}_3^E = \left\{ \{\mathsf{f}, \mathsf{i}, \mathsf{t}\}, \stackrel{*}{\neg}, \stackrel{*}{\rightarrow}, \stackrel{*}{\vee}, \stackrel{*}{\wedge}, \stackrel{*}{\leftrightarrow}, \{\mathsf{t}\} \right\}$$

---

[12] When no ambiguity arises, we simplify, for example, $\mathfrak{M}\mathfrak{B}_3^{\mathfrak{E}}$ by writing simply $\mathfrak{B}_3^E$.

## 5. Many-valued logics

identifies (or does not distinguish) i and f, while its functions behave classically with regard to t and f.

**5.5.13.** To finish, we emphasize that $\mathfrak{B}_3^E$, as well as

$$\mathfrak{B}_3^I = \{\{\mathtt{f}, \mathtt{i}, \mathtt{t}\}, \neg, \rightarrow, \vee, \wedge, \leftrightarrow, \{\mathtt{t}\}\},$$

is a reduct matrix of the *Bochvar matrix*

$$\mathfrak{B}_3 = \left\{\{\mathtt{f}, \mathtt{i}, \mathtt{t}\}, \neg, \rightarrow, \vee, \wedge, \leftrightarrow, \overset{*}{\neg}, \overset{*}{\rightarrow}, \overset{*}{\vee}, \overset{*}{\wedge}, \overset{*}{\leftrightarrow}, \{\mathtt{t}\}\right\}.$$

This highlights the fact that Bochvar's is indeed a single 3-valued logical system, which we can designate as $B_3$. Nevertheless, one must keep in mind that only

$$E\left(\mathfrak{B}_3^E\right) = E\left(\mathfrak{M}_2\right).$$

### 5.5.2. Kleene's 3-valued logics

Kleene's 3-valued logics (Kleene, 1938; 1952) have a clearer mathematical motivation than both Łukasiewicz's and Bochvar's logics. More specifically, he conceived them in large measure with the aim of allowing the (logical) analysis of partial functions. For instance, given the property $P$ such that

$$P(x) \quad \text{iff} \quad 1 \leq \frac{1}{x} \leq 2, \quad x \in \mathbb{R}$$

we have it that

$$P(x) = \begin{cases} \mathtt{t} & \text{if } \frac{1}{2} \leq x \leq 1 \\ \mathtt{i} & \text{if } x = 0 \\ \mathtt{f} & \text{otherwise} \end{cases}$$

where i (originally: u) clearly means *undefined*, or *undetermined* in the sense that one cannot determine whether $P(x)$ is true or false, whether because there is no algorithm to that end, or even because it is inessential for the question at issue.

**5.5.14.** In Kleene (1938), he defined the truth tables for $K_3^S$ (the superscript will be accounted for shortly; we omit subscripts in the connectives) in the following way:

## 5.5. Finitely many-valued propositional logics

| $A$ | $\neg A$ | | $A\backslash B$ | $\wedge$ t | i | f | $\vee$ t | i | f | $\to$ t | i | f | $\leftrightarrow$ t | i | f |
|---|---|---|---|---|---|---|---|---|---|---|---|---|---|---|---|
| t | f | | t | t | i | f | t | t | t | t | i | f | t | i | f |
| i | i | | i | i | i | f | t | i | i | t | i | i | i | i | i |
| f | t | | f | f | f | f | t | i | f | t | t | t | f | i | t |

Inspection of the truth tables for the connectives of $K_3^S$ produces the following results:

**5.5.15. (Prop.)** $K_3^S$ connectives are normal, uniform, and K-regular.
*Proof:* Left as an exercise.

**5.5.16. (Prop.)** Given the matrix

$$\mathfrak{K}_3^S = \{\{\mathtt{f}, \mathtt{i}, \mathtt{t}\}, \neg, \to, \vee, \wedge, \leftrightarrow, \{\mathtt{t}\}\}$$

we have

$$E\left(\mathfrak{K}_3^S\right) = \emptyset.$$

*Proof:* Left as an exercise.

This obviously leads to undesirable results, as, for example, the classical tautologies $A \to A$ (cf. Prop. 3.5.3, T3) and $A \leftrightarrow A$ not being so in $K_3$. In Kleene (1952), these connectives are referred to as *strong*, and new, *weak*, connectives of implication, disjunction, and conjunction are introduced (the connectives of negation and equivalence remain unchanged).

**5.5.17.** The following are the truth tables for the connectives of conjunction, disjunction, and material implication of Kleene's weak 3-valued logical system, denoted by $K_3^W$:

| $\wedge_*$ | t | i | f | $\vee_*$ | t | i | f | $\to_*$ | t | i | f |
|---|---|---|---|---|---|---|---|---|---|---|---|
| t | t | i | f | t | t | i | t | t | t | i | f |
| i | i | i | i | i | i | i | i | i | i | i | i |
| f | f | i | f | f | t | i | f | f | t | i | t |

Just as with the strong connectives, the aim was set with regard to arithmetical propositional functions, with the difference that now recursion is the main property in consideration, i.e. Kleene had in mind an effective calculation procedure that would allow the truth status of arithmetical propositions to be decided upon; if one of the constituents of a complex formula is assigned the truth value i, then the algorithm simply fails by "contagion" of the indeterminacy (at any stage of the

## 5. Many-valued logics

computation),[13] as in $B_3^I$; the "K" in K-regularity (Def. 5.3.3) is now accounted for: in effect, it just refers to *K*leene.

We have the following important results–whose proofs we leave as exercises–with respect to $K_3^W$:

**5.5.18. (Prop.)** $K_3^W$ connectives are normal, not uniform, and K-regular.

By revisiting the truth tables and matrix of $B_3^I$, we have it that:

**5.5.19. (Prop.)**
$$E\left(\mathfrak{B}_3^I\right) = E\left(\mathfrak{K}_3^W\right) = \emptyset.$$

In fact:

**5.5.20. (Prop.)**
$$\tilde{\heartsuit}_{B_3^I} = \tilde{\heartsuit}_{K_3^W}.$$

### 5.5.3. Finn's 3-valued logic

Finn (1972) carried on Bochvar's work (cf. Section 5.5.1), now differentiating between *propositional variables*, which can take any of the three truth values, and *sentential variables*, which can only take `truth` or `falsity` as valuations. In any case, the truth tables for the primitive connectives ($\neg, \wedge, \rightarrow$) are exactly those of Bochvar's.

**5.5.21.** The matrix of Finn's three-valued logical system $F_3$ is:

$$\mathfrak{F}_3 = (\{\mathtt{f},\mathtt{i},\mathtt{t}\}, \neg, \wedge, \rightarrow, +, -, \vee, \supset, \#, \{\mathtt{t}\})$$

**5.5.22. (Def.)** The secondary connectives of $F_3$ are defined as follows:

$$
\begin{aligned}
+A &:= A \rightarrow A \\
-A &:= +\neg A \\
A \vee B &:= \neg(\neg A \wedge \neg B) \\
A \supset B &:= \neg(A \wedge \neg B) \\
\#A &:= \neg(+A \vee \neg A) \\
A \leftrightarrow B &:= (A \rightarrow B) \wedge (B \rightarrow A)
\end{aligned}
$$

---

[13] That is to say that Kleene was not interested in "mending" the problem of the absence of tautologies and contradictions in his 3-valued logic; he was interested rather in recursion issues, what makes his system potentially relevant for computability theory.

## 5.5. Finitely many-valued propositional logics

$F_3$ has been fully axiomatized, but with a rather long list of axiom schemata (23, in total). Below we provide some of the axiom schemata of $F_3$ as an exercise.

### 5.5.4. Logics of nonsense: the 3-valued logics of Halldén, Åqvist, Segerberg, and Piróg-Rzepecka

Although at first sight *nonsense* might appear as too philosophical a concept, one of the interesting aspects of the so-called *logics of nonsense* is a mathematical treatment of nonsense or meaninglessness as a "third truth value." As a matter of fact, the logics of nonsense we shall approach now are, like Finn's 3-valued logic, to some extent continuations of Bochvar's 3-valued system. For uniformity, we shall always designate the "third truth value" as i.

We begin with Halldén's (1949) notion of "logical nonsense," which he concretized in the three-valued system that we shall denote as $Hn_3$. The first interesting aspect of this logic is the fact that Halldén actually considers i to be a designated value, along with truth. The (apparent) oddity lies in that Halldén actually considered the truth value "meaningless" or "nonsense" as simply "neither true nor false," a value to be assigned to propositions such as "The man with a hundred hairs is (not) bald."[14] However, Halldén introduces a new unary connective, +, such that $+P$ is read "$P$ is *meaningful*." This connective allows us to state simply of a meaningless proposition that it is false, i.e. not meaningful; in particular, "$\neg + P$" states that $P$ is meaningless.

**5.5.23.** The matrix of Halldén's 3-valued system $Hn_3$ is:

$$\mathfrak{H}n_3 = (\{f, i, t\}, +, \neg, \wedge, -, \rightarrow, \leftrightarrow, \{i, t\})$$

**5.5.24.** The following are the truth tables for the primitive connectives of $Hn_3$:

| $p$ | $+p$ |
|---|---|
| t | t |
| i | f |
| f | t |

| $p$ | $\neg p$ |
|---|---|
| t | f |
| i | i |
| f | t |

| $\wedge$ | t | i | f |
|---|---|---|---|
| t | t | i | f |
| i | i | i | i |
| f | f | i | f |

---

[14] The motivation here is clearly the sorites paradox (or *phalakros*–bald, in Greek–*paradox*, to be more precise). See, e.g., Bergmann (2008) for details on this paradox and related work in many-valued logics.

## 5. Many-valued logics

**5.5.25.** Note that the truth tables for negation and conjunction in Hn3 not involving the connective + are exactly those of $B_3^I$ (and thus of $K_3^W$), and the same holds for the remaining connectives of $O_L$. Just as in the case of these logics, if i is considered a designated value along with t, then formulae not involving + are tautologies if they are tautologies in CPL, and are contradictions if they are so in CPL. In particular, PEM is valid in Hn3, as by including i in the set $D$, Halldén requires of a valid proposition that it *not be false*, not that it must be true. And in effect, $P \vee \neg P$, though not always true in Hn3, is never false in this system.

**5.5.26.** Thus, if anything, Halldén's system preserves *non-falsity*, but this does not equate with *truth*. In particular, the MP rule is now restricted, as from $(P \vee \neg P) \to +P$ we cannot in fact infer that $+P$ if $P$ is meaningless.

**5.5.27. (Def.)** Halldén's secondary connectives–among which there is $-$, dual to $+$–are defined from the primitives in the standard way; $-$ is defined as:

$$-\phi := \neg + \phi$$

Clearly, "$-\phi$" states that $\phi$ is meaningless.

**5.5.28.** The following axiom schemata imposed on propositional variables, together with the rules MP and SUB, constitute an adequate axiom system for Hn3:

$(\mathcal{H}n1)$ $(\neg p \to p) \to p$
$(\mathcal{H}n2)$ $p \to (\neg p \to q)$
$(\mathcal{H}n3)$ $(p \to q) \to ((q \to r) \to (p \to r))$
$(\mathcal{H}n4)$ $+p \leftrightarrow +\neg p$
$(\mathcal{H}n5)$ $+(p \wedge q) \leftrightarrow (+p \wedge +q)$
$(\mathcal{H}n6)$ $p \to +p$

Setting out from Halldén's work, Åqvist (1962) decided to concentrate on the normativity of scientific statements: a statement that is valued as *normative* is according to him neither true nor false, i.e. is assigned the truth value i. We shall designate Åqvist's three-valued system as An3.

**5.5.29.** The matrix of Åqvist's system An3 is:

$$\mathfrak{An3} = (\{\mathtt{f,i,t}\}, \#, \neg, \vee, \wedge, \to, \leftrightarrow, F, L, M, \{\mathtt{t}\})$$

**5.5.30.** The following are the truth tables for the primitive connectives of An3:

## 5.5. Finitely many-valued propositional logics

| $p$ | $\#p$ |
|---|---|
| t | t |
| i | f |
| f | f |

| $p$ | $\neg p$ |
|---|---|
| t | f |
| i | i |
| f | t |

| $\vee$ | t | i | f |
|---|---|---|---|
| t | t | t | t |
| i | t | i | i |
| f | t | i | f |

The tables for negation and disjunction are the same as in L$_3$. In effect, An$_3$ can be considered as a variant of this system, as well as of K$_3^S$.

**5.5.31. (Def.)** Three new unary operators (**F**, **L**, and **M**) are introduced in An$_3$ and they are read and defined as follows:

$$\begin{aligned}
\mathbf{F}A \quad & (\text{``}A \text{ is false''}) & := \quad & \#\neg A \\
\mathbf{L}A \quad & (\text{``}A \text{ is meaningful''}) & := \quad & \#A \vee \mathbf{F}A \\
\mathbf{M}A \quad & (\text{``}A \text{ is meaningless''}) & := \quad & \neg \mathbf{L}A
\end{aligned}$$

The remaining secondary connectives of An$_3$, to wit, $\wedge$, $\rightarrow$, and $\leftrightarrow$ are defined in the standard way.

Another system inspired by Halldén's is Segerberg's (1965) three-valued logic. Actually, there are three different three-valued calculi here that we shall denote by Sn$_3^1$, Sn$_3^2$, and Sn$_3^3$.

**5.5.32.** The following are the matrices for Sn$_3^1$, Sn$_3^2$, and Sn$_3^3$:[15]

$$\mathfrak{Sn_3}^1 = (\{\mathtt{f},\mathtt{i},\mathtt{t}\}, \#, \neg, \wedge, \|, =, -, \vee, \rightarrow, \leftrightarrow, \{\mathtt{i},\mathtt{t}\})$$

$$\mathfrak{Sn_3}^2 = (\{\mathtt{f},\mathtt{i},\mathtt{t}\}, +, \wedge, \{\mathtt{i},\mathtt{t}\})$$

$$\mathfrak{Sn_3}^3 = (\{\mathtt{f},\mathtt{i},\mathtt{t}\}, \neg, \wedge, \{\mathtt{t}\})$$

**5.5.33.** The truth tables for the primitive connectives of Sn$_3^1$ are as follows:

| $p$ | $\#p$ |
|---|---|
| t | t |
| i | f |
| f | f |

| $p$ | $\neg p$ |
|---|---|
| t | f |
| i | i |
| f | t |

| $\wedge$ | t | i | f |
|---|---|---|---|
| t | t | i | f |
| i | i | i | i |
| f | f | i | f |

As can be seen, the connective # behaves exactly as in An$_3$ and the connective $\wedge$ does so exactly as in Hn$_3$. The connectives $\vee$, $\rightarrow$, and $\leftrightarrow$

---

[15] For Sn$_3^2$ and Sn$_3^3$ we provide only the *minimal matrix*, the matrix containing only the primary connectives.

## 5. Many-valued logics

are defined as usually. However, $\text{Sn}_3^1$ has three further secondary unary connectives.

**5.5.34. (Def.)** The remaining secondary unary connectives of $\text{Sn}_3^1$, to wit, $\|$, $=$, and $-$, are read and defined as:

$$\begin{aligned}
\| A \quad &(\text{"}A\text{ is false"}) & :=&\quad \#\neg A \\
= A \quad &(\text{"}A\text{ makes sense"}) & :=&\quad \neg(\neg\# A \wedge \neg\#\neg A) \\
- A \quad &(\text{"}A\text{ makes no sense"}) & :=&\quad \neg\# A \wedge \neg\#\neg A
\end{aligned}$$

**5.5.35.** The truth tables for the primitive connectives of $\text{Sn}_3^2$ are as follows:

| $p$ | $+p$ |
|---|---|
| t | i |
| i | f |
| f | t |

| $\wedge$ | t | i | f |
|---|---|---|---|
| t | t | i | f |
| i | i | i | i |
| f | f | i | f |

**5.5.36.** The truth tables for the primitive connectives of $\text{Sn}_3^3$ are as follows:

| $p$ | $\neg p$ |
|---|---|
| t | i |
| i | f |
| f | t |

| $\wedge$ | t | i | f |
|---|---|---|---|
| t | t | i | f |
| i | i | i | f |
| f | f | f | f |

We conclude this brief elaboration on the logics of nonsense with yet another three-valued logical system inspired by Halldén's first conception. This is the system elaborated on in Piróg-Rzepecka (1977) that we shall denote by $\text{Rn}_3$.

**5.5.37.** The matrix for $\text{Rn}_3$ is:

$$\mathfrak{Rn}_3 = (\{\mathtt{f}, \mathtt{i}, \mathtt{t}\}, \neg, \wedge, \rightarrow, \vee, \leftrightarrow, -, \#, \circ, \{\mathtt{t}\})$$

The primary connectives are $\neg$, $\wedge$, and $\rightarrow$. The two secondary binary connectives $\vee$ and $\leftrightarrow$ are standardly defined, and the author introduces a new secondary unary connective, $-$, and two new binary connectives, $\#$ and $\circ$. Interestingly, the major changes carried out by Piróg-Rzepecka fall on the connectives for material implication and material equivalence.

In particular, there is no assignment of the third truth value associated with these two connectives, which behave normally (cf. Def. 5.3.1) and decisively (Def. 5.3.4).

**5.5.38.** The following are the truth tables for the connectives of Rn$_3$:

| $p$ | $-p$ |
|---|---|
| t | f |
| i | t |
| f | t |

| $p$ | $\neg p$ |
|---|---|
| t | f |
| i | i |
| f | t |

| $\wedge$ | t | i | f |
|---|---|---|---|
| t | t | i | f |
| i | i | i | i |
| f | f | i | f |

| $\vee$ | t | i | f |
|---|---|---|---|
| t | t | i | t |
| i | i | i | i |
| f | t | i | f |

| $\to$ | t | i | f |
|---|---|---|---|
| t | t | f | f |
| i | t | t | t |
| f | t | t | t |

| $\leftrightarrow$ | t | i | f |
|---|---|---|---|
| t | t | f | f |
| i | f | t | t |
| f | f | t | t |

**5.5.39. (Def.)** The new connectives of Rn$_3$ are defined as:

$$\begin{aligned} -A &:= A \to \neg A \\ A \# B &:= -A \to B \\ A \circ B &:= -(-A \# - B) \end{aligned}$$

**5.5.40. (Prop.)** In Rn$_3$, the following holds true:

$$(\leftrightarrow_\circ^\to) \qquad A \leftrightarrow B \quad \text{iff} \quad (A \to B) \circ (B \to A)$$

*Proof:* Left as an exercise.

### 5.5.5. Heyting's 3-valued logic

In intuitionistic logic, PEM is not a tautology, and neither is $\neg\neg A \to A$. The motivations for this are complex (see Heyting, 1956), but as far as we are concerned Heyting proposed a three-valued system that rejects these as tautologies (Heyting, 1930). We shall denote this system by G$_3$, because Gödel (1930) provided the axiomatization for intuitionistic logic (see Section 5.5.8). This system is interesting, because it shows that once one has rejected some classical tautologies or contradictions as

## 5. Many-valued logics

such, one can manipulate the behavior of the connectives of a language, so as to make them behave in the desired way. But this system is particularly interesting, because it so happened that it was in the search for an adequate axiom system for Heyting's logic (a search that ended in failure) that intuitionistic propositional logic was conceived. Not a lesser feat if we know that this logic has only the two classical truth values.

**5.5.41.** The matrix for Heyting's three-valued system $G_3$ is:

$$\mathfrak{G}_3 = (\{f, i, t\}, \neg, \wedge, \vee, \rightarrow, \{t\})$$

**5.5.42.** The truth tables for the connectives of $G_3$ are as follows:

| $p$ | $\neg p$ | | $p \backslash q$ | \multicolumn{3}{c}{$\wedge$} | \multicolumn{3}{c}{$\vee$} | \multicolumn{3}{c}{$\rightarrow$} |
|---|---|---|---|---|---|---|---|---|---|---|---|---|
| | | | | t | i | f | t | i | f | t | i | f |
| t | f | | t | t | i | f | t | t | t | t | i | f |
| i | f | | i | i | i | f | t | i | i | t | t | f |
| f | t | | f | f | f | f | t | i | f | t | t | t |

It is easy to verify that the impacting changes are to be found in the behavior of negation and material implication (see the characterizations of all the connectives in Section 5.5.8).

### 5.5.6. Reichenbach's 3-valued logic

Reichenbach's (1946) three-valued logic provides a good example of a logical system conceived with practical applications in mind. In effect, this logic was conceived with the field of quantum mechanics in mind, it being the case, to put it simply, that some statements cannot be said to be true or false according to quantum theory; Reichenbach's third value, together with a large set of connectives, would then still allow a characterization of quantum phenomena. We denote his system by $Q_3$. Reichenbach did not provide completeness or soundness proofs for $Q_3$, let alone an axiom system, but we leave it here due to the complexity of its matrix and truth-functional connectives.[16]

**5.5.43.** The following is the matrix for $Q_3$:

$$\mathfrak{Q}_3 = (\{f, i, t\}, \neg, -, \sim, \wedge, \vee, \rightarrow, \supset, \rightsquigarrow, \leftrightarrow, \leftrightsquigarrow, \{t\})$$

---

[16] For a discussion of the (lack of) merits of Reichenbach's logic with respect to quantum theory, see, e.g., Haack (1974).

## 5.5. Finitely many-valued propositional logics

**5.5.44.** The truth tables for the three different connectives for negation, to wit, *cyclic* ($\sim$), *diametrical* ($-$), and *full negation* ($\neg$), are as follows:

| $p$ | $\sim p$ |
|---|---|
| t | f |
| i | t |
| f | i |

| $p$ | $-p$ |
|---|---|
| t | f |
| i | i |
| f | t |

| $p$ | $\neg p$ |
|---|---|
| t | i |
| i | t |
| f | t |

**5.5.45.** The remaining connectives of $Q_3$ have the following truth tables:

| $\vee$ | t | i | f |
|---|---|---|---|
| t | t | t | t |
| i | i | i | i |
| f | t | i | f |

| $\wedge$ | t | i | f |
|---|---|---|---|
| t | t | i | f |
| i | i | i | f |
| f | f | f | f |

| $p\backslash q$ | $\rightarrow$ | | | $\supset$ | | | $\rightsquigarrow$ | | | $\leftrightarrow$ | | | $\leftrightsquigarrow$ | | |
|---|---|---|---|---|---|---|---|---|---|---|---|---|---|---|---|
| | t | i | f | t | i | f | t | i | f | t | i | f | t | i | f |
| t | t | f | f | t | i | f | t | i | f | t | f | f | t | i | f |
| i | t | t | t | t | t | i | i | i | i | f | t | f | i | t | i |
| f | t | t | t | t | t | t | i | i | i | f | f | t | f | i | t |

### 5.5.7. Belnap's 4-valued logic

In classical logic, contradictions so to say infect a knowledge base integrally. According to N. Belnap, this need not be so. Belnap (1977) conceived a 4-valued logical system whose main potential application was in tackling inconsistency in information originating in diverse sources, where the reasoning agent is a computer (system). We shall denote this system by $B_4$.

**5.5.46.** With this objective in mind, Belnap conceived the set $\mathbf{4} = \{\mathbf{n}, \mathbf{f}, \mathbf{t}, \mathbf{b}\}$ of the truth values *none*, *false*, *true*, and *both*, respectively (we abbreviate *none* as $\mathbf{n}$ and *both* as $\mathbf{b}$). Importantly, $\mathbf{4}$ is actually the power set $2^{W_2} = \{\emptyset, \{\mathbf{f}\}, \{\mathbf{t}\}, \{\mathbf{f}, \mathbf{t}\}\}$. It is this feature that makes of $\mathbf{4}$ a *de-facto* set of logical truth values. This truth-value set actually

## 5. Many-valued logics

constitutes the *approximation lattice* $\mathcal{A}4 = (\mathbf{4}, \sqsubseteq)$, so called because its partial order "approximates the information in" (see Fig. 5.5.1.1). Mathematically, an approximation lattice just is a complete lattice where $x \sqsubseteq y$ is read as "$x$ approximates $y$."

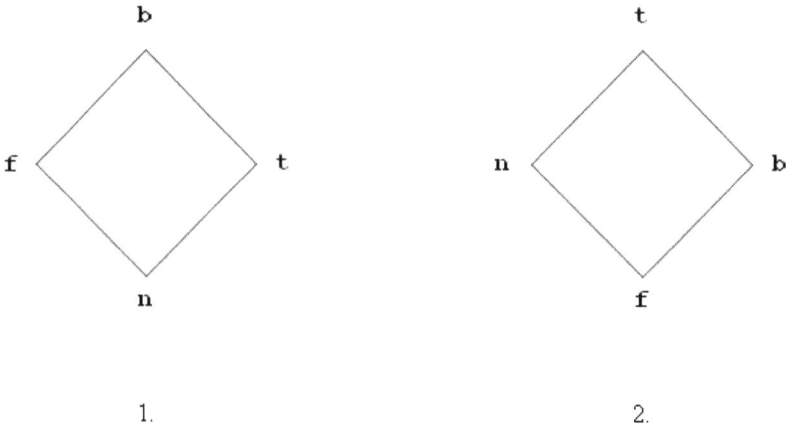

1.  2.

Figure 5.5.1.: The lattices $\mathcal{A}4$ (1.) and $\mathcal{L}4$ (2.).

**5.5.47.** The truth tables of $B_4$ actually also constitute a lattice, the *logical lattice* $\mathcal{L}4 = (\mathbf{4}, \leq)$ (Fig. 5.5.1.2). This lattice is called *logical*, because its ordering is in fact a logical ordering, with conjunction and disjunction as the binary operations of meet and join, respectively. Note that in this lattice we have the operations of join and meet as follows: $\mathbf{n} \vee \mathbf{b} = \mathbf{t}$ and $\mathbf{n} \wedge \mathbf{b} = \mathbf{f}$, respectively.

The *epistemic* rationale for the approximation lattice is as follows: **n**, the greatest lower bound (glb), is at the bottom because it gives no information at all, whereas **b**, the least upper bound (lub), is at the top because it gives too much (inconsistent) information. The "unabashedly" epistemic character of these truth values (cf. Belnap, 1977) is established as follows.

**5.5.48.** Of each atomic proposition $P$, which can be a predicate, it can be said that

1. $P$ is just told *true* (**t**);

2. $P$ is just told *false* (**f**);

3. $P$ is told *both true and false* (**b**);

4. $P$ is told *neither true nor false* (**n**).

## 5.5. Finitely many-valued propositional logics

With respect to 5.5.48.4, it is more correctly said of $P$ that it has an unknown epistemic status. It follows from this and the above that Belnap's truth values are more properly *told (to the computer) truth values*. Importantly, "told *true*" (**t**) is not the same as "true" (**t**). The computer is told *true* of a proposition $P$ just in case it has marked $P$ either with **t** or with **b**; the computer marks $P$ with **t** just in case it has been told *true* of $P$ and has not been told *false* of $P$. The same holds for the remaining truth values with a similar reasoning.

**5.5.49.** The following are the truth tables of B$_4$:

| ¬ | |
|---|---|
| b | b |
| t | f |
| f | t |
| n | n |

| ∧ | b | t | f | n |
|---|---|---|---|---|
| b | b | b | f | f |
| t | b | t | f | n |
| f | f | f | f | f |
| n | f | n | f | n |

| ∨ | b | t | f | n |
|---|---|---|---|---|
| b | b | t | b | t |
| t | t | t | t | t |
| f | b | t | f | n |
| n | t | t | n | n |

Negation may appear strange here, as both ¬**b** = **b** and ¬**n** = **n**, but Belnap accounts for this with the monotonic character of the approximation lattice: for a function to be monotonic, we must have it that if $x \sqsubseteq y$, then $f(x) \sqsubseteq f(y)$.

**5.5.50.** The semantics of this logic is summarily established as follows:

$$P \models_{B4} Q \quad \text{iff} \quad s(P) \leq s(Q).$$

where $s(\cdot)$ is a *set-up*, a mapping from all atomic formulae into **4** such that for any atomic formula $P$ we have $s(P) \in \mathbf{4}$. Thus, entailment in B$_4$ is the relation that holds just in case for any assignment of one of the four truth values to the variables, the value of $P$ does not exceed the value of $Q$. This is in agreement with the ordering of the logical lattice $\mathcal{L}4$.

A set-up represents not what is true, but what the computer has been told. This has epistemic implications that we shall not discuss here, namely insofar as complex formulae are concerned.

**5.5.51.** The matrix of B$_4$ is $\mathfrak{B}_4 = (\mathbf{4}, \neg, \wedge, \vee, D)$, where $D = \{\{\mathbf{t}\}\}$ or $D = \{\{\mathbf{t}\}, \{\mathbf{f}, \mathbf{t}\}\}$, depending on the applications in view. The latter set of designated values makes of B$_4$ a *relevance logic*, a logical system which stipulates that in order for a formula $P \rightarrow Q$ to be true there must be a relevant connection between $P$ and $Q$. This explains the absence of the connective $\rightarrow$ in B$_4$.

Finally, as expectable, quantification introduces some problems into the neat picture above (see Belnap, 1977).

## 5. Many-valued logics

### 5.5.8. The finitely $n$-valued logics of Post and Gödel

#### 5.5.8.1. Post logics

Formalizations of Post many-valued logics (Post, 1921) are especially important due to the many applications found for them, not the least because they are functionally complete. They were summarily introduced in Section 3.2.1, and we now complete in some further detail the discussion of these logics. We provide here a very succinct treatment of these logics. This succinctness may be misleading, as the Post finitely $n$-valued logics are indeed very complex logics, namely at the algebraic level (cf., e.g., Bolc & Borowik, 1992).[17]

The main feature that distinguishes Post logics from the other main many-valued logics is negation, which is ruled by the following truth table, according to the cyclic rotation function presented above (cf. Example 3.2.2).

**5.5.52.** Post's negation as rotation function is as follows:

| $A$ | $\neg_n A$ |
|---|---|
| 1 | 2 |
| 2 | 3 |
| 3 | 4 |
| $\vdots$ | $\vdots$ |
| $n-2$ | $n-1$ |
| $n-1$ | $n$ |
| $n$ | 1 |

Just as in Whitehead and Russell's *Principia mathematica* (Whitehead & Russell, 1910), negation forms with disjunction the pair of primitive connectives. The latter is in Post's systems $P_n$ defined as, for finite $n$, $t_i \vee_n t_j = max(t_i, t_j)$, and the remaining connectives are then defined as follows.

**5.5.53. (Def.)** The connectives $\wedge_n$, $\rightarrow_n$, and $\leftrightarrow_n$ are defined as follows in Post's $n$-valued systems:

---

[17] Note, however, that Post provided his non-standard logics $P_n$ with a semantical interpretation that does not interpret propositions but rather sets of propositions, which throws shadow on their status as propositional logics (see Rescher, 1969, p. 54-5; Malinowski, 1993, p. 46).

## 5.5. Finitely many-valued propositional logics

$$A \wedge_n B := \neg_n(\neg_n A \vee_n \neg_n B)$$
$$A \rightarrow_n B := \neg_n A \vee_n B$$
$$A \leftrightarrow_n B := (A \rightarrow_n B) \wedge_n (B \rightarrow_n A)$$

**5.5.54. (Def.)** For a given $n \geq 2$ and the set of linearly ordered objects $P_n = \{t_1, ..., t_n\}$, the algebraic structure

$$\mathfrak{P}_n = (\{t_1..., t_n\}, \neg_n, \vee_n)$$

is called an *n-valued Post algebra*, and the matrix $\mathfrak{MP}_n$ naturally associated with it

$$\mathfrak{MP}_n = (\{t_1, ..., t_n\}, \neg_n, \vee_n, \{t_n\})$$

is the (basic) *n-valued Post matrix*.

This is the minimal matrix for $P_n$. The functional definitions for the operations $\neg_n$ and $\vee_n$ were already given in Example 3.2.2. Other connectives of $P_n$ are defined as follows (the definition of $\leftrightarrow_n$ is left as an exercise).

**5.5.55. (Def.)** Let $t_i = i$. The connectives $\wedge_n$ and $\rightarrow_n$ have the following functional definitions:

1. $i \wedge j = min(i, j)$

2. $i \rightarrow j = \begin{cases} n & \text{if } i \leq j \\ j & \text{otherwise} \end{cases}$

**5.5.56.** An axiom system for $P_n$ can be found in Bolc & Borowik (1992).

Post also considered a family of $n$-valued logics with two parameters, $P_n^r$, in which the set $\{r, r+1, ..., n-1\}$ is treated as designated. We briefly elaborate on the Post logics $P_n^r$ drawing closely on Bolc & Borowik (1992) and using a notation that is also very commonly used for Post algebras.

**5.5.57. (Def.)** Let $e_0, e_1, ..., e_{n-1}$ be the chain of logical values, $e_0 = 0, e_1 = 1, ..., e_{n-1} = n-1$. The logical connectives in $P_n^r$, which include another unary connective $(-)$ for negation, are interpreted in terms of functions defined on the set $\{0, 1, ..., n-1\}$ as follows:

1. $\neg x = \begin{cases} x - 1 & \text{if } x \neq 0 \\ n - 1 & \text{otherwise} \end{cases}$

2. $-x = n - 1 - x$

3. $x \vee y = max(x, y)$

187

## 5. Many-valued logics

4. $x \wedge y = min(x, y)$

5. $x \to y = \begin{cases} n-1 & \text{when } x \leq y \\ y & \text{when } x > y \text{ and } x \geq r \\ n-x+y-1 & \text{when } x > y \text{ and } x < r \end{cases}$

The negation connective $-$ is identical to Lukasiewicz negation, and if $n-1$ is the unique designated value, then Post implication coincides with Lukasiewicz implication.

**Example 5.5.1.** We show two tables for Post implication in $P_5^3$ and $P_5^1$ (designated values are marked with *):

| $\to$ | 0 | 1 | 2* | 3* | 4* |
|---|---|---|---|---|---|
| 0 | 4 | 4 | 4 | 4 | 4 |
| 1 | 3 | 4 | 4 | 4 | 4 |
| 2* | 0 | 1 | 4 | 4 | 4 |
| 3* | 0 | 1 | 2 | 4 | 4 |
| 4* | 0 | 1 | 2 | 3 | 4 |

| $\to$ | 0 | 1 | 2 | 3 | 4* |
|---|---|---|---|---|---|
| 0 | 4 | 4 | 4 | 4 | 4 |
| 1 | 3 | 4 | 4 | 4 | 4 |
| 2 | 2 | 3 | 4 | 4 | 4 |
| 3 | 1 | 2 | 3 | 4 | 4 |
| 4* | 0 | 1 | 2 | 3 | 4 |

### 5.5.8.2. Gödel logics

Gödel (1930) defined another finitely $n$-valued logic, denoted by $G_n$, by means of the following matrix.

**5.5.58.** The following is the matrix for Gödel's $n$-valued logics:

$$\mathfrak{G}_n = \{\{0, 1, ..., n-1\}, \neg, \wedge, \vee, \to, \leftrightarrow, \{n-1\}\}$$

**5.5.59. (Def.)** The connectives of $G_n$ are defined as follows:

1. $\neg x = \begin{cases} n-1 & \text{if } x = 0 \\ 0 & \text{if } x \neq 0 \end{cases}$

2. $x \wedge y = min(x, y)$

3. $x \vee y = max(x, y)$

4. $x \to y = \begin{cases} n-1 & \text{if } x \leq y \\ y & \text{if } x > y \end{cases}$

5. $x \leftrightarrow y = \begin{cases} n-1 & \text{if } x = y \\ min(x, y) & \text{if } x \neq y \end{cases}$

**Example 5.5.2.** The truth tables for the connectives $\neg$, $\wedge$, $\vee$, and $\rightarrow$ of $G_3$ are shown in 5.5.42.

$G_3$ is a particularly interesting Gödel finitely $n$-valued logic, as its matrix provides a basis for an axiomatization of propositional intuitionistic logic (see Section 5.5.5).

**5.5.60.** The following axiom schemata for $G_3$, together with the rule MP, constitute an axiom system for the propositional intuitionistic logic:

$(\mathcal{G}_3 1)$    $(A \wedge B) \rightarrow A$
$(\mathcal{G}_3 2)$    $(A \wedge B) \rightarrow B$
$(\mathcal{G}_3 3)$    $A \rightarrow (A \vee B)$
$(\mathcal{G}_3 4)$    $B \rightarrow (A \vee B)$
$(\mathcal{G}_3 5)$    $A \rightarrow (B \rightarrow A)$
$(\mathcal{G}_3 6)$    $A \rightarrow (\neg A \rightarrow B)$
$(\mathcal{G}_3 7)$    $A \rightarrow (B \rightarrow (A \wedge B))$
$(\mathcal{G}_3 8)$    $(A \rightarrow \neg B) \rightarrow (B \rightarrow \neg A)$
$(\mathcal{G}_3 9)$    $(A \rightarrow (B \rightarrow C)) \rightarrow ((A \rightarrow B) \rightarrow (A \rightarrow C))$
$(\mathcal{G}_3 10)$    $(A \rightarrow C) \rightarrow ((B \rightarrow C) \rightarrow ((A \vee B) \rightarrow C))$
$(\mathcal{G}_3 11)$    $(\neg A \rightarrow B) \rightarrow (((B \rightarrow A) \rightarrow B) \rightarrow B)$

## 5.6. Fuzzy logics

Fuzzy formalisms (e.g., fuzzy sets), by and large constituting and/or contributing to *fuzzy logic* (FL), were conceived to formalize *vagueness* in natural language (Zadeh, 1965; 1975). FL is relevant in the context of many-valued logics in that, independently of the linguistic and philosophical problems it originally aimed at tackling (e.g., the sorites paradox; vague predicates, such as *young, fat, red*, etc.), it can be conceived as an extension of classical logic by allowing any truth value between 0 (**f**) and 1 (**t**).

**5.6.1. (Prop.)** Let it be the case that for any truth value $x \in W_{FL}$ we have it that $0 \le x \le 1$, i.e. $x \in [0,1]$. Then, $|W_{FL}| = \aleph_1$, by the property of the power of the continuum and by the continuum hypothesis.

*Proof:* Left as an exercise.

**5.6.2. (Def. )** *Validity in FL* – Let the set of designated values be defined as $D_\varepsilon = \{x | x \ge \varepsilon, \varepsilon \in (0,1]\}$. Then, we say that an inference

## 5. Many-valued logics

$X \models_\varepsilon \phi$ is *valid* iff for all $\chi \in X$ and all valuations *val*, if $val(\chi) \geq \varepsilon$, then $val(\phi) \geq \varepsilon$.

Each logic defined in this way is obviously a many-valued logic. In fact, it is governed by the definition of logical consequence that follows from the definition above:

**5.6.3. (Prop.)** *Entailment in FL* – For all $\varepsilon \in (0,1]$,

$$X \models \phi \quad \text{iff} \quad X \models_\varepsilon \phi.$$

*Proof:* Left as an exercise.

Because FL aims at expressing the continuous nature of some natural language properties, the set of FL truth values is the interval $[0, 1]$, and it is immediately obvious that we can conceive FL (or a subset thereof) as either identical to, or a variation of, Ł$_{\aleph_1}$ (see, e.g., Bellman & Zadeh, 1977). However, it would be wrong to equate FL with Ł$_{\aleph_1}$ (or simply Ł$_\aleph$). Indeed, Ł$_\aleph$ is but one of several logics known as *t-norm fuzzy logics*, the most important class of FLs. It is thus convenient to start by approaching the notion of t-norms and the *basic logic* (BL) whose construction it immediately allows.

**5.6.4. (Def.)** A *t-norm* is a function $\star : [0,1]^2 \longrightarrow [0,1]$ such that for all $x, y, z \in [0, 1]$, $\star$ is continuous on $[0, 1]^2$, the following properties hold:[18]

| | | |
|---|---|---|
| (TN1) | $x \star y = y \star x$ | *Commutativity* |
| (TN2) | $(x \star y) \star z = x \star (y \star z)$ | *Associativity* |
| (TN3) | $x \leq y$ implies $x \star z \leq y \star z$ | *Monotonicity* |
| (TN4) | $1 \star x = x$ | *Identity* |
| (TN5) | $0 \star x = 0$ | *Annihilating element* |

**Example 5.6.1.** For $x, y \in [0, 1]$, we have the following *fundamental t-norms*:

1. Łukasiewicz t-norm: $x \star_L y := max(0, x + y - 1)$

---

[18] $\star$ is continuous on $[0,1]^2$ if for all $x, y \in [0, 1]$, if $\lim_{i \to \infty} x_i = x$, for $x_i \geq 0$ a sequence, then
$$\lim_{i \to \infty} (x \star y) = x \star y.$$
This is an additional requirement for the FLs here approached, as there are non-continuous t-norms.

2. Gödel t-norm: $x \star_G y := min(x, y)$

3. Product t-norm: $x \star_\Pi y := x \cdot y$

In semantical terms, a t-norm is the interpretation for a specific kind of conjunction, &, known as *strong conjunction*, such that $x \star y = x \,\&\, y$. Then, a t-norm FL is determined by the selection of a continuous t-norm function for the conjunction connective &, with the remaining connectives defined in terms of it in the following way:

**5.6.5. (Def.)** For all $x, y, z \in [0, 1]$:

1. $x \,\&\, y \leq z$ iff $y \leq x \rightarrow z$ (the residuum of the t-norm)

2. $x \rightarrow y = 1$ iff $x \leq y$

3. $\neg x = x \rightarrow 0$

4. $x \wedge y = x \,\&\, (x \rightarrow y)$ (weak conjunction)

5. $x \vee y = ((x \rightarrow y) \rightarrow y) \wedge ((y \rightarrow x) \rightarrow x)$

It can further be shown that $x \wedge y = min(x, y)$ and $x \vee y = max(x, y)$. Importantly, if we select 1 as the single designated truth value, so that $D = \{1\}$, then any continuous t-norm defines a *continuum-valued logic* $L_\star$. In particular, the logic BL (*basic logic*) is characterized by the important property that any of its theorems is logically true in all $L_\star$ logics.

**5.6.6. (Def.)** *Basic logic (BL)* – The following axiom system, together with the rule MP, defines the logic BL for the formulae $A, B, C$ and the constant 0 denoting zero:

$(\mathcal{BL}1)$    $(A \rightarrow B) \rightarrow ((B \rightarrow C) \rightarrow (A \rightarrow C))$
$(\mathcal{BL}2)$    $(A \,\&\, B) \rightarrow A$
$(\mathcal{BL}3)$    $(A \,\&\, B) \rightarrow (B \,\&\, A)$
$(\mathcal{BL}4)$    $(A \,\&\, (A \rightarrow B)) \rightarrow (B \,\&\, (B \rightarrow A))$
$(\mathcal{BL}5)$    $(A \rightarrow (B \rightarrow C)) \rightarrow ((A \,\&\, B) \rightarrow C)$
$(\mathcal{BL}6)$    $((A \,\&\, B) \rightarrow C) \rightarrow (A \rightarrow (B \rightarrow C))$
$(\mathcal{BL}7)$    $((A \rightarrow B) \rightarrow C) \rightarrow (((B \rightarrow A) \rightarrow C) \rightarrow C)$
$(\mathcal{BL}8)$    $0 \rightarrow A$

## 5. Many-valued logics

The continuous t-norms in Example 5.6.1 are called fundamental because every continuous t-norm is a combination of them (Hájek, 1998). The connectives for material implication and negation (as residuum and pre-complement, respectively) derived from them behave in the defined ways below, giving origin to the several named FLs.

**5.6.7. (Def.)** $Ł_{\aleph}$ – For the Łukasiewicz t-norm, it can be easily shown that for all $x, y \in [0, 1]$:

$$x \to y = min\,(1, 1 - x + y)$$

$$\neg x = 1 - x$$

This logic is the *Łukasiewicz fuzzy logic* $Ł_{\aleph}$.

**5.6.8.** By adding to the axiom schemata of BL the axiom schema

$$\neg\neg A \to A$$

we obtain an adequate axiom system for the Łukasiewicz t-norm.

**5.6.9. (Def.)** $G_{\aleph}$ – For the Gödel t-norm, we have it that for all $x, y \in [0, 1]$:

$$x \to y = \begin{cases} 1 & \text{if } x \leq y \\ y & \text{if } x > y \end{cases}$$

$$\neg x = \begin{cases} 1 & \text{if } x = 0 \\ 0 & \text{if } x > 0 \end{cases}$$

This logic is known as *Gödel fuzzy logic*.[19]

**5.6.10.** By adding to BL the axiom schema

$$A \to (A \,\&\, A)$$

we obtain an adequate axiom system with respect to the Gödel t-norm.

**5.6.11. (Def.)** $\Pi_{\aleph}$ – For the product t-norm, it can be easily shown that for all $x, y \in [0, 1]$:

$$x \to y = \begin{cases} 1 & \text{if } x \leq y \\ y/x & \text{if } x > y \end{cases}$$

$$\neg x = \begin{cases} 1 & \text{if } x = 0 \\ 0 & \text{if } x > 0 \end{cases}$$

---

[19] Note that, in $G_{\aleph}$, & behaves exactly as ∧.

This logic is the *product (fuzzy) logic*.

**5.6.12.** If we add the following axiom schemata to BL,

$$\neg\neg C \to (((A\,\&\,C) \to (B\,\&\,C)) \to (A \to B))$$

$$(A \wedge \neg A) \to 0$$

we obtain an adequate axiom system with respect to the product t-norm.

**5.6.13. (Prop.)** BL-algebras, and the special cases thereof, viz., MV-algebras, G-algebras, and $\Pi$-algebras provide algebraic semantics for BL, $Ł_\aleph$, $G_\aleph$, and $\Pi_\aleph$, respectively.

*Proof:* Left as an exercise.

We close this section with some important properties of these logical systems that can be easily proved (as an exercise):

**5.6.14. (Prop.)** Let fuzzy validity be defined as in Definition 5.6.2 and let fuzzy entailment be as in Proposition 5.6.3. Then:

$$\top(Ł_\aleph), \top(G_\aleph), \top(\Pi_\aleph) \subsetneq \top(K)$$

$$\bot(Ł_\aleph), \bot(G_\aleph), \bot(\Pi_\aleph) \subsetneq \bot(K)$$

**5.6.15. (Prop.)** The propositional connectives of these logics are normal for the truth values 0 and 1.

**5.6.16. (Prop.)** If $X \models_{Ł_\aleph} \phi, X \models_{G_\aleph} \phi, X \models_{\Pi_\aleph} \phi$, then $X \models \phi$, but the converse does not hold.

**5.6.17. (Prop.)** Every degree-entailment, and every $n$-degree-entailment in these logics (when applicable; cf. Def.s 5.2.12 and 5.2.14) is a classical entailment, but the converse does not hold.

## 5.7. Quantification in many-valued logics

Given the plethora above of many-valued logics, it is understandably problematic to come up with a general method of quantification that suits them all, namely one that generalizes quantification in CL to the many-valued case. This is particularly important if general automated reasoning methods are to be devised for many-valued logics; because reasoning methods are already very developed for classical logic it is

## 5. Many-valued logics

highly desirable to make quantified many-valued formulae "behave classically." While the notions of *generalized* and *distribution quantifiers* allow us to achieve this aim for the finitely many-valued logics, the infinitely many-valued logics present specific problems. We thus discuss both cases separately.

### 5.7.1. Quantification in finitely many-valued logics

Recall from Definition 3.2.17 that a logical matrix $\mathfrak{M}$ for a propositional language $\mathsf{L} = (F, O)$ is a triple $\mathfrak{M} = (W, Fop, D)$ where $W$ is the set of truth values, $|W| \geq 2$, $Fop = \{f_{o_1}, ..., f_{o_n}\}$ is the set of interpretation functions, and $D \subset W, D \neq \emptyset$ is the set of designated values. We can extend this propositional matrix to a first-order one by adding to it a non-empty domain of quantification $\mathscr{D}$ and a set $Fq = \{f_{\blacklozenge_i} | \blacklozenge_i \in Q\}$ for $Q$ the set of quantifiers and where $f_{\blacklozenge_i}$ is a mapping from non-empty subsets of $W$ into $W$. (We abbreviate $f_{\blacklozenge_i}$ as $\tilde{\blacklozenge}_i$ ). The structure $\mathfrak{M}^* = (\mathscr{D}, W, Fop, Fq, D)$ is a logical matrix for the FO language $\mathsf{L}^*$. But this can be simplified as follows:

**5.7.1. (Def.)** A FO matrix for $\mathsf{L}^*$ is a triple $\mathfrak{M}^* = \left( W, \tilde{\heartsuit}, \tilde{\blacklozenge} \right)$ where $W$ is a set of truth values and $\tilde{\heartsuit}, \tilde{\blacklozenge}$ are as in Definition 2.1.8.

**5.7.2. (Def.)** *Generalized quantifiers* – Given a set of truth values $W_n = \{v_0, ..., v_{n-1}\}$, every binary connective $\heartsuit$ whose truth table $\tilde{\heartsuit}$ is commutative, associative and idempotent induces a quantifier $\blacklozenge^\heartsuit$, understood as "generalized $\heartsuit$", that is given by

$$\tilde{\blacklozenge}^\heartsuit(\{v_i\}) = v_i$$

$$\tilde{\blacklozenge}^\heartsuit(\{v_1, ..., v_k\}) = \tilde{\heartsuit}\left( v_1, \tilde{\heartsuit}\left( v_2, ..., \tilde{\heartsuit}(v_{k-1}, v_k)... \right) \right)$$

iff

$$\tilde{\blacklozenge}(\{v_i\}) = v_i$$

and for all $X, Y \neq \emptyset$, with $X \cup Y \subseteq W$ and $X \cap Y = \emptyset$, we have

$$\tilde{\blacklozenge}(X \cup Y) = \tilde{\blacklozenge}\left( X \cup \{\tilde{\blacklozenge}(Y)\} \right).$$

Then, the inducing connective is given by

$$\tilde{\heartsuit}(v_i, v_j) = \tilde{\blacklozenge}(\{v_i, v_j\}).$$

Binary connectives with the above properties can be identified with a finite partial ordering $\leq_\heartsuit$ of the upper semi-lattice $\mathcal{L}_\vee = (W_n, \leq_\heartsuit)$ for

## 5.7. Quantification in many-valued logics

which all non-empty sets have a supremum by the definition

$$v_i \leq v_j \quad \text{iff} \quad \widetilde{\heartsuit}(v_i, v_j) = sup_{\leq_\heartsuit}\{v_i, v_j\} = v_j$$

such that for some set $V \subseteq W_n$ we have $\blacklozenge^\heartsuit(V) = sup_{\leq_\heartsuit}(V)$. Do the same for the lower semi-lattice $\mathcal{L}_\wedge = (W_n, \leq_\heartsuit)$ with the same ordering such that we have $\blacklozenge^\heartsuit(V) = inf_{\leq_\heartsuit}(V)$. Let now $W_n$ be totally ordered and define the truth tables for the classical connectives that are commutative, associative, and idempotent, i.e. $\wedge$ and $\vee$, as:

$$\widetilde{\wedge}(v_i, v_j) = v_{min(i,j)} \qquad \widetilde{\vee}(v_i, v_j) = v_{max(i,j)}$$

Then, the classical quantifiers $\forall$ and $\exists$ can be defined as *generalizations* of $\wedge$ and $\vee$, respectively, so that we have for all non-empty $V \subseteq W_n$

$$\widetilde{\forall}(V) = v_{min\{i \mid v_i \in V\}}$$

and

$$\widetilde{\exists}(V) = v_{max\{i \mid v_i \in V\}}.$$

In an equivalent formulation, both $\wedge$ and $\forall$ / both $\vee$ and $\exists$ are based on the same semi-lattices.

Recall now also Definition 2.2.11.5, according to which the value of $\blacklozenge_i x$ is determined by the set of substitution instances of a formula $\phi$ given some valuation *val* and an interpretation $\mathcal{I}$, i.e., $val_\mathcal{I}(\blacklozenge_i x \phi(x)) = \blacklozenge_i(\Delta_{\mathcal{I},x}\phi(x))$ where $\Delta_{\mathcal{I},x}\phi(x) = \{val_{\mathcal{I}_a^x}\phi(a) \mid a \in \mathscr{D}\}$ is the distribution of $\phi(x)$ in $\mathcal{I}$ with respect to $x$ and $\mathcal{I}_a^x$ is the interpretation identical to $x$ when setting $\varpi(x) = a$. We can now define this distribution via the function $\blacklozenge_i$ as $val_\mathcal{I}(\blacklozenge_i x \phi(x)) = \blacklozenge_i(\{val_{\mathcal{I}_a^x}(\phi(a)) \mid a \in \mathscr{D}\})$ for a a constant such that we have $val_{\mathcal{I}_a^x}(\varpi(x) = a)$. Validity remains as in Definition 5.2.15 above. In fact, we are only interested in two quantifiers, $\forall$ and $\exists$, for which, in classical terms, we have the following:

**5.7.3. (Prop.)** Given a finite domain $\mathscr{D} = \{a_1, ..., a_n\}$ of nominal constants ascribed to objects, it is the case that

$$\forall x \phi(x) \equiv_\mathscr{D} \phi[x/a_1] \wedge ... \wedge \phi[x/a_n] = \phi(a_1) \wedge ... \wedge \phi(a_n)$$

$$\exists x \phi(x) \equiv_\mathscr{D} \phi[x/a_1] \vee ... \vee \phi[x/a_n] = \phi(a_1) \vee ... \vee \phi(a_n)$$

where $\equiv_\mathscr{D}$ designates formula equivalence under any interpretation in $\mathscr{D}$.

*Proof:* Left as an exercise.

With respect to Proposition 5.7.3, we say that $\forall$ *distributes over* $\wedge$ and $\exists$ *distributes over* $\vee$.

## 5. Many-valued logics

Definition 5.7.2 is naturally generalized to the many-valued logics in which both conjunction and disjunction are commutative, associative, and idempotent operations. Given that in many-valued logics the truth values come in degrees or in an ordered sequence, an easy way to generalize Definition 5.7.2 to them is as in Examples 5.7.1-2.

**Example 5.7.1.** For the logical systems $L_n^*$ and $P_n^*$, $n$ is finite, designating the FO extensions of $L_n$ and $P_n$, respectively, for any interpretation $\mathcal{I} = \mathcal{I}_a^x$ in a domain $\mathscr{D}$:

$$val_\mathcal{I}(\forall x A(x)) = min\{val_{\mathcal{I}_a^x}(A(a)) | a \in \mathscr{D}\}$$

$$val_\mathcal{I}(\exists x A(x)) = max\{val_{\mathcal{I}_a^x}(A(a)) | a \in \mathscr{D}\}$$

Contrast with the following Example, provided here for comparison.

**Example 5.7.2.** For the logical system $L_\aleph^*$, we introduce the quantifiers $\forall$ and $\exists$ in the following way for any interpretation $\mathcal{I} = \mathcal{I}_a^x$ in a domain $\mathscr{D}$:

$$val_\mathcal{I}(\forall x A(x)) = inf\{val_{\mathcal{I}_a^x}(A(a)) | a \in \mathscr{D}\}$$

$$val_\mathcal{I}(\exists x A(x)) = sup\{val_{\mathcal{I}_a^x}(A(a)) | a \in \mathscr{D}\}$$

Yet another way to generalize the classical quantifiers to many-valued logics is by means of descriptions.

**Example 5.7.3.** For Bochvar's FO logical system, and abbreviating $val_{\mathcal{I}_a^x}(\phi)$ as $/\phi/$, we define $\forall$ and $\exists$ as:

$$/\forall x A(x)/ = \begin{cases} t & \text{when} \quad /A(a)/ = t \text{ for every } a \in \mathscr{D} \\ i & \text{when} \quad /A(a)/ = i \text{ for some } a \in \mathscr{D} \\ f & \text{otherwise} \end{cases}$$

$$/\exists x A(x)/ = \begin{cases} f & \text{when} \quad /A(a)/ = f \text{ for every } a \in \mathscr{D} \\ i & \text{when} \quad /A(a)/ = i \text{ for some } a \in \mathscr{D} \\ t & \text{otherwise} \end{cases}$$

All the above follows the generalization proposed by Mostowski (1957). While in Examples 5.7.1 and 5.7.3 (i.e. for finitely many-valued logics) problems may arise in the new interpretation of the quantifiers, adequate axiomatic systems can be created for them as extensions of axiom systems for propositional calculi.

## 5.7. Quantification in many-valued logics

| | | |
|---|---|---|
| 1. | $\forall x P(x) \to P(a)$ | $\mathscr{L}_3^*6,\ \forall x A(x) / \forall x P(x),\ x/a$ |
| 2. | $\forall x (\neg P(x)) \to \neg P(a)$ | $\mathscr{L}_3^*6,\ \forall x A(x) / \forall x (\neg P(x)),\ x/a$ |
| 3. | $\neg\neg \forall x (\neg P(x)) \to \neg P(a)$ | DN (2) |
| 4. | $P(a) \to \neg \forall x (\neg P(x))$ | GConP (3) |
| 5. | $\forall x P(x) \to \neg \forall x (\neg P(x))$ | HS (1, 4) |
| 6. | $\forall x P(x) \to \exists x P(x)$ | QD$_\exists$ (5) |

Figure 5.7.1.: Proof of $\vdash_{\mathrm{L}_3^*} \forall x P(x) \to \exists x P(x)$ in $\mathscr{L}_3^*$.

**5.7.4.** It suffices to add to a suitable axiom system of this kind (e.g., the axiom system of Hilbert & Bernays, 1934; 1939) the following axioms $\mathcal{A}^*$ and $\mathcal{A}^{**}$, conditions QD$_{\exists,\forall}$, and the rule GEN$'$ (Malinowski, 1993):

| | | |
|---|---|---|
| ($\mathcal{A}^*$) | $\forall x (B \to A) \to (B \to \forall x A)$ | Condition: $x$ is not free in $B$ |
| ($\mathcal{A}^{**}$) | $\forall x A(x) \to A(y)$ | Condition: the substitution $A_y^x$ is permissible, i.e. $y$ is not to be bound in any place where $x$ is free in $A(x)$ |
| | $\forall x A(x) \to A(a)$ | Specification of $\mathcal{A}^{**}$ for constants |
| (QD$_{\exists,\forall}$) | $\exists x A = \neg \forall x (\neg A)$, $\forall x A = \neg \exists x (\neg A)$ | Property of mutual definability of the quantifiers |
| (GEN$'$) | $A(x)/\forall y A(y)$ | *Generalization rule* |

**Example 5.7.4.** If to $\mathscr{L}_3$ we add $\mathcal{A}^*$ and $\mathcal{A}^{**}$, relabeling the whole list of six axiom schemata respectively as $\mathscr{L}_3^*$1-6, and we further add to these the generalization rule GEN$'$, we obtain $\mathscr{L}_3^*$, an adequate proof calculus for L$_3^*$.

**Example 5.7.5.** We prove in $\mathscr{L}_3^*$ the theorem $\vdash_{\mathrm{L}_3^*} \forall x P(x) \to \exists x P(x)$ (see Fig. 5.7.1). For the derived rules used, see Example 5.4.3.

## 5. Many-valued logics

But Mostowski (1957) did not invent the wheel as far as generalized quantifiers are concerned. As we are mostly interested in conjunctive and disjunctive normal forms (see Part III), the pioneering work of J. B. Rosser and A. R. Turquette deserves some detailed consideration. Starting from the intuition that quantifiers can be treated as functions on the set of pairs $(x, A)$, $x \in Vi$ and $A$ is a formula (e.g., $\forall x A(x) = \blacklozenge_1(x, A); \exists x A(x) = \blacklozenge_2(x, A))$, Rosser & Turquette (1952) elaborated on a general theory of quantification for finitely many-valued logics. We next present some of the most important definitions and results of this theory.

**5.7.5. (Def.)** A *Rosser-Turquette (RT) generalized quantifier* is a formula of the form

$$\blacklozenge_i(x_1, ..., x_{m_i}, A_1, ..., A_{t_i})$$

where $x_1, ..., x_{m_i}$ are nominal variables and $A_1, ..., A_{t_i}$ are formulae composed of predicates, variables (propositional or nominal), and propositional connectives.

It is immediately obvious that a RT generalized quantifier may take several variables and formulae. The idea is to valuate the $\blacklozenge_i$ functions by means of an interpretation assigning to formulae values from the set $\{1, 2, ..., n\}$ (cf. Def. 5.2.5) and where the many-valued quantifiers are constructed by means of the connectives between the formulae.

To that end, Rosser and Turquette came up with the convenient notion of *partial normal form* (PaNF) both with respect to connectives and quantifiers.

**5.7.6. (Def.)** In the former case, for a given connective $\heartsuit$ (abbreviating $\heartsuit_i$), the $i$-th PaNF is a signed formula expression schema $E_i \equiv \heartsuit(A_1, ..., A_n)$ containing only (signed) formulae $A_1, ..., A_n$ and in CNF.[20]

**Example 5.7.6.** The following right sides of the equalities are the PaNFs for the connectives $\neg$ and $\vee$ of Ł$_3$:

---

[20] Rosser and Turquette used subscripts to indicate the truth value of a formula (e.g., $p_1$ reads "$p$ takes the value 1") and did not refer to this as "signed formulae" or "signed formula expressions."

## 5.7. Quantification in many-valued logics

$$\begin{aligned}
\{\mathtt{t}\}\,[\neg A]) &= \{\mathtt{f}\}\,[A] \\
\{\mathtt{i}\}\,[\neg A] &= \{\mathtt{i}\}\,[A] \\
\{\mathtt{f}\}\,[\neg A] &= \{\mathtt{t}\}\,[A] \\
\{\mathtt{t}\}\,[A \vee B] &= \{\mathtt{t}\}\,[A] \vee \{\mathtt{t}\}\,[B] \\
\{\mathtt{i}\}\,[A \vee B] &= (\{\mathtt{i}\}\,[A] \vee \{\mathtt{i}\}\,[B]) \wedge (\{\mathtt{f}\}\,[A] \vee \{\mathtt{i}\}\,[A]) \\
& \quad \wedge (\{\mathtt{f}\}\,[B] \vee \{\mathtt{i}\}\,[B]) \\
\{\mathtt{f}\}\,[A \vee B]) &= \{\mathtt{f}\}\,[A] \wedge \{\mathtt{f}\}\,[B]
\end{aligned}$$

The PaNFs for the remaining connectives of $L_3$ (and their negations; e.g., $\neg(\{\mathtt{f}\}\,[A \vee B]) = \neg(\{\mathtt{f}\}\,[A]) \vee \neg(\{\mathtt{f}\}\,[B])$ can now be easily completed, as it is evident that PaNFs are the *exhaustive* enumeration of the interpretations for the connectives; moreover, they are *mutually exclusive*, in the sense that under any given interpretation exactly one of the PaNFs is valuated as true.

We can now provide the following definition:

**5.7.7. (Def.)** (Zach, 1993). A signed formula expression schema $E$ is the *i-th partial form* of $(\blacklozenge x)\, A(x)$ iff

1. the signed atoms in $E$ belong to $\{u_{ji}\,(A\,(t_j))\,|\,1 \leq j \leq p, 1 \leq i \leq n\} \cup \{w_{ji}\,(A\,(c_j))\,|\,1 \leq j \leq q, 1 \leq i \leq n\}$, where $c_i$ are Skolem constants and $t_i$ term variables.

2. for every $E'$ a pre-instance[21] of $E$ and every interpretation $\mathcal{I}$:

    a) if for all $a_1, ..., a_q \in \mathscr{D}$ there are $e_1, ..., e_p \in \mathscr{D}$ such that

    $$\models_\mathcal{I} E\,\{t_1/e_1, ..., t_p/e_p, c_1/a_1, ..., c_q/a_q\}$$

    then $val_\mathcal{I}\,(\blacklozenge x A'(x)) = \mathtt{v}_i$.

    b) if for all $e_1, ..., e_p \in \mathscr{D}$ there are $a_1, ..., a_q \in \mathscr{D}$ such that

    $$\not\models_\mathcal{I} E\,\{t_1/e_1, ..., t_p/e_p, c_1/a_1, ..., c_q/a_q\}$$

    then $val_\mathcal{I}\,(\blacklozenge x A'(x)) \neq \mathtt{v}_i$

    where $A'$ is the instance of $A$ determined by $E'$.

---

[21] Given a set of formulae $F_{L^*}$, a *pre-instance* $E'$ of a schema $E$ is a formula of $F_{L^*}$ containing occurrences of the free variables and of the term variables of $E$. For instance, a pre-instance of the formula $A\,(c,t)$ is $(P\,(c,a) \wedge Q\,(t)) \to P\,(t,c)$.

## 5. Many-valued logics

**Example 5.7.7.** The PaNFs for the quantifiers $\forall$ and $\exists$ of Ł3 are as follows:

$$\begin{aligned}
\mathtt{t}\,[\forall x A(x)] &= \mathtt{t}\,[A(c)] \\
\mathtt{i}\,[\forall x A(x)] &= \mathtt{i}\,[A(t)] \wedge (\mathtt{i}\,[A(c)] \vee \mathtt{t}\,[A(c)]) \\
\mathtt{f}\,[\forall x A(x)] &= \mathtt{f}\,[A(t)] \\
\mathtt{t}\,[\exists x A(x)] &= \mathtt{t}\,[A(t)] \\
\mathtt{i}\,[\exists x A(x)] &= \mathtt{i}\,[A(t)] \wedge (\mathtt{f}\,[A(c)] \vee \mathtt{i}\,[A(c)]) \\
\mathtt{f}\,[\exists x A(x)] &= \mathtt{f}\,[A(c)]
\end{aligned}$$

Informally, an expression of the form $\forall x A(x)$ stands for the SFE resulting from $A(x)$ by replacing $x$ with some Skolem constant, and an expression of the form $\exists x A(x)$ stands for the SFE resulting from the substitution of $x$ with some variable term.

**5.7.8. (Prop.)** The $i$-th normal form of $\blacklozenge x A(x)$ is a PaNF if it is in CNF.

*Proof:* Left as an exercise.

Finally:

**5.7.9. (Prop.)** (Rosser & Turquette, 1952). For any formula of the form $\forall x \phi$, suppose that $\phi$ has only free occurrences of the individual variable $z$, and let $\phi_r(z)$ denote the $r$-th PaNF of $\phi$; then, the PaNFs $N_r$ of $\forall x \phi$ satisfy standard conditions iff the expression

$$\forall z\,(\phi_1(z) \vee ... \vee \phi_s(z)) \equiv (N_1 \vee ... \vee N_s)$$

is provable in the ordinary two-valued predicate calculus.

By "satisfying standard conditions," it is meant that $\forall x \phi$ takes a designated value iff $\phi$ always takes a designated value (cf. Def. 5.2.5).

Importantly, this broadened definition of standard conditions allows for the axiomatization of the standard $n$-valued predicate calculi in the following way.

**5.7.10.** We obtain an axiom system for the standard $n$-valued FO calculi by augmenting $\mathcal{A}$1-7 (in 5.2.8) with the two axiom schemata $\mathcal{A}^{*-**}$ of 5.7.4 (which we can relabel as $\mathcal{A}$8-9), and by further adding the generalization rule GEN and the following axiom schema:

$$(\mathcal{A}10) \quad N_r\,(\blacklozenge_i\,(x_1, ..., x_{m_i}, \phi_1, ...\phi_{t_i})) \rightarrow$$
$$J_r\,(\blacklozenge_i\,(x_1, ..., x_{m_i}, \phi_1, ...\phi_{t_i})) \quad \text{where } 1 \leq r \leq n,$$
$$1 \leq i \leq c$$

## 5.7. Quantification in many-valued logics

Given the signed formula expression schema $E_i \equiv \heartsuit_i^n(A_1, ..., A_n)$, it is now easy to see how the PaNF strategy applies to the $\blacklozenge_i$ functions as defined above, in order to specify the truth-functional behavior of quantifiers in many-valued logics. First, recall from Definition 2.1.8.3 that a truth function for a quantifier $\blacklozenge_i$ is a mapping from non-empty sets of truth values to truth values, i.e. $\blacklozenge_i = (2^W - \emptyset) \longrightarrow W$. Recall now also Definition 2.2.11.5.

Let us now retake the distribution subject:

**5.7.11. (Def.)** Let a formula $\psi = \blacklozenge_i x \phi(x)$ be given; the range of $\blacklozenge_i$ is the set of truth values that describes the situation in which ground instances of $x$ take exactly the values in this set under a given interpretation $\mathcal{I}$. Let $\Delta \subseteq W, \Delta \neq \emptyset$; then, $\psi$ takes the truth values $\blacklozenge_i(\Delta)$ if *for every* $v_j \in W$ it is the case that there is *some* $v_j \in \Delta$ iff there is some $a \in \mathcal{D}$ such that $val_\mathcal{I}(\phi(a)) = v_j$. We say that $\Delta$ is the *distribution* of $\psi$, and we call a like generalized quantifier a *distribution quantifier*.

**Example 5.7.8.** The truth functions for the distribution quantifiers $\tilde{\forall}$ and $\tilde{\exists}$ of $L_3^*$ are as follows:

$$
\begin{array}{rclcrcl}
\tilde{\forall}(\{t\}) &=& t & \quad & \tilde{\exists}(\{t\}) &=& t \\
\tilde{\forall}(\{i,t\}) &=& i & & \tilde{\exists}(\{i,t\}) &=& t \\
\tilde{\forall}(\{f,t\}) &=& f & & \tilde{\exists}(\{f,t\}) &=& t \\
\tilde{\forall}(\{f,i,t\}) &=& f & & \tilde{\exists}(\{f,i,t\}) &=& t \\
\tilde{\forall}(\{i\}) &=& i & & \tilde{\exists}(\{i\}) &=& i \\
\tilde{\forall}(\{f,i\}) &=& f & & \tilde{\exists}(\{f,i\}) &=& i \\
\tilde{\forall}(\{f\}) &=& f & & \tilde{\exists}(\{f\}) &=& f \\
\end{array}
$$

In terms of the inverse image of $\tilde{\blacklozenge}$, i.e. $\tilde{\blacklozenge}^{-1}$, we have the following:

$$
\begin{array}{rcl}
\tilde{\forall}^{-1}(t) &=& \{\{t\}\} \\
\tilde{\forall}^{-1}(i) &=& \{\{i\},\{i,t\}\} \\
\tilde{\forall}^{-1}(f) &=& \{\{f\},\{f,i\},\{f,t\},\{f,i,t\}\} \\
\tilde{\exists}^{-1}(t) &=& \{\{t\},\{f,t\},\{i,t\},\{f,i,t\}\} \\
\tilde{\exists}^{-1}(i) &=& \{\{i\},\{f,i\}\} \\
\tilde{\exists}^{-1}(f) &=& \{\{f\}\} \\
\end{array}
$$

## 5.7.2. Quantification in fuzzy logics

Some critical difficulties arise with relation to quantification in infinitely many-valued logics. For instance, in case the set $\mathscr{D}$ is infinite, the set $\{val_{\mathcal{I}_a^x}(A(a)) \,|\, a \in \mathscr{D}\}$ may not contain a least or a greatest element, and the operations $min$ and $max$ are thus not possible. A problematic case is Ł$_{\aleph_0}^*$, given that the set of truth values of this logic is not closed under the operations of conjunction and disjunction. As for Ł$_{\aleph_1}^*$, it has been shown (Scarpelini, 1962) that, if the quantifiers $\forall$ and $\exists$ are introduced as in Example 5.7.2, its predicate calculus is not axiomatizable.

With respect to quantification for t-norm fuzzy logics, it both is mathematically complex and confronts us often with unaxiomatizable systems. Nevertheless, we do indeed have well-behaving first-order FLs. In effect, and to begin with, in order to obtain L$_\star^*$, denoting first-order L$_\star$ (cf. Def. 5.6.5), we add to L$_\star$ a domain of quantification $\mathscr{D}$ and define $\forall$ and $\exists$ as the operations $inf$ and $sup$ (cf. Example 5.7.2).

**5.7.12. (Def.)** We obtain an axiom system for first-order BL, denoted by BL*, by adding to the axioms and rules of $\mathcal{BL}$ (Def. 5.6.6), for $B$ denoting some closed formula in which $x$ is not free, the following axiom schemata $\mathcal{BL}$*1-5 and rule for universal generalization GEN$\forall$:

$(\mathcal{BL}^*1)$ $\quad \forall x A(x) \to A(a)$
$(\mathcal{BL}^*2)$ $\quad A(a) \to \exists x A(x)$
$(\mathcal{BL}^*3)$ $\quad \forall x (B \to A) \to (B \to \forall x A)$
$(\mathcal{BL}^*4)$ $\quad \forall x (A \to B) \to (\exists x A \to B)$
$(\mathcal{BL}^*5)$ $\quad \forall x (B \vee A) \to (B \vee \forall x A)$

(GEN$\forall$) $\quad$ If $\vdash A_x^a$, then $\vdash \forall x A(x)$

However, only G$_\aleph^*$ has been provided with an adequate axiomatization, by adding to $\mathcal{BL}$1-8 and $\mathcal{BL}$*1-2 the axiom of Proposition 5.6.10, together with the rule MP (Hájek, 1997). With respect to Ł$_\aleph^*$, there is a first-order generalization of a Pavelka-style system[22] for Łukasiewicz fuzzy propositional logic (Novák, 1990; Novák, Perfilieva, & Močkoř, 1999). There is at present no adequate axiom system for $\Pi_\aleph^*$. Just as in the case of their propositional systems, BL*, Ł$_\aleph^*$, G$_\aleph^*$, and $\Pi_\aleph^*$ find an appropriate

---

[22] In a Pavelka-style system (Pavelka, 1979), besides the formulae $\phi$ one also considers their membership degrees to some fuzzy set $\Sigma^\sim$, it being the case that the membership degrees $\Sigma^\sim(\phi)$ are in fact the truth degrees, which form a residuated lattice $\mathcal{L} = (A, \cap, \cup, \star, \to, 0, 1)$. Bergmann (2008) provides various examples of Pavelka-style approaches to both propositional and predicate Łukasiewicz fuzzy logics.

## 5.7. Quantification in many-valued logics

algebraic semantics in BL-algebras, MV-algebras, G-algebras and Π-algebras, respectively (Esteva et al., 2003).

Thus, the specification of an interpretation for first-order FLs is particularly important.

**5.7.13. (Def.)** A triple $\mathcal{I} = (\mathscr{D}, \Theta, \varpi)$ is an interpretation for a first-order FL if the following conditions hold:

1. $\mathscr{D} \neq \emptyset$ is a domain;

2. $\Theta : \mathscr{D}^n \longrightarrow [0, 1]$ is a mapping such that for every predicate $P$ with arity $n > 0$ we have $val_\mathcal{I}(P(x_1, ..., x_n)) \in [0, 1]$;

3. $\varpi$ is an assignment of a member of $\mathscr{D}$ to each individual constant $a$ such that $\mathcal{I}(a) \in \mathscr{D}$.

It is important to remark that what is at stake in Definition 5.7.13 is the *degree of membership* of n-tuples of members of the domain $\mathscr{D}$ to a particular (fuzzy) set. This degree ranges from 1, for absolute membership, to 0, for definite non-membership.

**5.7.14. (Def.)** Given a triple $\mathcal{I} = (\mathscr{D}, \Theta, \varpi)$, the truth conditions for atomic formulae are defined in $Ł_\aleph^*$ as

$$val_\mathcal{I}(P(t_1, ..., t_n)) = \Theta(P)\left(val_{\mathcal{I}^*}(t_1), ..., val_{\mathcal{I}^*}(t_n)\right)$$

where the interpretation $\mathcal{I}^*$ is $\mathcal{I}$ or $\mathcal{I}_a^x$ for $t_i$ a constant or a variable, respectively.

**Example 5.7.9.** Let $N$ be the set of large digits and let $\mathscr{D} = \{0, 1, 2, 3, 4, 5, 6, 7, 8, 9\}$. We abbreviate $val_\mathcal{I}(N(x))$ as $/N(x)/$. Then, we have, for instance, $/N(0)/ = 0$, $/N(2)/ = 0.1$, $/N(5)/ = 0.5$, $/N(7)/ = 0.8$, and $/N(9)/ = 1$.

**5.7.15.** Let $val_{\mathcal{I}_a^x}(\phi) = /\phi/$ for some interpretation $\mathcal{I}$. Given a domain $\mathscr{D}$, the truth conditions for FOL (atomic) formulae in $Ł_\aleph^*$ are as:

1. $/\neg P/ = 1 - /P/$

2. $/P \wedge Q/ = min(/P/, /Q/)$

3. $/P \vee Q/ = max(/P/, /Q/)$

4. $/P \rightarrow Q/ = min(1, 1 - /P/ + /Q/)$

5. $/P \leftrightarrow Q/ = min(1, 1 - /P/ + /Q/, 1 - /Q/ + /P/)$

## 5. Many-valued logics

6. $/P\&Q/ = max\,(0, /P/ + /Q/ - 1)$

7. $/P\nabla Q/ = min\,(1, /P/ + /Q/)$

8. $/\forall xP(x)/ = glb\,\{/P(a)/\,|a \in \mathscr{D}\}$

9. $/\exists xP(x)/ = lub\,\{/P(a)/\,|a \in \mathscr{D}\}$

**5.7.16. (Prop.)** Let $z, a_x, b_x \in [0,1]$. The following equalities, where $a_x$ denotes the substitution of $x$ by $a$, hold for the FLs:

1. $glb\,\{x \pm z | x \in X\} = glb\,(X) \pm z$

2. $lub\,\{x \pm z | x \in X\} = lub\,(X) \pm z$

3. $glb\,\{-x | x \in X\} = -lub\,(X)$

4. $lub\,\{-x | x \in X\} = -glb\,(X)$

5. $glb\,\{z - x | x \in X\} = z - lub\,(X)$

6. $lub\,\{z - x | x \in X\} = z - glb\,(X)$

7. $lub\,\{a_x | x \in X\} \leq lub\,\{b_x | x \in X\}$ if $a_x \leq b_x$ for all $x \in X$

8. $glb\,\{max\,(z, x)\,| x \in X\} = max\,(z, glb\,(X))$

9. $glb\,\{z \star x | x \in X\} = z \star glb\,(X)$

10. $lub\,\{a_x | x \in X\} \star lub\,\{b_x | x \in X\} \leq lub\,\{a_x \star b_x | x \in X\}$

**Example 5.7.10.** Let the domain $\mathscr{D} = \{C, S, B, F, W\}$ be given, the members of which are the animals in *Schubert's steamroller* (cf. Example 4.6.13): *C*aterpillar, *S*nail, *B*ird, *F*ox, and *W*olf, respectively. Define the binary relation $MS \subseteq \mathscr{D} \times \mathscr{D}$, "animal $x$ is much smaller than animal $y$," as in Figure 5.7.2.

| MS | C | S | B | F | W |
|---|---|---|---|---|---|
| C | 0 | 0.5 | 0.8 | 1 | 1 |
| S | 0 | 0 | 0.7 | 1 | 1 |
| B | 0 | 0 | 0 | 0.9 | 1 |
| F | 0 | 0 | 0 | 0 | 0.5 |
| W | 0 | 0 | 0 | 0 | 0 |

Figure 5.7.2.: A fuzzy binary relation.

## 5.7. Quantification in many-valued logics

Clearly, $MS$ is a *fuzzy binary relation*, as it corresponds to the mapping $MS : \mathscr{D}^2 \longrightarrow [0,1]$. Let now $Ms(x,y)$ be the *binary fuzzy predicate* corresponding to the relation $MS$, i.e. $Ms : \mathscr{D}^2 \longrightarrow [0,1]$, and let further $MS$ establish the lowest possible truth value for the predicate $Ms(x,y)$. Then, and abbreviating $val_{\mathcal{I}}(\phi)$ as $/\phi/$ for any formula $\phi$, we have for instance $/\forall x \forall y\,((C(x) \wedge S(y)) \rightarrow Ms(x,y))/ = 1$ in $Ł_{\aleph}^*$. In effect, by 5.7.15.8, $/\forall x C(x)/ = 0$ and $/\forall y S(y)/ = 0$, as the truth degree to which any $x$ or $y$ can belong to the sets $C$ and $S$ respectively is the *minimum* degree of membership in the interval $[0,1]$ of any object to any set. In turn, the valuation of the conjunction $C(x) \wedge S(y)$ is itself also 0 by 5.7.15.2, and finally $/(C(x) \wedge S(y)) \rightarrow Ms(x,y)/ = 1$, because 0.5 is the minimum truth degree for the predicate $Ms(x,y)$ when $x \in C$ and $y \in S$, and by 5.7.15.4 we have $min(1, 1 - 0 + 0.5) = 1$.

**5.7.17.** Let $val_{\mathcal{I}_a^x}(\phi) = /\phi/$ for some interpretation $\mathcal{I}$. Given a domain $\mathscr{D}$, the truth conditions for FOL (atomic) formulae in $G_{\aleph}^*$ are:

1. $/\neg P/ = \begin{cases} 1 & \text{if } /P/ = 0 \\ 0 & \text{otherwise} \end{cases}$

2. $/P \wedge Q/ = /P \& Q/ = min(/P/, /Q/)$

3. $/P \triangledown Q/ = max(/P/, /Q/)$

4. $/P \rightarrow Q/ = \begin{cases} 1 & \text{if } /P/ \leq /Q/ \\ /Q/ & \text{otherwise} \end{cases}$

5. $/P \leftrightarrow Q/ = \begin{cases} 1 & \text{if } /P/ = /Q/ \\ min(/P/, /Q/) & \text{otherwise} \end{cases}$

6. $/\forall x P(x)/ = glb\,\{/P(a)/ | a \in \mathscr{D}\}$

7. $/\exists x P(x)/ = lub\,\{/P(a)/ | a \in \mathscr{D}\}$

**5.7.18.** Let $val_{\mathcal{I}_a^x}(\phi) = /\phi/$ for some interpretation $\mathcal{I}$. Given a domain $\mathscr{D}$, the truth conditions for FOL (atomic) formulae in $\Pi_{\aleph}^*$ are:

1. $/\neg P/ = \begin{cases} 1 & \text{if } /P/ = 0 \\ 0 & \text{otherwise} \end{cases}$

2. $/P \& Q/ = /P/ \cdot /Q/$

3. $/P \triangledown Q/ = /P/ + /Q/ - (/P/ \cdot /Q/)$

## 5. Many-valued logics

4. $/P \to Q/ = \begin{cases} 1 & \text{if } /P/ \leq /Q/ \\ \frac{/Q/}{/P/} & \text{otherwise} \end{cases}$

5. $/\forall x P(x)/ = glb\, \{/P(a)/ | a \in \mathscr{D}\}$

6. $/\exists x P(x)/ = lub\, \{/P(a)/ | a \in \mathscr{D}\}$

**5.7.19. (Prop.)** Let fuzzy validity be defined as in Definition 5.6.2 and let fuzzy entailment be as in Proposition 5.6.3. Then:

$$\top(Ł_\aleph^*), \top(G_\aleph^*), \top(\Pi_\aleph^*) \subsetneq \top(K_{pr})$$

$$\bot(Ł_\aleph^*), \bot(G_\aleph^*), \bot(\Pi_\aleph^*) \subsetneq \bot(K_{pr})$$

**5.7.20. (Prop.)** If $X \models_{Ł_\aleph^*} \phi, X \models_{G_\aleph^*} \phi, X \models_{\Pi_\aleph^*} \phi$, then $X \models \phi$, but the converse does not hold.

**5.7.21. (Prop.)** Every degree-entailment, and every $n$-degree-entailment in these logics (when applicable; cf. Def.s 5.2.12 and 5.2.14) is a classical entailment, but the converse does not hold.

We leave the proofs of the above propositions as exercises.

# Exercises

**Exercise 5.1.** Prove in $Ł_3$ the theorems in Example 5.4.2, i.e. show that they are indeed derived axioms.

**Exercise 5.2.** For the connectives $\neg$ and $\to$, give their truth tables in the logical systems $Ł_4$, $Ł_5$, and $Ł_8$.

**Exercise 5.3.** With functional completeness in view, we can expand $Ł_3$ with the two additional connectives $\&$ and $\triangledown$, called *bold* (or *strong*) *conjunction* and *bold* (or *strong*) *disjunction*, respectively. Their truth tables are as follows:

| &  | t | i | f |
|----|---|---|---|
| t  | t | i | f |
| i  | i | f | f |
| f  | f | f | f |

| $\triangledown$ | t | i | f |
|-----------------|---|---|---|
| t               | t | t | t |
| i               | t | t | i |
| f               | t | i | f |

1. Show that with these additional connectives $Ł_3$ is indeed functionally complete.

2. How do these additional connectives impact on the set of tautologies of $Ł_3$? (Hint: Check the classical tautologies in Proposition 3.5.3 with the additional connectives.)

**Exercise 5.4.** Let us consider the theorem proven in Example 5.4.4 as a derived axiom (no. 1) of $\mathscr{L}_\aleph$. Prove the remaining derived axioms of $\mathscr{L}_\aleph$ listed below:

1. $P \to ((P \to Q) \to Q)$
2. $\neg P \to (P \to Q)$
3. $((P \to Q) \to R) \to (\neg P \to R)$
4. $(((Q \to R) \to (P \to R)) \to S) \to ((P \to Q) \to S)$
5. $(P \to (Q \to R)) \to ((S \to Q) \to (P \to (S \to R)))$
6. $(P \to (Q \to R)) \to (Q \to (P \to R))$
7. $P \to P$
8. $\neg\neg P \to P$
9. $P \to \neg\neg P$

## 5. Many-valued logics

**Exercise 5.5.** Let $\mathfrak{K}_3^S = \{\{\mathtt{f},\mathtt{i},\mathtt{t}\}, \neg, \wedge, \{\mathtt{t}\}\}$ be the minimal matrix of $\mathrm{K}_3^S$. Define the remaining connectives in terms of the primitive ones.

**Exercise 5.6.** Check whether one can define the connectives of $\mathrm{B}_3^I$, $\mathrm{K}_E^S$, and Ł3 by means of the connectives of $\mathrm{B}_3^E$.

**Exercise 5.7.** Prove the following propositions:

1. (*Modified deduction-detachment theorem*) $X, A \vdash_{\text{Ł3}} B$ iff $X \vdash_{\text{Ł3}} A \rightarrow (A \rightarrow B)$.

2. Every quasi-entailment in Ł3, $\mathrm{K}_3^S$, $\mathrm{B}_3^I$, and $\mathrm{B}_3^E$ is a classical entailment.

3. (*Quasi-deduction theorem*) If $A_1, ..., A_n \models B$, then $A_1, ..., A_{n-1} \models A_n \rightarrow_{\text{KS}} B$.

4. Every degree-entailment of a formula $\phi$ by a set of formulae $X$ that holds in Ł3, $\mathrm{K}_3^S$, $\mathrm{B}_3^I$, or $\mathrm{B}_3^E$ is a classical entailment, but not vice-versa.

5. If $X$ is inconsistent, then $X \vdash_{\text{BE}} \neg\phi$.

**Exercise 5.8.** Build the truth tables for the following formulae:

1. $A \rightarrow_{\text{Ł3}} A$

2. $A \rightarrow_{\text{BI}} (\neg_{\text{BI}} A \wedge_{\text{BI}} B)$

3. $A \rightarrow_{\text{Ł3}} \neg_{\text{Ł3}} A$

4. $(\phi \rightarrow_{\text{BE}} \phi) \rightarrow_{\text{BE}} \psi$

5. $\phi \rightarrow_{\text{KS}} \psi$

6. $P \rightarrow_{\text{KW}} \neg_{\text{KW}} (P \vee_{\text{KW}} Q)$

7. $(P \wedge_{\text{KS}} Q) \rightarrow_{\text{KS}} (P \vee_{\text{KS}} Q)$

8. $(P \leftrightarrow_{\text{Q3}} Q) \wedge_{\text{Q3}} (P \rightarrow_{\text{Q3}} \neg_{\text{Q3}} Q)$

9. $P \nabla_{\text{Ł3}} \neg_{\text{Ł3}} P$

10. $\neg_{\text{Ł3}} (A \wedge_{\text{Ł3}} \neg_{\text{Ł3}} A)$

11. $P \to_{F3} P$

12. $P \supset_{F3} P$

13. $\#_{F3} (A \vee_{F3} \neg_{F3} A)$

14. $-_{F3} (P \wedge_{F3} \neg_{F3} P)$

15. $((A \to_{Hn3} B) \wedge_{Hn3} \neg_{Hn3} B) \to_{Hn3} \neg_{Hn3} A$

16. $\phi \leftrightarrow_{Hn3} (\neg_{Hn3} \neg_{Hn3} \phi)$

17. $\mathbf{F} \left( A \vee_{\mathring{A}n3} \neg_{\mathring{A}n3} A \right)$

18. $\mathbf{L}P \vee_{\mathring{A}n3} \mathbf{M} \left( \#_{\mathring{A}n3} P \right)$

19. $\#_{Sn_3^1} \left( A \vee_{Sn_3^1} \neg_{Sn_3^1} A \right)$

20. $\neg_{Sn_3^1} \left( Q \wedge_{Sn_3^1} \neg_{Sn_3^1} Q \right)$

21. $\|_{Sn_3^1} \left( A \vee_{Sn_3^1} \neg_{Sn_3^1} A \right)$

22. $=_{Sn_3^1} \left( A \vee_{Sn_3^1} \neg_{Sn_3^1} A \right)$

23. $-_{Sn_3^1} \left( Q \wedge_{Sn_3^1} \neg_{Sn_3^1} Q \right)$

24. $\left( -_{Rn3} \to_{BE3} \left( Q \wedge_{Sn_3^1} \neg_{Sn_3^1} Q \right) \right)$

25. $\sim_{Q3} (P \vee_{Q3} \neg_{Q3} P)$

26. $P \vee_{Q3} -_{Q3} P$

27. $P \vee_{Q3} \sim_{Q3} P$

28. $((A \rightsquigarrow_{Q3} B) \wedge_{Q3} A) \rightsquigarrow_{Q3} B$

29. $((A \supset_{Q3} B) \wedge_{Q3} A) \supset_{Q3} B$

30. $\sim_{Q3} (P \nabla_{L3} -_{Rn3} P)$

**Exercise 5.9.** Check whether the formulae in Exercise 5.4 have equivalent formulae in the different finite propositional many-valued logics of Sections 5.4-5.

## 5. Many-valued logics

**Exercise 5.10.** Check whether the classical tautologies (Prop. 3.5.3) are so in Ł3 and in the different finite propositional many-valued logics of Section 5.5 (except for G3).

**Exercise 5.11.** Given the primitive connectives of F3, produce the truth tables for its secondary connectives.

**Exercise 5.12.** Verify whether the following formulae are tautologies in F3:

1. $(p \wedge q) \rightarrow p$
2. $p \leftrightarrow \neg\neg p$
3. $\neg \# p$
4. $\# p \rightarrow \#(p \vee q)$
5. $(+p \wedge \# q) \rightarrow \neg(p \rightarrow q)$

**Exercise 5.13.** Check whether $\mathcal{H}n$1-6 (cf. Prop. 5.5.28) are valid formulae in $Sn_3^2$.

**Exercise 5.14.** Verify whether the following formulae provide an axiom system for Rn3:

1. $(p \rightarrow q) \rightarrow ((q \rightarrow r) \rightarrow (p \rightarrow r))$
2. $p \rightarrow (q \rightarrow p)$
3. $((p \rightarrow q) \rightarrow p) \rightarrow p$
4. $(p \wedge q) \rightarrow p$
5. $(p \wedge q) \rightarrow q$
6. $p \rightarrow (q \rightarrow (p \wedge q))$
7. $p \rightarrow (\neg p \rightarrow q)$
8. $\neg(\neg p) \rightarrow p$
9. $p \rightarrow \neg(\neg p)$
10. $p \rightarrow ((q \rightarrow \neg q) \rightarrow \neg(p \rightarrow q))$

11. $\neg(p \wedge q) \to \neg(q \wedge p)$

12. $\neg(p \wedge q) \to ((p \to \neg p) \to \neg p)$

13. $(p \wedge \neg q) \to \neg(p \wedge q)$

14. $(\neg p \wedge \neg q) \to \neg(p \wedge q)$

**Exercise 5.15.** Check the validity of the formulae above in Ł$_3$ and in the finitely many-valued systems of Section 5.5 (except Rn$_3$).

**Exercise 5.16.** Show why the following classical tautologies fail to be so in G$_3$:

1. PEM
2. $\neg\neg A \to A$

**Exercise 5.17.** Check whether the following are tautologies or contradictions in G$_3$:

1. $A \to \neg\neg A$
2. $\neg\neg A \vee \neg A$
3. PNC
4. $A \to A$
5. $A \leftrightarrow \neg A$
6. $(A \wedge B) \wedge \neg(A \vee B)$

**Exercise 5.18.** Show informally that B$_4$ is not a generalization of classical logic.

**Exercise 5.19.** Characterize each many-valued logical system in Section 5.5 according to their structural properties.

**Exercise 5.20.** Consider the finitely $n$-valued logics P$_n$ and P$_n^r$.

1. Provide the general truth tables for the connectives $\neg, -, \to, \wedge, \vee$.

2. Define the connective $\leftrightarrow$ in functional terms.

3. Give the truth table for the connective $\rightarrow$ in $P_6^3$.

**Exercise 5.21.** Given the characterization of the connective $\leftrightarrow$ in $G_3$ (Def. 5.5.59.5), give its truth table.

**Exercise 5.22.** Check whether the following arguments are valid in the FLs studied above:

1. $P \rightarrow \neg P / \neg P$

2. $P \vee P / P$

**Exercise 5.23.** For each of the following formulae in $L_3^*$ find an interpretation $\mathcal{I}$ that satisfies it:

1. $\forall x \, (P(x) \rightarrow \neg Q(x))$

2. $\forall x \forall y \, (R(x,y) \vee R(y,x))$

3. $\forall x \forall y \, (\neg R(x,y) \vee R(y,x))$

4. $\exists x \exists y \, (R(x,y) \wedge R(y,x))$

5. $\forall x \exists y \, (P(x) \rightarrow R(x,y))$

6. $\forall x \forall y \, ((Q(x) \wedge \neg Q(y)) \rightarrow R(x,y))$

**Exercise 5.24.** Verify which of the following formulae are valid in $L_3^*$ by means of the proof system $\mathscr{L}_3^*$:

1. $(\forall x P(x) \vee \exists x Q(x)) \rightarrow \exists x \, (P(x) \wedge Q(x))$

2. $\forall x \, (\phi) \leftrightarrow \neg \exists x \, (\neg \phi)$

3. $\forall x \forall y \, (P(x) \wedge \neg P(y)) \rightarrow \neg P(y)$

**Exercise 5.25.** Evaluate in $G_\aleph^*$ and $\Pi_\aleph^*$ the formula of Example 5.7.10: $\forall x \forall y \, ((C(x) \wedge S(y)) \rightarrow Ms(x,y))$.

**Exercise 5.26.** Determine the truth conditions for the connective $\leftrightarrow_{\Pi_\aleph}$.

**Exercise 5.27.** Given an interpretation $\mathcal{I}$ with domain $\mathscr{D} = \{a, b, c\}$ and a valuation $val_\mathcal{I}$ such that $P(a) = 1$ (abbreviating $val_\mathcal{I}(P(a)) = 1$), $P(b) = 0$, $P(c) = 0.5$, $Q(a) = 0.8$, $Q(b) = 0.9$, $Q(c) = 1$, $R(a) = 0.5$, $R(b) = 0.3$, $R(c) = 1$ $S(a,b) = 0.2$, $S(b,c) = 0.8$, $S(a,c) = 1$, $T(a) = 0.8$, $T(b) = 0.4$, and $T(c) = 0$, valuate in Ł$_\aleph^*$, G$_\aleph^*$, and Π$_\aleph^*$ (when applicable) the following formulae:

1. $P(a) \wedge Q(b)$
2. $P(b) \nabla T(b)$
3. $P(c) \to Q(a)$
4. $T(b) \& P(a)$
5. $R(b) \to S(c)$
6. $(R(a) \wedge S(a,c)) \to R(c)$
7. $P(a) \wedge \neg P(a)$
8. $P(b) \vee \neg P(b)$
9. $S(a,b) \to (T(a) \vee \neg T(c))$
10. $(S(b,c) \wedge \neg S(a,c)) \to R(c)$
11. $(P(b) \leftrightarrow \neg Q(b)) \wedge (\neg P(b) \leftrightarrow Q(b))$
12. $P(a) \& R(b)$
13. $P(b) \& \neg P(b)$
14. $(P(a) \nabla \neg P(a)) \wedge (P(a) \& \neg P(a))$

**Exercise 5.28.** Create a finitely many-valued propositional logic over L and state the rationale(s) for it. Of course, you need to specify the connectives of your logic, and build truth tables for them that are in accord with the rationale(s).

**Exercise 5.29.** Prove (Complete the proof of) the propositions and theorems in this Chapter that were left without a proof (with a sketchy proof, respectively).

**Exercise 5.30.** Determine the truth value of the following sentences in all the finitely many-valued logics in this Section:

## 5. Many-valued logics

1. This sentence is false.
2. You belong to this club iff you do not belong to this club.
3. You can only eat jam every other day, but not today.
4. This is not a sentence.
5. I am a liar.
6. I didn't see nobody. [Too direct a translation of Spanish *No vi a nadie.*]

# Part III.

# REFUTATION CALCULI FOR MANY-VALUED LOGICS

# 6. The signed SAT for many-valued logics

The satisfiability problem for many-valued logics is known as the *MV-SAT*. Decades of research into this problem from the perspective of automated deduction have produced fundamental results, both theoretical and practical, and we discuss them here. One of the main positive results is the *signed SAT for many-valued logics*, or *signed MV-SAT*, the development or adaptation of signed formalisms for tackling the MV-SAT. In this Chapter, we begin by discussing *general theoretical aspects* (Section 6.1), and then give the *transformations* that many-valued formulae are required to undergo so as to be amenable to a *computational approach* to the signed MV-SAT (Section 6.2).

## 6.1. From the MV-SAT to the signed MV-SAT

Recall what was said in the Introduction (Section 1.2) and in Section 4.1 on the satisfiability problem. The SAT for some formula $\phi$ in a many-valued logic–the MV-SAT–can be expressed as follows:

**6.1.1. (Def.)** *MV-SAT* – Given a many-valued logic $Ł_n$ with a set of truth values $W_n$ and a set of designated values $D_k \subset W_n$, $0 < k < n$, for an arbitrary formula $\phi \in F_{Ł_n}$ we ask whether there is at least one interpretation $\mathcal{I}$ for which $val_\mathcal{I}(\phi) = x \in D$.

**Example 6.1.1.** Let the logic $Ł_3$ be given. Recall that the matrix of $Ł_3$ is:

$$\mathfrak{M}Ł_3 = \left( \left\{ 0, \frac{1}{2}, 1 \right\}, \neg, \rightarrow, \vee, \wedge, \leftrightarrow, \{1\} \right).$$

Then, the MV-SAT for a formula $\phi \in F_{Ł_3}$ is expressed as "Is it ever the case that $val_\mathcal{I}(\phi) = 1$ in $Ł_3$?"

**6.1.2. (Def.)** Given the duality between validity and unsatisfiability (cf. Prop. 3.2.14), we say that a formula $\phi$ is *D-valid* iff it is not $\overline{D}$-*satisfiable*. Or, if we define the sets $D^+$ and $D^-$ (cf. Def. 5.2.10), such that $W = D^+ \cup D^-$ and $D^+ \cap D^- = \emptyset$, we say that $\phi$ is $D^+$-*valid* iff it is not $D^-$-*satisfiable*.

Put like this, it is obvious that we are reducing the notion of many-valued satisfiability to the Boolean satisfiability problem. This can be adequately formalized by means of *signed logic* (see Section 2.5). By always "marking" a formula in a many-valued logic with the truth value(s) that it does or may take (its *sign*), we obtain a formalism that allows us to generalize classical notions such as validity and satisfiability to the many-valued logics with computational advantages. As a matter of fact, any formula in classical logic is *signed* in the sense that we have $P$ iff we do not have $\neg P$, which, given the classical truth-value set $W_2 = \{0, 1\}$, equates with *signing* the formulae $(\neg) P$ as $\{0, 1\} [P]$ or $\{0, 1\} [\neg P]$. Usually, we have it that $P$ corresponds to $\{1\}[P]$ and $\neg P$ corresponds to $\{0\}[P]$.[1]

Recall now the contents of Section 2.5.

**6.1.3. (Prop.)** Let $A$ be the signed formula

$$A = (\uparrow i_1 [P_1] \wedge ... \wedge \uparrow i_k [P_k]) \rightarrow \uparrow j [Q].$$

Then, by the deduction theorem (Theorems 3.2.5-6), an interpretation $\mathcal{I}$ satisfies $A$ iff it does not satisfy one of $\uparrow i_1 [P_1] \wedge ... \wedge \uparrow i_k [P_k]$ or it satisfies $\uparrow j [Q]$. Thus, the signed formula $A$ is equivalent to the Horn formula $B = \overline{\uparrow i_1} [P_1] \vee ... \vee \overline{\uparrow i_k} [P_k] \vee \uparrow j [Q]$.

*Proof:* Left as an exercise.

This equivalence shows how signed logic naturally generalizes standard CPL; in effect, by Theorem 3.2.6, $\uparrow j [Q]$ is a logical consequence of $(\uparrow i_1 [P_1] \wedge ... \wedge \uparrow i_k [P_k])$ iff $\neg ((\uparrow i_1 [P_1] \wedge ... \wedge \uparrow i_k [P_k]) \rightarrow \uparrow j [Q])$ is unsatisfiable, i.e. iff $(\uparrow i_1 [P_1] \wedge ... \wedge \uparrow i_k [P_k]) \not\models \uparrow j [Q]$. By the classical definition of material implication (cf. 2.4.7.1), which is straightforwardly generalized to signed logic, we have

$$\overline{(\uparrow i_1 [P_1] \wedge ... \wedge \uparrow i_k [P_k])} \vee \uparrow j [Q].$$

Applying negation to this formula, we have

$$\overline{\overline{(\uparrow i_1 [P_1] \wedge ... \wedge \uparrow i_k [P_k])} \vee \uparrow j [Q]} = \uparrow i_1 [P_1] \wedge ... \wedge \uparrow i_k [P_k] \wedge \overline{\uparrow j} [Q].$$

By working with regular signed formulae, i.e. formulae signed with regular signs (cf. Def. 2.5.2), as above, we are actually working with sets of truth values, instead of individual truth values, as signs. This

---
[1] See Section 7.2 for further discussion on this topic.

## 6.1. From the MV-SAT to the signed MV-SAT

has computational advantages, and we shall henceforth work mostly with (not necessarily regular) signs $S$ that are sets of truth values.[2]

**6.1.4. (Def.)** Let $\mathcal{I} = (\mathscr{D}, \Theta, \varpi)$ be an interpretation for $\mathsf{L}^*$ and $W$. Then, given a set $S \subseteq W$ and some formula $\phi$, the *semantics of signed formulae* is given by a valuation $sval_{\mathcal{I}}$ assigning either $\mathtt{f}$ or $\mathtt{t}$ to each signed formula $S[\phi]$ defined as:

1. $sval_{\mathcal{I}}(S[\phi]) = \mathtt{t}$ iff $val_{\mathcal{I}}(\phi) \in S$
2. $sval_{\mathcal{I}}(\top) = \mathtt{t}$ and $sval_{\mathcal{I}}(\bot) = \mathtt{f}$
3. $sval_{\mathcal{I}}(\neg \phi) = \mathtt{t}$ iff $sval_{\mathcal{I}}(\phi) = \mathtt{f}$
4. $sval_{\mathcal{I}}(\phi \wedge \psi) = \mathtt{t}$ iff $sval_{\mathcal{I}}(\phi) = \mathtt{t}$ and $sval_{\mathcal{I}}(\psi) = \mathtt{t}$
5. $sval_{\mathcal{I}}(\phi \vee \psi) = \mathtt{t}$ iff $sval_{\mathcal{I}}(\phi) = \mathtt{t}$ or $sval_{\mathcal{I}}(\psi) = \mathtt{t}$
6. $sval_{\mathcal{I}}(\phi \to \psi) = \mathtt{t}$ iff $sval_{\mathcal{I}}(\phi) = \mathtt{f}$ or $sval_{\mathcal{I}}(\psi) = \mathtt{t}$
7. $sval_{\mathcal{I}}(\forall x \phi(x)) = \mathtt{t}$ iff $sval_{\mathcal{I}_a^x}(\phi(a)) = \mathtt{t}$ for all $a \in \mathscr{D}$
8. $sval_{\mathcal{I}}(\exists x \phi(x)) = \mathtt{t}$ iff $sval_{\mathcal{I}_a^x}(\phi(a)) = \mathtt{t}$ for some $a \in \mathscr{D}$

In terms of the *S-satisfiability* (henceforth: *satisfiability*)[3] relation, we then have:

**6.1.5. (Def.)** Given a sign $S \subseteq W$, a signed formula $S[\phi]$ *is satisfied* exactly by the interpretations $\mathcal{I}$ such that $val_{\mathcal{I}}(\phi) \in S$, and we write

$$\models_{\mathcal{I}} S[\phi] \quad \text{iff} \quad val_{\mathcal{I}}(\phi) \in S.$$

Otherwise, $S[\phi]$ is *unsatisfiable*. A signed formula $S[\phi]$ is *valid* if it is satisfied by all interpretations $\mathcal{I}$ such that $val_{\mathcal{I}}(\phi) \in S$, and we write simply $\models S[\phi]$. Otherwise, $S[\phi]$ is *invalid*. A signed formula $S[\phi]$ is a *tautology* iff every interpretation satisfies $\phi$ and $S = D^{(+)} \subseteq W$.

**6.1.6. (Def.)** An interpretation $\mathcal{I}$ satisfies a signed literal $S[P]$ iff there is some valuation $val_{\mathcal{I}}$ such that $val_{\mathcal{I}}(P) \in S$, and we write

$$\models_{\mathcal{I}} S[P] \quad \text{iff} \quad val_{\mathcal{I}}(P) \in S.$$

---

[2] Actually, note that we began precisely by considering sets of truth values as signs and individual truth values as a special case thereof (cf. Def.s 2.5.1-2).

[3] Evidently, we need not specify that this is *S-satisfiability* (or *S-validity*, *S-tautology*, etc.) if it is clear from the context that we are working with signed logic, as the classical notions of satisfiability and validity generalize very naturally to signed logic. This is particularly so when we have sets $D, \overline{D}$, in which case we are operating with Boolean valuations.

Otherwise, $S\,[P]$ is *unsatisfiable*. An interpretation *satisfies* a signed clause iff it satisfies at least one of its signed literals, and it satisfies a signed CNF formula if it satisfies all its clauses.[4] A signed CNF formula is *satisfiable* iff there exists at least one interpretation that satisfies all its signed clauses; otherwise, it is *unsatisfiable*. Two signed clauses (signed CNF formulae) are *equivalent* iff they are satisfied by the same interpretations. They are said to be *satisfiability-equivalent*, or *equisatisfiable*, iff they are both either satisfiable or unsatisfiable.

**6.1.7. (Prop.)** The signed empty clause is always unsatisfiable and the signed empty CNF formula is always satisfiable.

*Proof:* Left as an exercise.

It should be clear that the semantics of signed CNF formulae is classical *above the literal level*, reason why classical proof procedures can be naturally generalized to them, sufficing for that end to take special care at the literal level. In particular, this means that we can reduce the problem of deciding the satisfiability of formulae of a finitely many-valued logic to the problem of deciding the satisfiability of formulae in signed CNF.

**6.1.8. (Def.)** The problem of deciding the satisfiability of formulae in signed CNF/DNF is called *the signed satisfiability problem*, or *signed SAT*.

It is important to remark at the outset that by working in the framework of signed logic, we avoid technically difficult and often "shaky" distinctions between object- and metalanguages or levels. By merely associating a unique ordering to a finitely many-valued truth-value set $W_n$, we can, given any many-valued logic $L_n$, introduce a new logic $L_n^S$ whose atoms are signed formulae, and we can then work directly in this logic. This commonly extends well to the first-order case, and many inference techniques can be naturally generalized to such logics (Murray & Rosenthal, 1993). This entails that we can apply the well-known techniques of classical logic in the study of the not so well-known many-valued logics. (Of course, one can also argue, as Murray & Rosenthal (1993) do, that human reasoning is "essentially classical." See above Sections 5.2.1-2.)

---

[4] We concentrate here on CNFs, because signed resolution (Chapter 8) requires the transformation of many-valued formulae into signed CNFs. Nevertheless, we treat the aspects of DNFs that are important for the CNFs themselves, as well as their general features that are of import for signed tableaux calculi (Chapter 7). In any case, recall that every formula in CNF has an equivalent DNF and vice-versa (cf. Prop. 2.4.16).

## 6.2. From many-valued formulae to signed formulae

### 6.2.1. General notions and definitions

Although there are no "natural" clause forms in the many-valued logics, there are procedures to transform and translate many-valued formulae into clause forms. This loss of "naturalness" is compensated for by the computational advantages of working with clause forms, in particular when combining clause forms with signed logic.[5] In effect, the following results are well known:

**6.2.1. (Prop.)** Any finitely many-valued formula can be translated into a satisfiability-equivalent signed CNF formula in polynomial time, and every signed CNF formula can be translated into a satisfiability-equivalent regular CNF formula with an arbitrary total ordering on $W$ also in polynomial time.

*Proof:* Left as an exercise.

See, for example, Beckert, Hähnle, & Manyà (2000) for a more comprehensive elaboration.[6]

The following theorem is fundamental in that it "licenses" us to work with sets as signs in the context of refutation methods for the MV-SAT, which requires the use of DNF and/or CNF representations:

**Theorem 6.2.1.** *(Hähnle, 1991) For $W$ finite, $n = |W|$, let $\mathscr{S} \subseteq (2^W - \emptyset)$ be a family of truth values satisfying the condition*

$$(\#) \quad \forall i \in W, \exists S_1, ..., S_k \in \mathscr{S} \text{ such that } \bigcap_{j=1}^{k} S_j = \{i\}.$$

*If $\phi = S\left[\heartsuit(\phi_1, ..., \phi_m)\right]$, for $m \geq 1$ and $S \in \mathscr{S}$, is a signed formula from an $n$-valued logic $\mathrm{L}_n$ equipped with a function $\tilde{\heartsuit}$ such that $S \cap rg\left(\tilde{\heartsuit}\right) \neq \emptyset$, then there are numbers $M_1, M_2 \leq n^m$, index sets $I_1, ..., I_{M_1}, J_1, ..., J_{M_2} \subseteq \{1, ..., m\}$, and signs $S_{rs}, S_{kl} \in \mathscr{S}$ with $1 \leq r \leq M_1, 1 \leq k \leq M_2$ and*

---

[5] However, it must be remarked that clause-based reasoning applied to many-valued logics is independent of the logics the clause forms were obtained from. In this context, the distinction that Baaz, Fermüller, & Salzer (2001) establish between internal and external proof procedures is relevant. According to this distinction, both clausal and signed logics are external formalisms with respect to the many-valued logics for which they cater.

[6] Matters are more complex in the case of the infinitely many-valued logics. See Section 8.3.

## 6. The signed SAT for many-valued logics

$s \in I_r, l \in J_k$ such that

$$\phi \text{ is satisfiable}$$

$$iff$$

$$\bigvee_{r=1}^{M_1} \left( \bigwedge_{s \in I_r} S_{rs}[\phi_s] \right) \text{ is satisfiable}$$

$$iff$$

$$\bigwedge_{k=1}^{M_2} \left( \bigvee_{l \in J_k} S_{kl}[\phi_l] \right) \text{ is satisfiable.}$$

*Proof:* A proof can be found in, e.g., Carnielli (1987). **QED**

In Theorem 6.2.1, the first expression is a *signed DNF representation*, and the second is a *signed CNF representation*.[7]

Salzer and colleagues (e.g., Salzer, 2000; Baaz, Fermüller, & Salzer, 2001) provide transformation procedures for signed formulae that allow a formalization of many-valued semantics; this, in turn, allows these logics to be reasoned upon classically, namely by means of the following equivalences.

**6.2.2. (Prop.)** The following equivalences are valid:

1. $\{\}\,[\phi] \equiv \square \equiv \bot; W\,[\phi] \equiv \top$

2. $\neg (S\,[\phi]) \equiv \overline{S}\,[\phi] = (W - S)\,[\phi]$

3. $S_1\,[\phi] \vee S_2\,[\phi] \equiv (S_1 \cup S_2)\,[\phi]$, in particular $\{v_1\}\,[\phi] \vee ... \vee \{v_n\}\,[\phi] \equiv \{v_1, ..., v_n\}\,[\phi]$

4. $S_1\,[\phi] \wedge S_2\,[\phi] \equiv (S_1 \cap S_2)\,[\phi]$, in particular $(\{v_1\}\,[\phi] \wedge \{v_2\}\,[\phi]) = |$ for $v_1 \neq v_2$

5. $\forall x\,(S_1\,[\phi]) \wedge \exists x\,(S_2\,[\phi]) \equiv \forall x\,((S_1 \cap S_2)\,[\phi])$

6. $\exists x\,(S_1\,[\phi]) \vee \forall x\,(S_2\,[\phi]) \equiv \exists x\,(S_1\,[\phi]) \vee \forall x\,((S_1 \cup S_2)\,[\phi])$

---

[7] Note that by speaking of *representation* we indicate clearly that we are at the metalanguage level. Moreover, in order to guarantee the existence of DNF/CNF representations $\mathscr{S}$ has to satisfy condition #, i.e. enough signs must be available.

## 6.2. From many-valued formulae to signed formulae

*Proof:* Left as an exercise.

Negations can be wholly eliminated from SFEs (cf. Def. 2.5.5) by 6.2.2.2, and 6.2.2.3 allows the elimination of all non-singleton signs by the introduction of disjunctions. Equivalences 6.2.2.5-6 express the fact that given $\forall x \, (S_1 \, [\phi])$ truth values not occurring in $S_1$ can be removed or added at will in certain other formulae, and in the presence of $\exists x \, (S_1 \, [\phi])$ truth values occurring in $S_1$ can undergo the same operations.

It is thus obvious that, for any arbitrary connective and distribution quantifier, their semantics can be expressed as SFEs.

**Example 6.2.1.** For a number of many-valued logics (e.g., Gödel and Łukasiewicz logics), given two formulae $A$ and $B$, conjunction and disjunction can be defined by

$$A \wedge B = min \, (A, B)$$

and

$$A \vee B = max(A, B)$$

where *min* and *max* refer to the truth values of $A$ and $B$, i.e.

$$\widetilde{\wedge} \, (\mathbf{v}_i, \mathbf{v}_j) = \mathbf{v}_{min(i,j)}$$

and

$$\widetilde{\vee} \, (\mathbf{v}_i, \mathbf{v}_j) = \mathbf{v}_{max(i,j)}.$$

In terms of SFEs, we then have

$$\{\mathbf{v}_i\} \, [\phi_1 \wedge \phi_2] \equiv \{\mathbf{v}_j | j \geq i\} \, [\phi_1] \wedge \{\mathbf{v}_j | j \geq i\} \, [\phi_2] \wedge (\{\mathbf{v}_i\} \, [\phi_1] \vee \{\mathbf{v}_i\} \, [\phi_2])$$

and

$$\{\mathbf{v}_i\} \, [\phi_1 \vee \phi_2] \equiv \{\mathbf{v}_j | j \leq i\} \, [\phi_1] \wedge \{\mathbf{v}_j | j \leq i\} \, [\phi_2] \wedge (\{\mathbf{v}_i\} \, [\phi_1] \vee \{\mathbf{v}_i\} \, [\phi_2]).$$

Then, for a three-valued logic with $W_3 = \{\mathtt{f}, \mathtt{i}, \mathtt{t}\}$ and a universal quantifier $\forall^*$ whose truth function for $\Delta \subseteq W_3$, $\Delta \neq \emptyset$, is

$$\widetilde{\forall^*} \, (\Delta) = \begin{cases} \mathtt{t} & \text{if } \Delta = \{\mathtt{t}\} \\ \mathtt{f} & \text{if } \mathtt{f} \in \Delta \text{ or } \mathtt{i} \in \Delta \end{cases}$$

## 6. The signed SAT for many-valued logics

the following SFE is obtained as a specification of its semantics:

$$[\{\mathtt{t}\}\,[\forall^* x A\,(x)] \equiv \forall x\,(\{\mathtt{t}\}\,[A\,(x)])] \wedge [\{\mathtt{f}\}\,[\forall^* x A\,(x)] \equiv \exists x\,(\{\mathtt{f},\mathtt{i}\}\,[A\,(x)])].$$

**6.2.3. (Def.)** Given a pair $(\phi, \Phi)$, where $\phi$ is a signed formula and $\Phi$ is a SFE, $\phi \Longrightarrow \Phi$ is a *transformation rule*.[8] A transformation rule is *correct* iff $\phi \equiv \Phi$ is valid. A transformation rule is said to be *reducing* (i.e. it is a *reduction rule*) iff every many-valued formula in $\Phi$ is a proper subformula of the many-valued formula in $\phi$.

**6.2.4. (Def.)** Let $RT$ be a set of transformation rules. $RT$ is said to be *complete* if for every signed formula $S\,[\phi]$ there is a reduction rule in $RT$ that is applicable to the formula, unless $\phi$ is atomic. In other words, $RT$ is complete if it contains rules $S\,[\heartsuit_i\,(\phi_1, ..., \phi_n)] \Longrightarrow \Phi$ and $S\,[\blacklozenge_i x\,(\phi)] \Longrightarrow \Phi$ for every $n$-ary $i$-th connective $\heartsuit_i$, every $i$-th quantifier $\blacklozenge_i$, and every sign $S \neq \emptyset$.

Obviously, given a complete set $RT$ of reduction rules, we can transform any signed formula into any equivalent SFE consisting solely of atomic signed formulae. Although singleton sets suffice for completeness, the consideration of sets of signs with cardinality greater than one is computationally advantageous. In particular, and as seen above, working only with the sets $S$ and $\overline{S}$ allows us, so to say, to work classically.

We now provide definitions of the general transformation rules. We give examples from the more familiar classical logic as a warming-up for the more complex scenario for the many-valued logics.

**6.2.5. (Def.)** For an $n$-ary connective $\heartsuit$, a *propositional transformation rule* is an expression of the form

$$S\,[\heartsuit\,(A_1, ..., A_n)] \Longrightarrow \bigodot_i \bigoplus_j S_{ij}\,[A'_{ij}]$$

where each $A'_{ij} \in \{A_1, ..., A_n\}$ and the symbols $\odot, \oplus$ are duals and stand for either of $\wedge, \vee$. In effect, and specifying, given a connective $\heartsuit$ and a sign $S$, we have it that a SFE $\Phi$ of a formula $S\,[\heartsuit\,(A_1, ..., A_n)]$ is called a

1. CNF for $\heartsuit$ and $S$ iff $\Phi = \bigwedge_i \left( \bigvee_j S_{ij}\,[A_{ij}] \right)$ and $S\,[\heartsuit\,(A_1, ..., A_n)] \equiv \Phi$ is valid;

---

[8]The distinction between signed formula and SFE may be marked by a special notation (e.g., Baaz, Fermüller, & Salzer, 2001), a strategy we find superfluous, namely because there is no difference whatsoever as regards the behavior of the connectives ($\wedge$ and $\vee$) and, as for the quantifiers, we know we are working with distribution quantifiers.

2. DNF for $\heartsuit$ and $S$ iff $\Phi = \bigvee_i \left( \bigwedge_j S_{ij} [A_{ij}] \right)$ and $S [\heartsuit (A_1, ..., A_n)] \equiv \Phi$ is valid.

**Example 6.2.2.** For $W_2 = \{\mathtt{f}, \mathtt{t}\}$ and the connectives $\neg$, $\wedge$ and $\vee$, the following is a (sub)set of correct reduction rules for CPL:

$$
\begin{array}{rl}
(\wedge_\mathtt{t}) & \{\mathtt{t}\} [A \wedge B] \implies \{\mathtt{t}\} [A] \wedge \mathtt{t} [B] \\
(\wedge_\mathtt{f}) & \{\mathtt{f}\} [A \wedge B] \implies \{\mathtt{f}\} [A] \vee \mathtt{f} [B] \\
(\neg_\mathtt{t}) & \{\mathtt{t}\} [\neg A] \implies \{\mathtt{f}\} [A] \\
(\neg_\mathtt{f}) & \{\mathtt{f}\} [\neg A] \implies \{\mathtt{t}\} [A]
\end{array}
$$

**Example 6.2.3.** Given the set of reduction rules in Example 6.2.2, the signed formula $\{\mathtt{f}\} [A \wedge \neg (B \wedge \neg C)]$ undergoes the following transformations:

$$
\begin{array}{rll}
\{\mathtt{f}\} [A \wedge \neg (B \wedge \neg C)] & \equiv & \{\mathtt{f}\} [A] \vee \{\mathtt{f}\} [\neg (B \wedge \neg C)] \quad \text{by } \wedge_\mathtt{f} \\
& \equiv & \{\mathtt{f}\} [A] \vee \{\mathtt{t}\} [B \wedge \neg C] \quad \text{by } \neg_\mathtt{f} \\
& \equiv & \{\mathtt{f}\} [A] \vee \{\mathtt{t}\} [B] \wedge \{\mathtt{t}\} [\neg C] \quad \text{by } \wedge_\mathtt{t} \\
& \equiv & \{\mathtt{f}\} [A] \vee \{\mathtt{t}\} [B] \wedge \{\mathtt{f}\} [C] \quad \text{by } \neg_\mathtt{t}
\end{array}
$$

**6.2.6. (Def.)** For a quantifier $\blacklozenge$, a *quantifier transformation rule* is an expression of the *general form*

$$
S [\blacklozenge x A(x)] \implies \bigodot_i \left( \bigoplus_j \blacklozenge x \left( \bigoplus_k S_{ijk} [A(x)] \right) \right)
$$

where the symbols $\bigodot, \bigoplus$ are duals and stand for either of $\bigwedge, \bigvee$. In effect, and specifying, given a quantifier $\blacklozenge$ and a sign $S$, we have it that a SFE $\Phi$ of a formula $S [\blacklozenge x A(x)]$ is called a

1. CNF for $\blacklozenge = \forall, \exists$ and $S$ iff $\Phi = \bigwedge_i \left( \bigvee_j \blacklozenge x (S_{ij} [A(x)]) \right)$ and $S [\blacklozenge x A(x)] \equiv \Phi$ is valid;

2. DNF for $\blacklozenge = \forall, \exists$ and $S$ iff $\Phi = \bigvee_i \left( \bigwedge_j \blacklozenge x (\bigwedge_k S_{ijk} [A(x)]) \right)$ and $S [\blacklozenge x A(x)] \equiv \Phi$ is valid.[9]

---

[9]In 6.2.6.1, we do not have $\Phi = \bigwedge_i \left( \bigvee_j \blacklozenge x (\bigvee_k S_{ijk} [A(x)]) \right)$, because $\bigvee_k S_{ijk} [A(x)]$ can be simplified to $S_{ij} [A(x)]$ where $S_{ij} = \bigcup_k S_{ijk}$.

## 6. The signed SAT for many-valued logics

**Example 6.2.4.** The following are the CNF[10] correct reduction rules for the universal and existential quantifiers of Ł$_3^*$ and P$_3^*$ (cf. Example 5.7.7; see Example 6.2.7 for a step-by-step computation):

$(\forall_\mathtt{t})$ $\quad \{\mathtt{t}\}[\forall x A(x)] \Longrightarrow \forall x(\{\mathtt{t}\}[A(x)])$
$(\forall_\mathtt{i})$ $\quad \{\mathtt{i}\}[\forall x A(x)] \Longrightarrow \exists x(\{\mathtt{i}\}[A(x)]) \wedge \forall x(\{\mathtt{i},\mathtt{t}\}[A(x)])$
$(\forall_\mathtt{f})$ $\quad \{\mathtt{f}\}[\forall x A(x)] \Longrightarrow \exists x(\{\mathtt{f}\}[A(x)])$

$(\exists_\mathtt{t})$ $\quad \{\mathtt{t}\}[\exists x A(x)] \Longrightarrow \exists x(\{\mathtt{t}\}[A(x)])$
$(\exists_\mathtt{i})$ $\quad \{\mathtt{i}\}[\exists x A(x)] \Longrightarrow \exists x(\{\mathtt{i}\}[A(x)]) \wedge \forall x(\{\mathtt{f},\mathtt{i}\}[A(x)])$
$(\exists_\mathtt{f})$ $\quad \{\mathtt{f}\}[\exists x A(x)] \Longrightarrow \forall x(\{\mathtt{f}\}[A(x)])$

### 6.2.2. Transformation rules for many-valued connectives

**6.2.7. (Prop.)** Let $\heartsuit$ be an $n$-ary connective and let $i$ be a fixed truth value. By examination of the truth table for $\heartsuit$, a SFE $E$ containing only signed formulae can be constructed from $\{j[A_k] \,|\, 1 \leq k \leq n, j \in W\}$ and such that $i[\heartsuit(A_1, ..., A_n)] \equiv E$. In fact, two (non-unique) SFEs $C$ and $D$ can be constructed such that

1. $C, D$ do not contain $\neg$;

2. $C, D$ contain only literals from the set $\{j[A_k] \,|\, 1 \leq k \leq n, j \in W\}$;

3. $C$ is in CNF and $D$ is in DNF;

4. $C, D$, and $E$ are equivalent as Boolean formulae, and hence they are all equivalent to $i[\heartsuit(A_1, ..., A_n)]$.

*Proof:* In effect, for any formula $A$ and any truth value $i$, we have it that

$$(\S) \qquad \neg(i[A]) \equiv \bigvee_{j \neq i} j[A].$$

Thus, and by applying De Morgan's law DM$_\vee$, for any SFE $E$ an equivalent SFE $C$ that is in CNF and does not contain the connective $\neg$ can be found. We show how in the following Propositions 6.2.8-9. **QED**

---
[10] Actually, *cnf*; see below.

## 6.2. From many-valued formulae to signed formulae

In particular, for a sign $S$ we obtain a CNF representation by duality: we compute a DNF representation for $\overline{S}$ and then replace $\neg \bigvee \wedge ... S'...$ with $\bigwedge \bigvee ... \overline{S'}...$ by applying De Morgan's laws. We make this process more precise. We draw here mostly on Baaz, Fermüller, & Salzer (2001).

**6.2.8.** *Obtaining DNFs for many-valued connectives* – For an $n$-ary connective $\heartsuit$ and some sign $S$, we define

$$DNF_\heartsuit(S) := \bigvee_{\substack{v_1, ..., v_n \in W \\ \widetilde{\heartsuit}(v_1, ..., v_n) \in S}} \left( \bigwedge_{i=1}^{n} \{v_i\}[A_i] \right).$$

$DNF_\heartsuit(S)$ is maximal, in the sense that every DNF that is equivalent to $S[\heartsuit(A_1, ..., A_n)]$ has at most as many disjuncts as it, their number being bounded by $|W|^n$. Normal forms bounded by $|W|^{n-1}$ have fewer disjuncts and conjuncts. In effect, we have

$$dnf_\heartsuit(S) := \bigvee_{v_1, ..., v_{n-1} \in W} \left( \bigwedge_{i=1}^{n-1} \{v_i\}[A_i] \wedge \{v_n | \widetilde{\heartsuit}(v_1, ..., v_n) \in S\}[A_n] \right)$$

which in fact corresponds to the direct reading off a truth table $\widetilde{\heartsuit}$ as, for arbitrary truth values $u, w \in W$,

$$\{u\}[\heartsuit(A, B)] \equiv \bigvee_{w \in W} \left( \{w\}[A] \wedge \{w' | \widetilde{\heartsuit}(w, w') = u\}[B] \right).$$

**6.2.9.** *Obtaining CNFs for many-valued connectives* – Given § and the equivalences $\neg DNF_\heartsuit(\overline{S}) \equiv CNF_\heartsuit(S)$ or $\neg dnf_\heartsuit(\overline{S}) \equiv cnf_\heartsuit(S)$ by Proposition 8.2.2.2, we can start with these equivalences and apply then De Morgan's law (cf. 2.4.7) and eliminate all negation signs, so that we obtain

$$CNF_\heartsuit(S) := \bigwedge_{\substack{v_1, ..., v_n \in W \\ \widetilde{\heartsuit}(v_1, ..., v_n) \in \overline{S}}} \left( \bigvee_{i=1}^{n} \overline{\{v_i\}}[A_i] \right)$$

and

$$cnf_\heartsuit(S) := \bigwedge_{v_1, ..., v_{n-1} \in W} \left( \bigvee_{i=1}^{n-1} \overline{\{v_i\}}[A_i] \vee \overline{\{v_n | \widetilde{\heartsuit}(v_1, ..., v_n) \in S\}}[A_n] \right)$$

6. The signed SAT for many-valued logics

corresponding to

$$\overline{\{u\}}\,[\heartsuit\,(A,B)] \equiv \bigwedge_{w \in W} \left(\overline{\{w\}}\,[A] \vee \left\{w' | \widetilde{\heartsuit}\,(w,w') \neq u\right\}[B]\right)$$

$$\equiv$$

$$\overline{\{u\}}\,[\heartsuit\,(A,B)] \equiv \bigwedge_{w \in W} \left(\overline{\{w\}}\,[A] \vee \left\{w' | \widetilde{\heartsuit}\,(w,w') = u\right\}[B]\right).$$

**Example 6.2.5.** Let $\heartsuit$ be implication in Ł$_3$. For two (atomic) formulae $A$ and $B$, we want to compute the CNF for $\{i\}\,[A \to_{Ł3} B]$. Thus, we begin by computing the DNF for $\{f,t\}\,[A \to_{Ł3} B]$:

$$\bigvee_{\substack{v_1, v_2 \in W_{Ł3} \\ v_1 \to v_2 \neq i}} (\{v_1\}\,[A] \wedge \{v_2\}\,[B])$$

This produces, by examination of the truth table, the DNF for $\overline{\{i\}}$ (we omit the subscript for the connectives henceforth):

$$\bigvee \begin{pmatrix} (\{t\}\,[A] \wedge \{t\}\,[B]) \\ (\{t\}\,[A] \wedge \{f\}\,[B]) \\ (\{i\}\,[A] \wedge \{t\}\,[B]) \\ (\{i\}\,[A] \wedge \{i\}\,[B]) \\ (\{f\}\,[A] \wedge \{t\}\,[B]) \\ (\{f\}\,[A] \wedge \{i\}\,[B]) \\ (\{f\}\,[A] \wedge \{f\}\,[B]) \end{pmatrix}$$

Given this DNF, the computation of the CNF for $\{i\}\,[A \to_{Ł3} B]$ is now an easy matter, and we obtain the formula:

$$\bigwedge \begin{pmatrix} (\{f,i\}\,[A] \vee \{f,i\}\,[B]) \\ (\{f,i\}\,[A] \vee \{i,t\}\,[B]) \\ (\{f,t\}\,[A] \vee \{f,i\}\,[B]) \\ (\{f,t\}\,[A] \vee \{f,t\}\,[B]) \\ (\{i,t\}\,[A] \vee \{f,i\}\,[B]) \\ (\{i,t\}\,[A] \vee \{f,t\}\,[B]) \\ (\{i,t\}\,[A] \vee \{i,t\}\,[B]) \end{pmatrix}$$

In the above, $dnf_\heartsuit(S)$ and $cnf_\heartsuit(S)$ optimize the process of obtaining minimal normal forms in many-valued logics.

**6.2.10. (Prop.)** Let $\heartsuit$ be any $n$-ary connective. Let further $E$ stand for any (a) $DNF_\heartsuit(S)$, (b) $dnf_\heartsuit(S)$, (c) $CNF_\heartsuit(S)$, or (d) $cnf_\heartsuit(S)$, where $S$ is a sign. Then, for every $\heartsuit$ and every sign $S$, the rule

$$S\left[\heartsuit\left(A_1, ..., A_n\right)\right] \Longrightarrow E$$

is correct and reducing. The number of disjuncts (conjuncts) in (a) and (c) [(b) and (d)] is bounded by $|W|^n$ and $|W|^{n-1}$, respectively.

*Proof:* Left as an exercise.

**Example 6.2.6.** For the connective $\rightarrow_{Ł3}$, by applying the minimization above we obtain the *cnfs* in Figure 6.2.1. We deleted literals of the forms $\{\}\,[\phi] \equiv \bot$ and disjunctions where one of the disjuncts is of the form $\{\mathtt{f},\mathtt{i},\mathtt{t}\}\,[\phi] \equiv \top$ (cf. Prop. 6.2.2.1). Further simplifications were, for instance, $\{\mathtt{i},\mathtt{t}\}\,[A] \wedge \{\mathtt{f},\mathtt{t}\}\,[A] \equiv \{\mathtt{t}\}\,[A]$ (cf. Prop. 6.2.2.4), which in turn allowed the elimination of the conjunct $\{\mathtt{f},\mathtt{i}\}\,[A]$ in the process of obtaining the *cnf* for $\{\mathtt{f}\}\,[A \rightarrow B]$.

The simplifications concretized in the *cnfs* can be made automatically by means of the system MULTlog (see Baaz et al., 1996).

### 6.2.3. Transformation rules for many-valued quantifiers

Given the elaboration above on transformation rules for many-valued connectives, the following definitions for many-valued quantifiers should be evident once some necessary supplementary definitions have been formulated.

Recall from Definitions 2.2.11.5 and 5.7.11 that the *distribution* of a quantified formula $\blacklozenge_i x \phi(x)$ is the set of truth values that is obtained by evaluating $\phi(x)$ for all possible values of $x$ given a domain $\mathscr{D}$: $\Delta_{\mathcal{I},x}(\phi(x)) = \{val_{\mathcal{I}_a^x}(\phi(a)) \,|\, a \in \mathscr{D}\}$. In other words, the distribution of $\phi(x)$ is the collection of all truth values that can be obtained by an evaluation of $\phi(x)$; then, $\Delta$ is the distribution of $\phi(x)$ iff $\forall x\,(\Delta\,[\phi(x)])$ and $\exists x\,(\{\mathbf{v}_i\}\,[\phi(x)])$, i.e. if the value assigned to $\phi(x)$ is in $\Delta$ for all $x$ and iff for every truth value $\mathbf{v}_i \in \Delta$ there is an $x$ such that the valuation of $x$ corresponds to $\mathbf{v}_i$ for all $\mathbf{v}_i \in \Delta$, that is, if there is a valuation function according to which a variable can be interpreted in $W$ (cf. the valuation for a quantifier $\forall^*$–$\widetilde{\forall}^*$– in Example 6.2.1).

**6.2.11.** *Obtaining DNFs for many-valued quantifiers* – For a distri-

## 6. The signed SAT for many-valued logics

$\{\mathtt{t}\}[A \to B] \implies \{\mathtt{i},\mathtt{t}\}[A] \vee \{\mathtt{f},\mathtt{i},\mathtt{t}\}[B]) \wedge (\{\mathtt{f},\mathtt{t}\}[A] \vee \{\mathtt{i},\mathtt{t}\}[B]) \wedge (\{\mathtt{f},\mathtt{i}\}[A] \vee \{\mathtt{t}\}[B])$
$\equiv (\{\mathtt{f},\mathtt{t}\}[A] \vee \{\mathtt{i},\mathtt{t}\}[B]) \wedge (\{\mathtt{f},\mathtt{i}\}[A] \vee \{\mathtt{t}\}[B])$
$\equiv (\{\mathtt{f}\}[A] \vee \{\mathtt{i},\mathtt{t}\}[B]) \wedge (\{\mathtt{f},\mathtt{i}\}[A] \vee \{\mathtt{t}\}[B])$

$\{\mathtt{i}\}[A \to B] \implies (\{\mathtt{i},\mathtt{t}\}[A] \vee \{\}[B]) \wedge (\{\mathtt{f},\mathtt{t}\}[A] \vee \{\mathtt{f},\mathtt{t}\}[B]) \wedge (\{\mathtt{f},\mathtt{i}\}[A] \vee \{\mathtt{i}\}[B])$
$\equiv \{\mathtt{i},\mathtt{t}\}[A] \wedge (\{\mathtt{f},\mathtt{t}\}[A] \vee \{\mathtt{f},\mathtt{t}\}[B]) \wedge (\{\mathtt{f},\mathtt{i}\}[A] \vee \{\mathtt{i}\}[B])$
$\equiv \{\mathtt{t}\}[A] \vee \{\mathtt{f}\}[B]) \wedge (\{\mathtt{i}\}[A] \vee \{\mathtt{i}\}[B])$

$\{\mathtt{f}\}[A \to B] \implies \{\mathtt{i},\mathtt{t}\}[A] \vee \{\}[B]) \wedge (\{\mathtt{f},\mathtt{t}\}[A] \vee \{\}[B]) \wedge (\{\mathtt{f},\mathtt{i}\}[A] \vee \{\mathtt{f}\}[B])$
$\equiv \{\mathtt{i},\mathtt{t}\}[A] \wedge \{\mathtt{f},\mathtt{t}\}[A] \wedge (\{\mathtt{f},\mathtt{i}\}[A] \vee \{\mathtt{f}\}[B])$
$\equiv \{\mathtt{t}\}[A] \wedge \{\mathtt{f}\}[B]$

Figure 6.2.1.: *cnf*s for the connective $\to_{L3}$.

## 6.2. From many-valued formulae to signed formulae

bution quantifier ♦ and a sign $S$, we have

$$DNF_\blacklozenge(S) := \bigvee_{\substack{\emptyset \subset \Delta \subseteq W \\ \widetilde{\blacklozenge}(\Delta) \in S}} \left( \forall x \left( \Delta \left[ A(x) \right] \right) \wedge \bigwedge_{v_i \in \Delta} \exists x \left( \{v_i\} \left[ A(x) \right] \right) \right).$$

$DNF_\blacklozenge(S)$ has one characteristic formula for each distribution $\Delta$ satisfying the condition $\widetilde{\blacklozenge}(\Delta)$. The bound for the number of disjuncts in $DNF_\blacklozenge(S)$ is now $2^{|W|}$, but, as in the case of the transformation rules for many-valued connectives, this bound can be lowered to a minimal bound of $2^{|W|-1}$ by applying the definition

$$dnf_{\blacklozenge,u}(S) := \bigvee_{\Delta \subseteq (W - \{u\})} \left( \forall x \left( \alpha_S(\Delta) \left[ A(x) \right] \right) \wedge \bigwedge_{v_i \in \beta_S(\Delta)} \exists x \left( \{v_i\} \left[ A(x) \right] \right) \right)$$

where $u \notin \Delta$ is an arbitrary truth value, $\alpha_S(\Delta)$ and $\beta_S(\Delta)$ are given in the table bellow, and $dnf_{\blacklozenge,u}(S)$ is obtained by merging the characteristic formulae for $\Delta$ and $\Delta \cup \{u\}$ in a single expression.

| $\widetilde{\blacklozenge}(\Delta)$ | $\widetilde{\blacklozenge}(\Delta \cup \{u\})$ | $\alpha_S(\Delta)$ | $\beta_S(\Delta)$ |
|---|---|---|---|
| $\notin S$ | $\notin S$ | $\{\}$ | $\{\}$ |
| $\notin S$ | $\in S$ | $\Delta \cup \{u\}$ | $\Delta \cup \{u\}$ |
| $\in S$ | $\notin S$ | $\Delta$ | $\Delta$ |
| $\in S$ | $\in S$ | $\Delta \cup \{u\}$ | $\Delta$ |

**6.2.12.** *Obtaining CNFs for many-valued quantifiers* – By applying De Morgan's law to $\neg DNF_\blacklozenge(\bar{S})$ and $\neg dnf_{\blacklozenge,u}(\bar{S})$ and eliminating all negations, we obtain the definitions

$$CNF_\blacklozenge(S) := \bigwedge_{\substack{\emptyset \subseteq \Delta \subseteq W \\ \widetilde{\blacklozenge}(\Delta) \in \bar{S}}} \left( \exists x \left( \overline{\Delta} \left[ A(x) \right] \right) \vee \bigvee_{v_i \in \Delta} \forall x \left( \overline{\{v_i\}} \left[ A(x) \right] \right) \right)$$

and

$$cnf_{\blacklozenge,u}(S) :=$$

## 6. The signed SAT for many-valued logics

$$\bigwedge_{\Delta \subseteq (W-\{\mathtt{u}\})} \left( \exists x \left( \overline{\alpha_{\overline{S}}(\Delta)} [A(x)] \right) \vee \bigvee_{v_i \in \beta_{\overline{S}}(\Delta)} \forall x \left( \overline{\{v_i\}} [A(x)] \right) \right).$$

**Example 6.2.7.** We consider the quantifier $\widetilde{\forall}$ in $Ł_3^*$ defined by $\widetilde{\forall}(\{\mathtt{t}\}) = \mathtt{t}$ and

$$\widetilde{\forall}(\Delta) = \begin{cases} \mathtt{f} & \text{for } \mathtt{f} \in \Delta \\ \mathtt{i} & \text{otherwise} \end{cases}.$$

In order to obtain the $cnf_{\forall,\mathtt{i}}\{\mathtt{i}\}$ we start by getting $\alpha_{\overline{\{\mathtt{i}\}}} = \alpha_{\{\mathtt{f},\mathtt{t}\}}$ and $\beta_{\overline{\{\mathtt{i}\}}} = \beta_{\{\mathtt{f},\mathtt{t}\}}$ (cf. the truth functions for the distribution quantifiers of $Ł_3^*$ in Example 5.7.8):

| $\Delta$ | $\widetilde{\forall}(\Delta)$ | $\widetilde{\forall}(\Delta \cup \{\mathtt{i}\})$ | $\alpha_{\{\mathtt{f},\mathtt{t}\}}(\Delta)$ | $\beta_{\{\mathtt{f},\mathtt{t}\}}(\Delta)$ |
|---|---|---|---|---|
| $\{\}$ | — | $\mathtt{i}$ | $\{\}$ | $\{\}$ |
| $\{\mathtt{f}\}$ | $\mathtt{f}$ | $\mathtt{f}$ | $\{\mathtt{f},\mathtt{i}\}$ | $\{\mathtt{f}\}$ |
| $\{\mathtt{t}\}$ | $\mathtt{t}$ | $\mathtt{i}$ | $\{\mathtt{t}\}$ | $\{\mathtt{t}\}$ |
| $\{\mathtt{f},\mathtt{t}\}$ | $\mathtt{f}$ | $\mathtt{f}$ | $\{\mathtt{f},\mathtt{i},\mathtt{t}\}$ | $\{\mathtt{f},\mathtt{t}\}$ |

We next compute $cnf_{\forall,\mathtt{i}}\{\mathtt{i}\}$:

$$\bigwedge \begin{pmatrix} \exists x \left( \overline{\{\}} [A(x)] \right) \\ \left( \exists x \left( \overline{\{\mathtt{f},\mathtt{i}\}} [A(x)] \right) \vee \forall x \left( \overline{\{\mathtt{f}\}} [A(x)] \right) \right) \\ \left( \exists x \left( \overline{\{\mathtt{t}\}} [A(x)] \right) \vee \forall x \left( \overline{\{\mathtt{t}\}} [A(x)] \right) \right) \\ \left( \exists x \left( \overline{\{\mathtt{f},\mathtt{i},\mathtt{t}\}} [A(x)] \right) \vee \forall x \left( \overline{\{\mathtt{f}\}} [A(x)] \right) \vee \forall x \left( \overline{\{\mathtt{t}\}} [A(x)] \right) \right) \end{pmatrix}$$

$$\equiv$$

$$\bigwedge \begin{pmatrix} (\exists x (\{\mathtt{t}\} [A(x)]) \vee \forall x (\{\mathtt{i},\mathtt{t}\} [A(x)])) \\ (\exists x (\{\mathtt{f},\mathtt{i}\} [A(x)]) \vee \forall x (\{\mathtt{f},\mathtt{i}\} [A(x)])) \\ (\forall x (\{\mathtt{i},\mathtt{t}\} [A(x)]) \vee \forall x (\{\mathtt{f},\mathtt{i}\} [A(x)])) \end{pmatrix}$$

By applying optimization rules (e.g., elimination of redundant formulae), we obtain the following minimal expression with only two conjuncts:

$$\exists x (\{\mathtt{i}\} [A(x)]) \wedge \forall x (\{\mathtt{i},\mathtt{t}\} [A(x)])$$

See Example 6.2.4 for the *cnfs* of the remaining signed quantified formulae of $Ł_3^*$.

## 6.2. From many-valued formulae to signed formulae

One should note that these transformation rules involving distribution quantifiers are quite general in the sense that the $CNF_\blacklozenge$ and the $cnf_{\blacklozenge,\mathtt{u}}$ are the same for finitely many-valued logics based on t-norms, as well as for the finitely-valued Post logics, given that they all share the same truth functions for the quantifiers $\forall$ and $\exists$ (Example 5.7.1). Thus, the $cnf_{\forall,\mathtt{i}}\{\mathtt{i}\}$ for the quantifier $\forall$ in Ł$_3^*$ defined as above is the same for the same quantifier in G$_3^*$ and in P$_3^*$. In effect, while these logical systems differ greatly at the propositional level (see above), they are all similar at the first-order level. Hence, it is obvious that the disjunctive and conjunctive normal forms as defined above are identical for Ł$_n^*$, G$_n^*$, and P$_n^*$ for some $i$, $3 \leq i \leq n$, $n$ is finite.

**6.2.13. (Prop.)** Let $\blacklozenge$ be any finitely many-valued distribution quantifier. Let further $E$ stand for any (a)$DNF_\blacklozenge(S)$, (b) $dnf_{\blacklozenge,\mathtt{u}}(S)$, (c) $CNF_\blacklozenge(S)$, or (d) $cnf_{\blacklozenge,\mathtt{u}}(S)$, where $S$ is a sign and $\mathtt{u}$ is a truth value. Then, for every $\blacklozenge$, every sign $S$, and every truth value $\mathtt{u}$ the rule

$$S\left[\blacklozenge x A(x)\right] \Longrightarrow E$$

is correct and reducing. The number of disjuncts (conjuncts) in (a) and (c) [(b) and (d)] is bounded by $2^{|W|}$ and $2^{|W|-1}$, respectively.

*Proof:* Left as an exercise.

### 6.2.4. Transformation rules and preservation of structure

The transformation rules above may corrupt the structure of the original formulae, a feature that is shared by many-valued logics and CL alike.

**6.2.14.** In fact, the transformation rules just elaborated on are merely *language preserving*, in the sense that to the original formulae no new symbols are added in the process of transformation. As an alternative, one may apply *structure preserving* transformation rules, which encode the structure of the original formulae by introducing predicate symbols. Therefore, we may call them *predicate introduction rules*, and they are as follows:

1. 
$$S\left[\phi\left[A(\vec{x})\right]\right] \Longrightarrow S\left[\phi\left[P(\vec{x})\right]\right] \wedge \bigwedge_{\mathtt{v}_i \in W} \forall \vec{x}\left(\{\mathtt{v}_i\}\left[A(\vec{x})\right] \equiv \{\mathtt{v}_i\}\left[P(\vec{x})\right]\right)$$

## 6. The signed SAT for many-valued logics

2.

$$S\left[\phi\left[A\left(\vec{x}\right)\right]\right] \Longrightarrow S\left[\phi\left[P\left(\vec{x}\right)\right]\right] \wedge \bigwedge_{v_i \in W} \forall \vec{x}\left(\{v_i\}\left[A\left(\vec{x}\right)\right] \rightarrow \{v_i\}\left[P\left(\vec{x}\right)\right]\right)$$

where $\phi$ is a many-valued formula, $A$ is a subformula contained in $\phi$ in one or more positions, and $\vec{x} = (x_1, x_2, ..., x_n)$ is the vector of free variables in $A$. $\phi[P(\vec{x})]$, where $P$ is a new predicate symbol occurring nowhere else in $\phi$, denotes the substitution by $P(\vec{x})$ in $\phi$ of one or more occurrences of $A$.

In 6.2.14.1, $(\{v_i\}[A(\vec{x})] \equiv \{v_i\}[P(\vec{x})])$ guarantees that in every interpretation satisfying this equivalence it is the case that, for all $x$, $P(\vec{x})$ and $A(\vec{x})$ evaluate to the same truth value $v_i \in W$. Note that 6.2.14.2 is obtained precisely by dropping one of the two logically equivalent conjuncts:

$$\bigwedge_{v_i \in W} \forall \vec{x}\left(\{v_i\}[A(\vec{x})] \rightarrow \{v_i\}[P(\vec{x})]\right) \wedge \bigwedge_{v_i \in W} \forall \vec{x}\left(\{v_i\}[P(\vec{x})] \rightarrow \{v_i\}[A(\vec{x})]\right)$$

Clearly, the introduction of new predicate symbols entails that both rules above are not strictly correct, but both sides of both rules are equivalent as far as satisfiability is concerned, namely with respect to resolution proof procedures. However, these rules produce a much lower number of clauses than the language preserving rules. In effect, a formula with $m$ occurrences of at most $n$-ary connectives and $l$ occurrences of quantifiers contains not more than $m|W|^n + l2^{|W|} + 1$ clauses. Again, we have to trade off between different properties of many-valued logics.

### 6.2.5. Translation to clausal form

As in CL, resolution proof procedures applied to the verification of the validity of a many-valued formula $\phi$ require that $\phi$ be translated into clausal form. Because resolution is a refutation method, the first step in the proof procedure is the transformation of $S[\phi] - \bar{S}[\phi]$, where $S$ is typically the set of designated values (i.e. $\phi$ is true if it takes one of the truth values in $S$), into a set of clauses by (1) eliminating many-valued connectives and quantifiers, (2) eliminating any existential quantifiers by replacing any existentially quantified variables by Skolem terms, and (3) obtaining CNFs by repeatedly applying distributivity.

However, before all this we have to compute the transformation rules for the logic at hand. These are relatively easy to perform–for low cardi-

## 6.2. From many-valued formulae to signed formulae

$$
\begin{aligned}
\{v_i\}[A \wedge B] &\implies (\{v_j\}[A] \wedge \{v_j\}[B]) \wedge (\{v_i\}[A] \vee \{v_i\}[B]), \forall j \geq i \\
\{v_i\}[A \vee B] &\implies (\{v_j\}[A] \wedge \{v_j\}[B]) \wedge (\{v_i\}[A] \vee \{v_i\}[B]), \forall j \leq i \\
\{\mathtt{t}\}[A \to B] &\implies \{v_j\}[A] \wedge \{v_k\}[B], \forall j < c \leq k, 1 \leq c \leq n-1 \\
\{\mathtt{f}\}[A \to B] &\implies \{\mathtt{t}\}[A] \wedge \{\mathtt{f}\}[B] \\
\{v_i^*\}[A \to B] &\implies \{v_k\}[A] \wedge \{v_s\}[B], k \neq r \text{ and} \\
& \quad (n-1)-(r-s) \neq i,\ 0 \leq r \leq n-1 \\
\{v_i\}[\neg A] &\implies \{v_{(n-1)-i}\}[A] \\
\{\mathtt{t}\}[\forall x A(x)] &\implies \forall x(\{\mathtt{t}\}[A(x)]) \\
\{\mathtt{f}\}[\forall x A(x)] &\implies \exists x(\{\mathtt{f}\}[A(x)]) \\
\{v_i^*\}[\forall x A(x)] &\implies \forall x(\{v_j\}[A(x)]) \wedge \exists x(\{v_i\}[A(x)]), \forall j \geq i \\
\{\mathtt{t}\}[\exists x A(x)] &\implies \exists x(\{\mathtt{t}\}[A(x)]) \\
\{\mathtt{f}\}[\exists x A(x)] &\implies \forall x(\{\mathtt{f}\}[A(x)]) \\
\{v_i^*\}[\exists x A(x)] &\implies \forall x(\{v_j\}[A(x)]) \wedge \exists x(\{v_i\}[A(x)]), \forall j \leq i
\end{aligned}
$$

Figure 6.2.2.: Transformation rules for signed formulae of $L_n$.

nality of $W$, of course–once we have the specifications of its connectives and quantifiers, as well as of other properties, such as the ordering in $W$ and the set $D$ of designated values. As a matter fact, these computations can be carried out in a wholly automated way by the MULtlog system.

**Example 6.2.8.** Figure 6.2.2 shows the transformation rules for the t-norm based logic $Ł_n^*$ where the primitive connectives are defined as (we mark $v_i$ with $*$ whenever $v_i \neq \mathtt{f}, \mathtt{t}$)

$$\widetilde{\neg}(v_i) = v_{(n-1)-i}$$

and

$$\widetilde{\to}(v_i, v_j) = \begin{cases} \mathtt{t} & \text{if } i \leq j \\ v_{(n-1)-(i-j)} & \text{otherwise} \end{cases}.$$

**Example 6.2.9.** Let the formula $A = \forall x P(x) \to \exists y P(y)$ be considered in $Ł_3^*$ as $\{\mathtt{i}\}[A]$. In Example 6.2.6, we obtained the result:

$$\{\mathtt{i}\}[A \to B] \equiv (\{\mathtt{t}\}[A] \vee \{\mathtt{f}\}[B]) \wedge (\{\mathtt{i}\}[A] \vee \{\mathtt{i}\}[B])$$

Thus, $A$ is equivalent to the formula:

$$(\{\mathtt{t}\}[\forall x P(x)] \vee \{\mathtt{f}\}[\exists y P(y)]) \wedge (\{\mathtt{i}\}[\forall x P(x)] \vee \{\mathtt{i}\}[\exists y P(y)])$$

We show (Fig. 6.2.3) the clause translation process, indicating by means of numbers the main operations carried out, and underlining the specific

## 6. The signed SAT for many-valued logics

parts upon which those operations fall in case the whole expression is not being operated upon.

The indicated operations are as follows:

(1) Rule $\forall_t$
(2) Rule $\forall_i$
(3) Distributivity law
(4) Rule $\exists_f$
(5) Rule $\exists_i$
(6) Contraction of disjunctions of formula expressions equal up to variable renaming
(7) Removal of the 3rd and 4th conjuncts, which are redundant in presence of the 2nd conjunct
(8) Skolemization and deletion of $\forall$

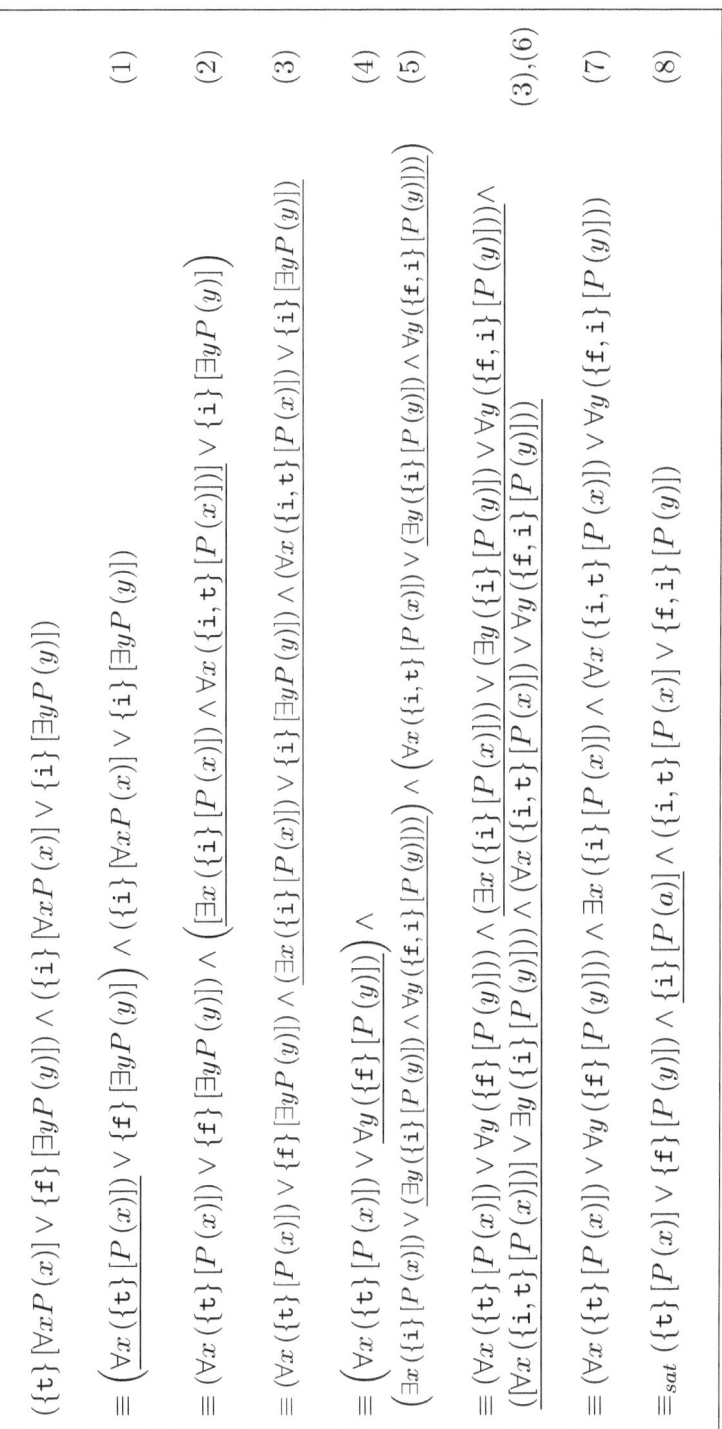

Figure 6.2.3.: Clause transformation process of a signed FO many-valued formula.

6. The signed SAT for many-valued logics

## Exercises

**Exercise 6.1.** Consider the formulae of Exercises 5.8.1-10 in Ł$_3$:

1. Obtain the SFEs of their DNFs.

2. Obtain the SFEs of their *dnfs*.

3. Obtain the SFEs of their CNFs.

4. Obtain the SFEs of their *cnfs*.

**Exercise 6.2.** Obtain the CNF representations of the formulae in Exercises 5.23-4.

**Exercise 6.3.** Write the transformation rules for the t-norm based $G_n^*$ (cf. Def. 5.6.9) that are different from those of the t-norm based $Ł_n^*$, to wit:

1. $\{t\}\,[\neg A]$

2. $\{f\}\,[\neg A]$

3. $\{v_i\}\,[\neg A]$, $v_i \neq f, t$

4. $\{v_i\}\,[A \to B]$, $v_i \neq f, t$

**Exercise 6.4.** Prove (Complete) the propositions and theorems in this Chapter that were left without a proof (with a sketchy proof, respectively).

# 7. Signed tableaux for the MV-SAT

Although somehow out of grace with the field of automated deduction, *signed tableaux* (abbreviating *signed analytic tableaux*) are quite adequate for tackling computationally the MV-SAT. After some introductory remarks (Section 7.1), we discuss both the *origins of signed tableaux in classical logic* (Section 7.2) and *Surma's important good-for-all algorithm* for the many-valued logics (Section 7.3). This preparatory work being done, we elaborate at length on *signed tableaux for the MV-SAT*, both for *the finitely* and *the infinitely many-valued logics*. This elaboration is carried out in different Sections (7.4-5), because the problems posed by them are rather distinct.

## 7.1. Introductory remarks

Needless to say, a tableaux calculus should be generic enough to constitute an adequate proof system for any many-valued logic. This, however, seems unlikely, as the infinitely many-valued logics pose specific problems. But not only these pose problems: the general tableau rule schema for finite tableaux (Fig. 7.1.1; $\Phi$ stands for the premise) indicates that the size (columns and rows) of the expansions can be quite large,[1] depending on the number of truth values of some many-valued logic $L_n$.

$$\frac{\Phi}{\begin{array}{ccc} \phi_{11} & \cdots & \phi_{m1} \\ \vdots & & \vdots \\ \phi_{1r_1} & \cdots & \phi_{mr_m} \end{array}}$$

Figure 7.1.1.: General tableau rule schema for finite tableaux.

Be it as it may, great progress has been achieved as far as the finitely

---
[1] Especially compared with the classical case, in which we have $m \leq 2$ and $r_i \leq 2$.

## 7. Signed tableaux for the MV-SAT

many-valued logics are concerned. Although there are several tableaux calculi for the MV-SAT (c.f., e.g., Hähnle, 1999), we shall restrict our discussion to the signed tableaux calculus. In our view, not only is this the most efficient tableaux application to the finitely many-valued logics, but also it is a very natural extension of the analytic tableaux method for classical logic. In effect, this is intrinsically a signed proof procedure (see Section 7.2). In particular, we focus on sets as signs, as this has significant computational advantages. In the case of infinitely many-valued logics, we elaborate on an approach that considers intervals as signs.

One of the first applications of analytic tableaux to many-valued logics was Surma's (1977) presentation of an algorithm claimed to provide a complete axiomatization of any finitely many-valued logic. Interestingly enough, Surma does not use the labels "analytic" or "tableaux" anywhere in this paper, though he clearly puts this method in relation with the Gentzen sequent calculus and he briefly refers the reader to Smullyan (1968)–albeit only with regard to Hintikka sets. In any case, Surma's method can actually be considered an instance of what we now call *signed tableaux*. In effect, axiomatization and a proof procedure based on the sequent and tableaux calculi are naturally associated in one and the same algorithm, at least for finitely many-valued propositional logics (e.g., Carnielli, 1985).

Surma's work inspired more recent progress in the subject, namely with work by Carnielli (1987, 1991), and the topic somehow crystallized in Hähnle (1999), after previous abundant work by Hähnle and colleagues. This work actually produced a tableaux-based theorem prover for FO finitely many-valued logics, $_3T^AP$,[2] implementable in Prolog (Hähnle, Beckert, & Gerberding, 1994). However, it is obvious that work in signed tableaux is somehow behind the "popularity" of signed resolution for many-valued logics. One of the reasons for this–besides the general popularity of resolution–might be the fact that the former requires a far more complex mathematical (indeed, algebraic) foundation than the latter. For instance, for the completeness of the signed resolution calculus, Herbrand semantics suffices, whereas we need an algebraic semantics for the completeness of the signed tableaux calculus (at least in this text; see Section 7.4).

An important note, however, is required before we move on. In what follows, we shall often *implicitly* assume that we are working with a homomorphism $hom\,(\mathfrak{L}, \mathfrak{U})$ such that the (power set) of the set $W_n$ of truth values is in fact, in the case of signed logic, the (power set) of

---

[2] An acronym for **3**-valued **ta**bleaux-based theorem **p**rover.

the set $E_n \subseteq \mathfrak{U}$.[3] Instead of calling this calculus a "logic over $\mathfrak{U}$," as Surma does (see Section 7.3), we simply talk of *signed logic (for* L*)*; in particular, we talk of the *signed tableaux calculus* for a many-valued logic $L_n$. Nevertheless, we do not wish to emphasize this, as it is the case that the analytic tableaux calculus presented in Chapter 4 just is, too, a calculus that can be isolated from classical logic proper, so that we have the pair K = $(\Vdash, \mathcal{TC})$, i.e. the pair constituted by the classical consequence relation and the analytic tableaux calculus (*ad-hocly* denoted by $\mathcal{TC}$) as *one adequate* proof system for this particular notion of logical consequence. In this sense, the former pair is a *different logical system* from, say, the pair K = $(\Vdash, \mathcal{LK})$ , where $\mathcal{LK}$ is also an adequate proof system for classical logic (see Gabbay, 1994). In any case, from the contents of Section 7.2 it should be obvious that any signed tableau for classical logic can be immediately converted into an unsigned tableau; this, however, requires a homomorphism $hom\,(\mathfrak{L}, \mathfrak{U})$, too, but it is more often than not implicit.

Be it as it may, on some occasions we shall need to make *explicit* what our algebraic basis for the signed tableaux calculus is, and we pave the way for this in Section 7.4.

## 7.2. Signed analytic tableaux for classical formulae

Recall from Definitions 2.5.1-2 that a signed formula is an expression of the form $i\,[\phi]$ where $i \in W_n$ is the sign of $\phi$. For $W_2$, there are only two ways to sign a formula: $\mathtt{t}\,[\phi]$ and $\mathtt{f}\,[\phi]$. Given a formula $\phi$ (and some model $\mathcal{M}$), $\mathtt{t}\,[\phi]$ asserts that the formula $\phi$ is true (in $\mathcal{M}$), and $\mathtt{f}\,[\phi]$ asserts that $\phi$ is false (in $\mathcal{M}$). This provides us with the root of the analytic tableaux proof calculus, as in order to refute $\mathtt{t}\,[\neg\phi]$ we start with $\mathtt{f}\,[\phi]$, i.e. we begin by assuming that $\phi$ is false. Indeed, a tableaux proof is always so if it is a countermodel $\overline{\mathcal{M}}$ to the assertion that $\phi$ is false in some semantics $\mathfrak{S}$ (cf. Section 4.5.1).

This simple procedure of signing a formula actually provides us with the syntactical means to start a tableaux proof machinery, namely by providing us with the rules of construction for tableaux.

**7.2.1. (Prop.)** Let $A, B$ be formulae. Let further a formula of the form $i\,[\phi]$, $i = \mathtt{f}, \mathtt{t}$, be a signed formula. Then, the following are rules of construction of tableaux for CPL:

---

[3]Alas, this is actually more complex; see Section 7.4. In particular, see Hähnle (2001) for the algebraic foundations of signed logic.

## 7. Signed tableaux for the MV-SAT

1.
$$(i\neg) \quad \frac{\mathtt{t}\,[\neg A]}{\mathtt{f}\,[A]} \quad \frac{\mathtt{f}\,[\neg A]}{\mathtt{t}\,[A]}$$

2.
$$(i\to) \quad \frac{\mathtt{t}\,[A \to B]}{\mathtt{f}\,[A] \mid \mathtt{t}\,[B]} \quad \frac{\mathtt{f}\,[A \to B]}{\mathtt{t}\,[A]\ \ \mathtt{f}\,[B]}$$

3.
$$(i\wedge) \quad \frac{\mathtt{t}\,[A \wedge B]}{\mathtt{t}\,[A]\ \ \mathtt{t}\,[B]} \quad \frac{\mathtt{f}\,[A \wedge B]}{\mathtt{f}\,[A] \mid \mathtt{f}\,[B]}$$

4.
$$(i\vee) \quad \frac{\mathtt{t}\,[A \vee B]}{\mathtt{t}\,[A] \mid \mathtt{t}\,[B]} \quad \frac{\mathtt{f}\,[A \vee B]}{\mathtt{f}\,[A]\ \ \mathtt{f}\,[B]}$$

*Proof:* The proof is by checking the truth tables. **QED**

**7.2.2.** The above rules correspond to the A and B expansion rules, corresponding to the $\alpha\beta$-classification, in Figure 7.2.1.

| $\alpha$ | $\alpha_1$ | $\alpha_2$ |
|---|---|---|
| $\mathtt{t}\,[A \wedge B]$ | $\mathtt{t}\,[A]$ | $\mathtt{t}\,[B]$ |
| $\mathtt{f}\,[A \vee B]$ | $\mathtt{f}\,[A]$ | $\mathtt{f}\,[B]$ |
| $\mathtt{f}\,[A \to B]$ | $\mathtt{t}\,[A]$ | $\mathtt{f}\,[B]$ |

| $\beta$ | $\beta_1$ | $\beta_2$ |
|---|---|---|
| $\mathtt{f}\,[A \wedge B]$ | $\mathtt{f}\,[A]$ | $\mathtt{f}\,[B]$ |
| $\mathtt{t}\,[A \vee B]$ | $\mathtt{t}\,[A]$ | $\mathtt{t}\,[B]$ |
| $\mathtt{t}\,[A \to B]$ | $\mathtt{f}\,[A]$ | $\mathtt{t}\,[B]$ |

Figure 7.2.1.: Signed tableaux expansion rules: $\alpha\beta$-classification.

Comparing Figure 7.2.1 with Figure 4.5.1, it is readily seen that the symbol for negation $\neg$ has been consistently replaced by the sign $\mathtt{f}$. It suffices to do the same in the case of the $\gamma\delta$-classification to obtain the expansion rules for signed quantified formulae (see Fig. 7.2.2).

**7.2.3.** The expansion rules for signed quantified formulae, corresponding to the $\gamma\delta$-classification, are shown in Figure 7.2.2.

| $\gamma$ | $\gamma(a)$ |
|---|---|
| $\mathtt{t}\,[\forall x A(x)]$ | $\mathtt{t}\,[A(a)]$ |
| $\mathtt{f}\,[\exists x A(x)]$ | $\mathtt{f}\,[A(a)]$ |

| $\delta$ | $\delta(a)$ |
|---|---|
| $\mathtt{t}\,[\exists x A(x)]$ | $\mathtt{t}\,[A(a)]$ |
| $\mathtt{f}\,[\forall x A(x)]$ | $\mathtt{f}\,[A(a)]$ |

Figure 7.2.2.: Signed tableaux expansion rules: $\gamma\delta$-classification.

Interestingly, this is in fact Smullyan's (1968) first approach to the analytic tableaux calculus, and he was inspired to do so by the work of

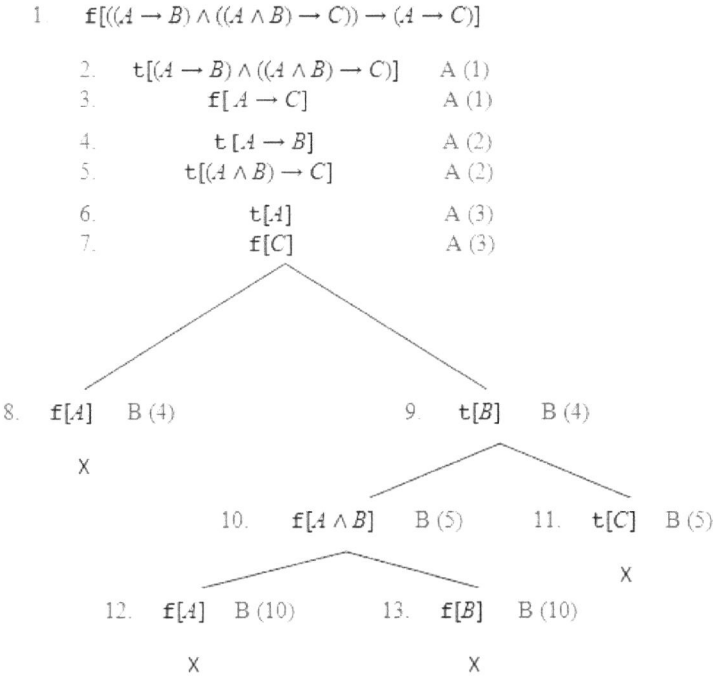

Figure 7.2.3.: A signed tableau proof of a CPL theorem.

the Polish logician Z. Lis (1960), who signed formulae by using the signs + and − for true and false formulae, respectively, in the context of the semantic tableaux calculus. If in the above 7.2.1-2 one replaces the signs f, t by −, +, respectively, one has recovered Lis' "expansion" rules.

**7.2.4. (Def.)** Let $\mathcal{T}_\phi^i$ be a tableau for the signed formula $i\,[\phi]$. We say that a branch $\mathcal{B} \in \mathcal{T}_\phi^i$ is closed if it contains some signed (sub)formula and its complement. $\mathcal{T}_\phi^i$ is closed if every branch of $\mathcal{T}_\phi^i$ is closed. By a proof of $\phi$ is meant a closed tableau for $\mathtt{f}\,[\phi]$.

**Example 7.2.1.** Figure 7.2.3 shows the tableau proof of the formula $((A \to B) \land (A \land B) \to C) \to (A \to C)$ (cf. Example 4.5.1). Compare with Figure 4.5.2.

## 7.3. Surma's algorithm

Although some important posterior work has been carried out on the

## 7. Signed tableaux for the MV-SAT

subject of signed tableaux (see above), we think that Surma's (1977) still provides the best *locus* to approach the subject. We next expound on the central aspects of Surma (1977) with adapted notation. Interestingly, he calls his "signed tableaux calculus" the "logic over $\mathfrak{U}$."

**7.3.1. (Def.)** Let the propositional language L be similar to the finite algebra of formulae $\mathfrak{L} = (F_L, o_1, ..., o_m)$, $\mathfrak{L}$ is a free algebra freely generated by the set of propositional variables $V_{\mathfrak{L}}$, and let the algebra $\mathfrak{U} = (G_n, f_1, ..., f_m)$ be similar to $\mathfrak{L}$ (cf. Def. 3.2.3). We define a tautology of L as $h(p) = n$ for an arbitrary formula $p \in F_L$ and for every $h \in hom(\mathfrak{L}, \mathfrak{U})$.

**7.3.2. (Def.)** Let now any formula of the form $a_i(p)$ be called a *signed formula* of $p$. For every $p \in V_L$, every $i \leq n$ unary propositional connective in the metalanguage of L, and every homomorphism $h \in hom(\mathfrak{L}, \mathfrak{U})$, we have

$$(\clubsuit) \quad h(a_i(p)) = \begin{cases} n & \text{for } h(p) = i \\ 1 & \text{otherwise} \end{cases}$$

where $a_i(p)$ actually expresses a kind of "negation" of $p$.

Obviously, this provides a signed logic basis for a refutation procedure. Now enters the tableaux part.

**7.3.3.** Let

$$i(\heartsuit, j_1, ..., j_m)$$

stand for the condition

$(\clubsuit\clubsuit)$ There exists a homomorphism $h$ of $\mathfrak{L}$ onto $\mathfrak{U}$ such that $h(p_k) = j_k$ for every $k \leq m$, and such that $f(j_1, ..., j_m) = i$.

This expresses the fact that, given a propositional formula $\phi$ with an $m$-ary connective $\heartsuit$ of the form $\heartsuit(p_1, ..., p_m)$, $\phi$ takes the value $i$ under a homomorphism $h$ and under the substitution $j_1$ for $p_1$, $j_2$ for $p_2$, ..., and $j_m$ for $p_m$.

**7.3.4.** Then, all the rules of *decomposition* of the propositional formulae from $\mathfrak{L}$ can be stated as

$$\frac{a_i(\heartsuit(p_1, ..., p_m))}{\bigvee \{a_{j_1}(p_1) \wedge ... \wedge a_{j_m}(p_m) \mid j_1, ..., j_m \leq n, i(\heartsuit, j_1, ..., j_m)\}}$$

where $i \leq n$ and $p_1, ..., p_m \in V$. This means that for every propositional connective $\heartsuit$ there are $n$ rules of decomposition, and with the help of these rules we can decompose any formula from $F_{\mathfrak{L}}$ into an at most $n^m$-adic tree.

**7.3.5.** It so happens that the trees are only of two shapes, to wit,

$$
\begin{array}{c}
p \\
| \\
q_1 \\
| \\
q_2 \\
\vdots \\
q_k
\end{array}
$$

and

which are symbolized respectively as

$$\frac{p}{q_1 \wedge \ldots \wedge q_k}$$

and

$$\frac{p}{q_1 \vee \ldots \vee q_k}.$$

The reader will readily recognize the essentials of both the analytic tableaux calculus (Section 4.5) and signed logic (Section 2.5).

**Example 7.3.1.** (Surma, 1977) Let $n = m = 2$. We define the algebra $\mathfrak{U} = (\{1, 2\}, \leftrightarrow)$ by:

$$(1 \leftrightarrow 1) = (2 \leftrightarrow 2) = 2$$

$$(1 \leftrightarrow 2) = (2 \leftrightarrow 1) = 1$$

Then, for any two homomorphisms $h_1, h_2 \in hom(\mathfrak{L}, \mathfrak{U})$ satisfying conditions ♣ and ♣♣, we have:

|       | $p$ | $a_1(p)$ | $a_2(p)$ |
|-------|-----|----------|----------|
| $h_1$ | 1   | 2        | 1        |
| $h_2$ | 2   | 1        | 2        |

And the two corresponding rules of decomposition are:

## 7. Signed tableaux for the MV-SAT

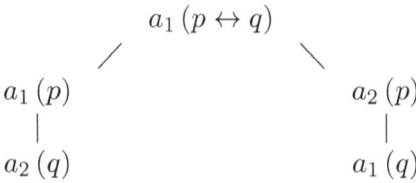

and

$$
\begin{array}{c}
a_2 \, (p \leftrightarrow q) \\
\diagup \qquad \diagdown \\
\begin{array}{c} a_1 \, (p) \\ | \\ a_1 \, (q) \end{array} \qquad \begin{array}{c} a_2 \, (p) \\ | \\ a_2 \, (q) \end{array}
\end{array}
$$

These rules can be rewritten in the form

$$\frac{a_1 \, (p \leftrightarrow q)}{(a_1 \, (p) \wedge a_2 \, (q)) \vee (a_2 \, (p) \wedge a_1 \, (q))}$$

and

$$\frac{a_2 \, (p \leftrightarrow q)}{(a_1 \, (p) \wedge a_1 \, (q)) \vee (a_2 \, (p) \wedge a_2 \, (q))}.$$

**7.3.6. (Def.)** Let the (at most) $n^m$-adic tree in 7.3.4 be called a *decomposition tree*. We say that a decomposition tree for a signed formula of the form $a_i \, (p)$ is *closed* if each branch of that tree has signed formulae of the form $a_j \, (q)$ and $a_k \, (q)$, where $j \neq k$ and $q$ is a subformula of the formula $p$.

Now, the reader should readily recognize that the pair $(a_j, a_k)$ is a contradictory pair, i.e. a branch is closed iff it contains a contradictory pair $(a_j, a_k)$ where $j \neq k$.

**Example 7.3.2.** Note that in Example 7.3.1 the decomposition tree is closed for $a_1 \, (p \leftrightarrow q)$. Indeed, in any of its two branches we have the contradictory pair $(a_1, a_2)$. On the contrary, the decomposition tree for $a_2 \, (p \leftrightarrow q)$ can be said to be *open*.

**7.3.7. (Def.)** We say that a formula $p \in F_L$ is *provable* (in the logic over $\mathfrak{A}$) if, for every $i < n$, the $i$ decomposition trees for the signed formulae $a_i \, (p)$ are closed.

**Example 7.3.3.** The decomposition tree for $a_1 (p \leftrightarrow q)$ in Example 7.3.1 shows that the formula $a_{n=2} (p \leftrightarrow q)$ is provable in the classical propositional logic.

In intuitive terms, Definition 7.3.7 means that in order to prove that formulae are theorems from a specific propositional logic we have to derive a contradiction for each and every one of the negations

$$a_1 (p), a_2 (p), ..., a_{n-1} (p)$$

of the formula $p$, which, in turn, entails $n - 1$ separate decomposition trees.

Clearly, and differently from Definition 5.2.5, the set $G_n = \{1, ..., n\}$ has here a linear ordering $\leq$ with $1 \leq i \leq n - 1$ referring to "falsity" and $n$ to truth. Now, let $G_n = \{1, 2\}$; then, we need only a single decomposition tree. This is the classical case for a closed tableau. But if we are dealing with a three-valued logical system, then we need already two decomposition trees.

**Example 7.3.4.** (Surma, 1977) We now provide a complete example for a formula in Ł$_3$. We shall be occupied with the connective $\rightarrow_{Ł3}$. According to Surma's algorithm, there are three rules of decomposition for this connective (we omit the subscript; compare with the truth table for $\rightarrow_{Ł3}$ in 5.4.3):

1.
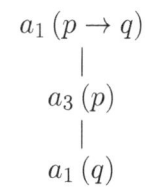

2.
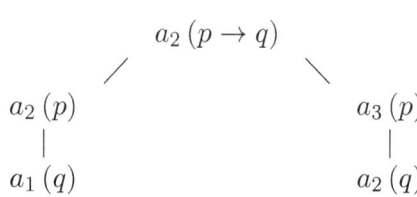

## 7. Signed tableaux for the MV-SAT

3.

$$\frac{a_3\,(p \to q)}{\bigvee \begin{pmatrix} (a_1\,(p) \wedge a_1\,(q)) \\ (a_2\,(p) \wedge a_2\,(q)) \\ (a_3\,(p) \wedge a_3\,(q)) \\ (a_1\,(p) \wedge a_2\,(q)) \\ (a_2\,(p) \wedge a_3\,(q)) \\ (a_1\,(p) \wedge a_3\,(q)) \end{pmatrix}}$$

It is now an easy matter to show that the formula $p \to_{L3} (q \to_{L3} p)$ is a theorem in $L_3$. Because we have $E_{L3} = \{1, 2, 3\}$, we need two decomposition trees, for $1\,(p \to_{L3} (q \to_{L3} p))$ and $2\,(p \to_{L3} (q \to_{L3} p))$, where $i = 1, 2, 3$ abbreviates $a_i$. We show both trees (we omit the subscript in the connective):

1.

$$1\,(p \to (q \to p))$$
$$|$$
$$3\,(p)$$
$$|$$
$$1\,(q \to p)$$
$$|$$
$$3\,(q)$$
$$|$$
$$1\,(p)$$
$$\times$$

2.

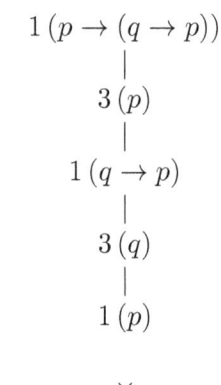

Both decomposition trees are closed. That is to say that the decomposition trees for $n-1$ are closed. Thus, the formula $p \to_{Ł3} (q \to_{Ł3} p)$ has been proven to be a theorem by means of Surma's algorithm, or, as he put it, in the logic over $\mathfrak{U}$.

Surma (1977) proved the completeness of his logic over $\mathfrak{U}$, and the calculus is also sound.

## 7.4. Signed tableaux for finitely many-valued logics

Just as for the classical tableaux calculus, we begin by giving a general algorithm for signed tableaux for many-valued logics (Algorithm 7.1); the different sub-calculi require the specifications to be elaborated on in this Chapter.

From Section 6.2, recall that Theorem 6.2.1 establishes the equisatisfiability condition for a DNF/CNF representation of a signed formula. As a matter of fact, this theorem directly gives rise to rules for signed tableaux and signed sequents. With respect to the former, the DNF representation of a signed formula just is the many-valued signed tableau rule. Putting Surma's results in more recent jargon (e.g., Hähnle, 1999), and letting $\overset{\bullet}{\mathcal{E}}_k$ denote the set of conjuncts occurring in each conjunction $\mathcal{E}_k$ whenever $\bigvee_{k=1}^{M} \mathcal{E}_k$ is a DNF representation of a signed formula:

**7.4.1. (Def.)** If $\bigvee_{r=1}^{M} \mathcal{E}_r$, for $\mathcal{E}_r = \bigwedge_{s \in I_r} i_{rs}[\phi_s]$ and $1 \leq r \leq M$, is a signed DNF representation of $\phi = i[\heartsuit(\phi_1, ..., \phi_m)]$, $m \geq 1$, then a *many-valued signed tableau rule* for $\phi$ is defined as

$$\frac{i[\heartsuit(\phi_1, ..., \phi_m)]}{\overset{\bullet}{\mathcal{E}}_1 \mid ... \mid \overset{\bullet}{\mathcal{E}}_M}$$

where the $\overset{\bullet}{\mathcal{E}}_r = \{i_{rs}[\phi_s] \mid s \in I_r\}$ are *expansions* of the rule.

**Example 7.4.1.** The following are the expansion rules for the formula $i[A \to B]$ in Ł$_3$:

| | | $\{t\}[A \to B]$ | | | |
|---|---|---|---|---|---|
| $\{t\}[A]$ | $\{i\}[A]$ | $\{i\}[A]$ | $\{f\}[A]$ | $\{f\}[A]$ | $\{f\}[A]$ |
| $\{t\}[B]$ | $\{t\}[B]$ | $\{i\}[B]$ | $\{t\}[B]$ | $\{i\}[B]$ | $\{f\}[B]$ |

7. *Signed tableaux for the MV-SAT*

---

**Algorithm 7.1** Signed tableaux proof procedure for MV-SAT

---

**Input:** A signed formula $S[A]$, where $|S| = i$ for $0 \leq i \leq |W_n|$

**Output:** A closed binary tree $\mathcal{T}^\times_{\overline{S}[A]}$ that constitutes a proof of validity of $S[A]$

---

1. Input $\overline{S}[A]$ in the root of a binary tree.

2. Apply the expansion rules (see below) to the signed formulae on the tree, thus adding new signed formulae and splitting branches until no more rules can be applied and the leaves are constituted by signed literals of the form $S[L]$.

3. Close contradictory branches, i.e. branches that contain a contradictory set

$$\perp_S = \left\{ S_1[L], ..., S_r[L] \,\Big|\, r \geq 1, \bigcap_{i=1}^{r} S_i = \emptyset \right\}.$$

4. Terminate successfully iff all branches are closed; unsuccessfully, otherwise.

---

## 7.4. Signed tableaux for finitely many-valued logics

- 
$$\frac{\{\mathtt{i}\}\,[A \to B]}{\{\mathtt{t}\}\,[A] \;\;|\;\; \{\mathtt{i}\}\,[A]}$$
$$\{\mathtt{i}\}\,[B] \;\;|\;\; \{\mathtt{f}\}\,[B]$$

- 
$$\frac{\{\mathtt{f}\}\,[A \to B]}{\{\mathtt{t}\}\,[A]}$$
$$\{\mathtt{f}\}\,[B]$$

Given Definition 7.4.1, we can then define branch closure as follows:

**7.4.2. (Def.)** Let $\mathcal{T}^i_\Phi$ be a signed tableau for a finite set of signed formulae $\Phi = \{i_1\,[\phi]\,,...,i_r\,[\phi]\}$ and let $\mathcal{B} \in \mathcal{T}^i_\Phi$ be one of its branches. $\mathcal{B}$ is a *closed* branch iff one of the following conditions hold:

1. there are signed formulae $i\,[\phi]\,, j\,[\phi]$ on $\mathcal{B}$ such that $i \neq j$, i.e. $i\,[\phi]$ and $j\,[\phi]$ are complementary;

2. there is a signed formula $\phi = i\,[\heartsuit\,(\phi_1, ..., \phi_m)]$, $m \geq 0$, on $\mathcal{B}$ such that $i$ does not occur in the range of $\heartsuit$, i.e. $\phi$ is self-contradictory.

Just as in the case of classical logic, for a many-valid logic a tree $\mathcal{T}^i_\Phi$ is closed iff all its branches are closed. The central theorem is obviously due to Surma (1977), and can be formulated as follows:

**Theorem 7.4.1.** *(Surma, 1977)* Let $\phi$ be a formula in a finitely n-valued logic and let $S \subseteq W_n$, $i \in S$. Then $\phi = i\,[\phi]$ is valid[4] iff there are $n-1$ closed signed tableaux for $\mathtt{v}_1, ..., \mathtt{v}_{n-1} \in \overline{S}$, $|\overline{S}| = n-1$.

*Proof:* The proof can be reconstructed from Surma's (1977) proof of completeness. **QED**

Despite the groundbreaking character of Surma's algorithm in the context of proof theory for finitely many-valued logics, the above calculus, further formalized in Carnielli (1987, 1991), is not efficient enough. Firstly, the expansion rules can be too complex (see Example 7.4.1, in particular the expansion rule for $\{\mathtt{t}\}\,[\phi \to_{\mathrm{L3}} \psi])$, actually introducing redundancy in the signed tableaux. Secondly, and most importantly, one needs to build $n-1$ tableaux for $n-1$ truth values different from $n$ (truth or designated value). In a 3-valued logic, this equates with building two tableaux for each formula one wishes to prove. If the formula is

---
[4] Another way to put this is: $\phi$ is S-valid.

## 7. Signed tableaux for the MV-SAT

complex, this can be indeed very hard work. Thus, the next progressive step should allow us to be able to prove the validity of a formula (in an arbitrary finitely many-valued logic) by generating a single tableau.

In the discussion above (Section 7.3), it will readily be realized that Surma's "negations" $a_1$ and $a_2$ in the logic Ł$_3$ are in fact the truth values f and i, respectively. This shows that Surma's algorithm can be said to prove a signed formula $n\,[\phi]$ if, given a set of truth values $W_n$, all the decomposition trees of $i\,[\phi]$ for all $i \leq (n-1)$ are closed. In fact, the expansion rules in Example 7.4.1 suggest immediately that one could work with sets of truth values rather than with individual truth values. Indeed, to express in Ł$_3$ the fact that a formula $\phi$ is valuated as *either* $val\,(\phi) = \text{f}$ *or* $val\,(\phi) = \text{i}$, we can simply write $\{\text{f},\text{i}\}\,[\phi]$, or, what is the same, $\overline{\{\text{t}\}}\,[\phi]$, as we have $val\,(\phi) \neq \text{t}$. In order to prove $\text{t}\,[\phi]$ in Ł$_3$, we thus must refute $\{\text{f},\text{i}\}\,[\phi]$.

This has been duly formalized (see below), and we restrict our further discussion of signed tableaux to the so-called *sets-as-signs tableaux*. We first approach the propositional case and then tackle the FO case. In both cases, we mostly draw on work developed by R. Hähnle and colleagues; although both cases are treated from different perspectives, they are both strongly algebra-based.[5]

### 7.4.1. Propositional signed tableaux

We follow here mostly R. Hähnle, who, via an "enriching of the syntax of signs" and "interpreting them semantically in a different way" (Hähnle, 1991), produced the signed tableaux calculus of Example 7.4.4. We leave formulations of the signed tableau rules according to Hähnle's method for the main propositional many-valued logics as exercises. We complement this approach with another somehow kindred formalization of a signed propositional calculus for the logical system MK (see Example 7.4.6). (It might be an interesting exercise for the reader to speculate on the semantics of this calculus.)

**7.4.4. (Def.)** The *base of signs* is defined as:

$$\mathscr{S} = \{\{k_1, ..., k_m\} \mid \{k_1, ..., k_m\} \in W_n\} = 2^{W_n}$$

**7.4.5. (Prop.)** For a logic L, the *set of signs* $\mathscr{S}_\text{L}$ will obey:

$$\{\{i\} \mid i \in W_n\} \subseteq \mathscr{S}_\text{L} \subseteq \mathscr{S}$$

---

[5]We dispense with the in-our-text redundant "sets-as-signs" label.

## 7.4. Signed tableaux for finitely many-valued logics

*Proof:* Left as an exercise.

**Example 7.4.2.** The set of signs $\mathscr{S}_{L3}$ of the logic $L_3$ can be selected as:
$$\mathscr{S}_{L3} = \{\{f\}, \{i\}, \{t\}, \{f, i\}\}$$
Clearly, the subset $\{f, t\}$ is of no practical interest if we want to prove (quasi-)tautologies; the same holds for $\{i, t\}$, as the only designated value in $L_3$ is $t$. On the other hand, $\{f, i, t\}$ just is a trivial sign, and $\emptyset$ indicates that no expansion rule can be defined with respect to a formula.

**7.4.6. (Def.)** *Algebra of signs* – Let $\mathfrak{U} = (G_n, f_1, ..., f_m)$ defined as in Definition 3.2.2. be similar to the algebra $\mathfrak{W} = \left(\mathscr{S}, f'_1, ..., f'_r\right)$. Then, $\mathfrak{W}$ is called an algebra of signs with operations defined as:

$$f'_i(S_1, ..., S_m) = \bigcup \{f_i(v_1, ..., v_m) \,|\, v_k \in S_k, 1 \leq k \leq m\}$$

For practical purposes, we make $G_n = \{v_0, ..., v_{n-1}\} = W_n$; the contents of Definition 3.2.2 hold. For each *set of signs* $S = \{v_1, ..., v_m\}$, we denote the set $\{(v_1, ..., v_m) \,|\, f(v_1, ..., v_m) \in S\}$ where $S$ is the *set of truth values* corresponding to $S = \{v_1, ..., v_m\}$ by $\mathcal{S}$.[6] Let now $\mathfrak{L}$ be an algebra of formulae similar to $\mathfrak{U}$. Then, $\mathfrak{W}$ defines the semantics of $\mathfrak{L}$ in terms of a set of signs $\mathscr{S}_L$. Another way to "license" this semantics is by seeing a propositional logic L as the triple $(\mathfrak{L}, \mathfrak{M}_\mathfrak{U}, \mathscr{S}_L)$ where $\mathfrak{L}$ is defined as in Definition 3.2.3, $\mathfrak{M}_\mathfrak{U} = (G_n, d_1, ..., d_k, f_1, ..., f_r)$, $k \leq |G_n|$ and all the $d_i$ are 0-ary operators on $G_n$, is an algebra similar to $\mathfrak{W}$ called a propositional matrix for $\mathfrak{L}$, and $\mathscr{S}_L$ is the set of signs for L.

**7.4.7. (Def.)** The triple $L = (\mathfrak{L}, \mathfrak{M}_\mathfrak{U}, \mathscr{S}_L)$ is a propositional logic.

Given a language $\mathfrak{L}$ (which we identify with its set of formulae), we shall refer to its set of signed formulae as $\mathfrak{L}^S \subseteq \mathfrak{L}$.

The above allows for a formalization of signed tableau rules for many-valued logics as follows:

**7.4.8. (Def.)** Let $L = (\mathfrak{L}, \mathfrak{M}_\mathfrak{U}, \mathscr{S}_L)$ be a many-valued logic whose formulae are of the form $\heartsuit(\phi_1, ..., \phi_m)$. Let further $f$ and $f'$ be the interpretation of $\heartsuit$ in $\mathfrak{M}_\mathfrak{U}$ and $\mathfrak{W}$, respectively. A *tableau rule* is a function $\pi: S \longrightarrow \heartsuit$ assigning to every signed formula $S[\phi] \in \mathfrak{L}^S$ a signed tree $\mathcal{T}^S$ with root $S[\heartsuit(\phi_1, ..., \phi_m)]$ and linear subtrees $S_1[\phi_{i_1}] \circ ... \circ S_t[\phi_{i_t}]$ such that $S_1, ..., S_t \in \mathscr{S}_L$, where $1 \leq t \leq m$ and $\mathcal{H}^S(\heartsuit(S_1, i_1), ..., (S_t, i_t))$,

---

[6] To be more rigorous, we should distinguish two sets, the set of signs and the set of truth values corresponding to it, but this brings further notational complexity in what is already a complex notation. Note that we are simply making both sets *coincide*, namely by a homomorphism $hom(\mathfrak{U}, \mathfrak{W})$.

## 7. Signed tableaux for the MV-SAT

stating that there is a homomorphism $h : \mathfrak{L} \to \mathfrak{W}$, holds for each of the subtrees. The root is called the *premise* and the subtrees are the *expansions*.

**7.4.9. (Def.)** The collection of such expansions is called a *conclusion* of a tableau rule if

1. for any $v_1, ..., v_m \in S$ there is an expansion $S_1 [\phi_{i_1}] \circ ... \circ S_t [\phi_{i_t}]$ with $v_{i_k} \in S_k$ for all $1 \leq k \leq t$, and

2. there is no set of expansions with fewer elements satisfying 1.

**7.4.10. (Prop.)** A tableau rule is *sound* iff there is a homomorphism $hom\,(\mathfrak{L}, \mathfrak{W})$ such that for $h \in hom\,(\mathfrak{L}, \mathfrak{W})$, $h$ satisfies the following conditions:[7]

1. $h(\phi_{i_k}) = S_k$ for $1 \leq k \leq t$;

2. $f'\left(S'_1, ..., S'_m\right) \subseteq S$ whenever $S'_{i_k} = S_k$ for all $1 \leq k \leq t$ and the other $S'_j$ are arbitrary;

3. for no $1 \leq k \leq t$ is there a $S'_k$ with $S'_k \supsetneq S_k$ that satisfies 1 and 2;

4. there is no $t'$ with $t' < t$ that satisfies 1 and 2.

*Proof:* Left as an exercise.

Intuitively, condition 7.4.9.1 specifies soundness of rules with respect to $\mathfrak{W}$: each truth table belonging to $S$ must be covered by an expansion. Condition 7.4.9.2 minimizes the number of expansions. As said above, $\mathfrak{W}$ defines the semantics of $\mathfrak{L}$; Proposition 7.4.10.1 simply formalizes this property. By Proposition 7.4.10.2 we are assured of completeness with respect to $\mathfrak{W}$. As for Propositions 7.4.10.3-4, the first makes $S_k$ maximal and the second minimizes the number of formulae in expansions. Obviously, Definition 7.4.9.1 and Propositions 7.4.10.1-2 suffice to assure us of the existence of sound and complete expansion rules.

Now, in order to guarantee that we are not missing any truth values, we need the following condition:

**7.4.11. (Def.)** *Complete set of signs* – Given an $n$-valued logic L, let $f$ be the interpretation of an $m$-ary connective $\heartsuit$. Let further $\mathscr{S}_L$ be the set of signs of L. We say that $\mathscr{S}_L$ is complete (with respect to L) iff for all connectives in L and all $S \in \mathscr{S}_L$ the following condition holds:

---

[7] That is, $\mathcal{H}^S\left(\heartsuit\left(S_1, i_1\right), ..., \left(S_t, i_t\right)\right)$ holds.

## 7.4. Signed tableaux for finitely many-valued logics

$$\forall v_1, ..., v_m \in \mathcal{S}, \exists S_1, ..., S_m \in \mathscr{S}_L \cup W \text{ such that } v_l \in S_l, 1 \leq l \leq m,$$

$$\text{and } f'(S_1, ..., S_m) \subseteq \mathscr{S}_L.$$

In what follows, keep this algebraic foundation in the back of your mind; when it is necessary to bring it back to the frontal cortex, we shall make this clear.

We now reformulate Definition 7.4.1 for the signed propositional tableaux calculus where sets instead of individual truth values are the only signs.

**7.4.12. (Def.)** If $\bigvee_{r=1}^{M} \left( \bigwedge_{s \in I_r} S_{rs} [\phi_s] \right)$ is a signed DNF representation of $\phi = S [\heartsuit (\phi_1, ..., \phi_m)]$, $m \geq 1$, then a *many-valued signed tableau rule* for $\phi$ is defined as

$$\frac{S [\heartsuit (\phi_1, ..., \phi_m)]}{\overset{\bullet}{\mathcal{E}}_1 \mid ... \mid \overset{\bullet}{\mathcal{E}}_M}$$

where the $\overset{\bullet}{\mathcal{E}}_r = \{S_{rs} [\phi_s] \mid s \in I_r\}$ are *expansions* of the rule.

More precisely, if $S_{rs}, S_{kl}$ are as in Theorem 6.2.1, then:

**7.4.13. (Prop.)** The following tableau rule is sound and complete for the connective in the premise provided that $S$ is complete with respect to $\mathscr{S}$:

$$\frac{S [\heartsuit (\phi_1, ..., \phi_m)]}{\begin{array}{c|c|c} \vdots \mid \vdots \mid & & \vdots \\ S_{1s} [\phi_s] \mid & \cdots & \mid S_{M_1 s} [\phi_s] \\ \vdots \mid \vdots \mid & & \vdots \end{array}}$$

*Proof:* Left as an exercise.

Note in Theorem 6.2.1 condition #; we are now considering only signs that satisfy this condition.[8]

**Example 7.4.3.** We obtain the signed tableau expansion rule for $\{\mathtt{f}, \mathtt{i}\} [A \to_{L3} B]$ with the method above. Revisiting the truth table for the connective $\to_{L3}$ (cf. 5.4.3 and Example 7.4.1), we see that $\{\mathtt{f}\} [A \to_{L3} B]$ is the case when we have $\{\mathtt{t}\} [A]$ and $\{\mathtt{f}\} [B]$; we have

---

[8] A stronger condition in Theorem 6.2.1 would be (cf. Prop. 7.4.5)

$$(\#') \quad \{\{i\} \mid i \in W\} \subseteq \mathscr{S}$$

Then, in Theorem 6.2.1 the condition $S \in \mathscr{S}$ can be relaxed to $S = S_1 \cup ... \cup S_k$ for $\{S_1, ..., S_k\} \subseteq \mathscr{S}$ by applying the theorem separately to each $S_i$. Given this "relaxed condition," we can then give representations of arbitrary $S \subseteq W$.

## 7. Signed tableaux for the MV-SAT

$\{i\}\,[A \to_{L3} B]$ when (i) either $\{i\}\,[A]$ and $\{f\}\,[B]$ or (ii) $\{t\}\,[A]$ and $\{i\}\,[B]$. Combining both runs as follows: we have $\{f,i\}\,[A \to_{L3} B]$ when (i) we have $\{t\}\,[A]$ *and* either $\{i\}\,[B]$ or $\{f\}\,[B]$, i.e. $\{f,i\}\,[B]$, (ii) *or* we have $\{f\}\,[B]$ *and* either $\{t\}\,[A]$ or $\{i\}\,[A]$, i.e. $\{i,t\}\,[A]$. The tableau rule is thus:

$$\frac{\{f,i\}\,[A \to_{L3} B]}{\begin{array}{c|c} \{i,t\}\,[A] & \{t\}\,[A] \\ \{f\}\,[B] & \{f,i\}\,[B] \end{array}}$$

In more formal terms, we appeal to the homomorphism $h : \mathfrak{L} \longrightarrow \mathfrak{W}$ (cf. Def. 7.4.8), according to which for $\{f\}\,[A \to_{L3} B]$ we have $h_1(A) = \{t\}$ and $h_1(B) = \{f\}$; for $\{i\}\,[A \to_{L3} B]$ we have $h_2(A) = \{i,t\}$ and $h_2(B) = \{f,i\}$, where $h_2(\phi) = h_2'(\phi) \cup h_2''(\phi)$, for $h_2'(A) = \{i\}$ and $h_2'(B) = \{f\}$, and $h_2''(A) = \{t\}$ and $h_2''(B) = \{i\}$. Add to this the fact that a sign constituted by a set of truth values has the metalogical structure $\bigvee (\bigwedge (\bigvee i\,[\phi]))$ for $i$ a truth value. That is to say that, as compared to a rule for a sign constituted by an individual truth value, we now have an additional level of nesting of connectives.

**Example 7.4.4.** Hähnle (1991) built a sound and complete 3-valued calculus, $\mathcal{L}_3$, with the tableaux method above. The matrix of L3 is

$$\mathcal{L}_3 = \{\{f,i,t\}, \neg, \sim, \odot, \wedge, \vee, \to, \{t\}\}$$

and the truth tables for the connectives are as follows:

| $p$ | $\sim p$ |
|---|---|
| t | f |
| i | t |
| f | t |

| $p$ | $\neg p$ |
|---|---|
| t | f |
| i | i |
| f | t |

| $p$ | $\odot p$ |
|---|---|
| t | t |
| i | t |
| f | f |

| $\wedge$ | | | |
|---|---|---|---|
| $p \backslash q$ | t | i | f |
| t | t | i | f |
| i | i | i | f |
| f | f | f | f |

| $\vee$ | | | |
|---|---|---|---|
| t | i | f | |
| t | t | t | |
| t | i | i | |
| t | i | f | |

| $\to$ | | | |
|---|---|---|---|
| t | i | f | |
| t | i | f | |
| t | t | t | |
| t | t | t | |

The signed tableau rules for the connectives $\neg, \wedge, \vee, \to$ are shown in Figure 7.4.1.

## 7.4. Signed tableaux for finitely many-valued logics

$$(\neg_{L3_{RE}}) \quad \frac{\{\mathbf{t}\}\,[\neg A]}{\{\mathbf{f}\}\,[A]} \quad \frac{\{\mathbf{i}\}\,[\neg A]}{\{\mathbf{i}\}\,[A]} \quad \frac{\{\mathbf{f}\}\,[\neg A]}{\{\mathbf{t}\}\,[A]} \quad \frac{\{\mathbf{f},\mathbf{i}\}\,[\neg A]}{\{\mathbf{i}\}\,[A]\,|\,\{\mathbf{t}\}\,[A]}$$

$$(\to_{L3_{RE}}) \quad \frac{\{\mathbf{t}\}\,[A\to B]}{\{\mathbf{f},\mathbf{i}\}\,[A]\,|\,\{\mathbf{t}\}\,[B]} \quad \frac{\{\mathbf{i}\}\,[A\to B]}{\{\mathbf{t}\}\,[A]} \quad \frac{\{\mathbf{f}\}\,[A\to B]}{\{\mathbf{t}\}\,[A]} \quad \frac{\{\mathbf{f},\mathbf{i}\}\,[A\to B]}{\{\mathbf{t}\}\,[A]}$$
$$\qquad\qquad\qquad\qquad\qquad\qquad\qquad \{\mathbf{i}\}\,[B] \qquad \{\mathbf{f}\}\,[B] \qquad \{\mathbf{f},\mathbf{i}\}\,[B]$$

$$(\land_{L3_{RE}}) \quad \frac{\{\mathbf{t}\}\,[A\land B]}{\{\mathbf{t}\}\,[A]} \quad \frac{\{\mathbf{i}\}\,[A\land B]}{\{\mathbf{i}\}\,[A]\,|\,\{\mathbf{i}\}\,[A]\,|\,\{\mathbf{t}\}\,[A]} \quad \frac{\{\mathbf{f}\}\,[A\land B]}{\{\mathbf{f}\}\,[A]\,|\,\{\mathbf{f}\}\,[B]} \quad \frac{\{\mathbf{f},\mathbf{i}\}\,[A\land B]}{\{\mathbf{f},\mathbf{i}\}\,[A]\,|\,\{\mathbf{f},\mathbf{i}\}\,[B]}$$
$$\qquad \{\mathbf{t}\}\,[B] \qquad \{\mathbf{t}\}\,[B]\,|\,\{\mathbf{i}\}\,[B]\,|\,\{\mathbf{i}\}\,[B]$$

$$(\lor_{L3_{RE}}) \quad \frac{\{\mathbf{t}\}\,[A\lor B]}{\{\mathbf{t}\}\,[A]\,|\,\{\mathbf{t}\}\,[B]} \quad \frac{\{\mathbf{i}\}\,[A\lor B]}{\{\mathbf{i}\}\,[A]\,|\,\{\mathbf{f},\mathbf{i}\}\,[A]} \quad \frac{\{\mathbf{f}\}\,[A\lor B]}{\{\mathbf{f}\}\,[A]} \quad \frac{\{\mathbf{f},\mathbf{i}\}\,[A\lor B]}{\{\mathbf{f},\mathbf{i}\}\,[A]}$$
$$\qquad\qquad\qquad \{\mathbf{f},\mathbf{i}\}\,[B]\,|\,\{\mathbf{i}\}\,[B] \quad \{\mathbf{f}\}\,[B] \quad \{\mathbf{f},\mathbf{i}\}\,[B]$$

Figure 7.4.1.: Signed tableau rules for some connectives of $L_3$.

## 7. Signed tableaux for the MV-SAT

The semantics of signed tableau rules for sets as signs thus established, we now elaborate on the closure rules for branches and trees.

**7.4.14. (Def.)** Given a finite set $\Phi^S$ of signed formulae, a many-valued signed tableau for $\Phi^S$, denoted by $\mathcal{T}^S_\Phi$, is a labeled directed tree in which the labels are the signed formulae $S_i[\phi] \in \Phi^S$, $1 \leq i \leq r$. A branch $\mathcal{B} \in \mathcal{T}^S_\Phi$ is said to be *complete* if all possible expansion rules have been applied, and $\mathcal{T}^S_\Phi$ is *complete* if every of its branches is complete.

The distinction between an analytic tableau for classical logic and a signed tableau for many-valued signed formulae falls on the definition of closure.

**7.4.15. (Def.)** Let $\mathcal{T}^S_\Phi$ be a signed tableau for a finite set of signed formulae $\Phi^S = \{S_1[\phi], ..., S_r[\phi]\}$ and let $\mathcal{B} \in \mathcal{T}^S_\Phi$ be one of its branches. $\mathcal{B}$ is a *closed* branch iff one of the following conditions holds:

1. there are signed formulae $S_1[\phi], ..., S_r[\phi]$ on $\mathcal{B}$ such that $\bigcap_{i=1}^{r} S_i = \emptyset$, i.e. $\mathcal{B}$ is *inconsistent*;

2. there is a signed formula $S[\heartsuit(\phi_1, ..., \phi_m)]$, $m \geq 0$, on $\mathcal{B}$ such that $S \cap rg(\tilde{\heartsuit}) = \emptyset$.

That is, $\mathcal{B}$ is a closed branch iff it contains a *contradiction set*

$$\perp_S = \left\{ S_1[\phi], ..., S_r[\phi] \,|\, r \geq 1, \bigcap_{i=1}^{r} S_i = \emptyset, S_i[\phi] \in \Phi^S \right\}$$

$$\cup \left\{ S[\phi] \,|\, \text{no rule defined for } S[\phi] \in \Phi^S \right\}.$$

Otherwise, $\mathcal{B}$ is *open*.

**7.4.16. (Def.)** A tree $\mathcal{T}^S_\Phi$ is *closed* iff each of its branches is closed. Otherwise, $\mathcal{T}^S_\Phi$ is said to be *open*.

**Example 7.4.5.** We wish to know whether it is the case that the formula $((A \to B) \wedge A) \to B$ (cf. Prop. 3.5.3, T5) is valid[9] in L$_3$ (Example 7.4.4). Recall that the set of designated values of the corresponding matrix is $D = \{t\}$, i.e. we wish to know whether the signed formula $\{t\}[((A \to B) \wedge A) \to B]$ is valid in L$_3$. Thus, we place the signed formula $\{f, i\}[((A \to B) \wedge A) \to B]$ at the root of the tree. Figure 7.4.2 shows the closed tree for this formula; note the complementary pairs $(\{t\}[A], \{f, i\}[A])$ and $(\{t\}[B], \{f, i\}[B])$. We conclude that the classical tautology known as *modus ponens* is a valid formula in L$_3$.

---

[9] Or $S$-valid.

## 7.4. Signed tableaux for finitely many-valued logics

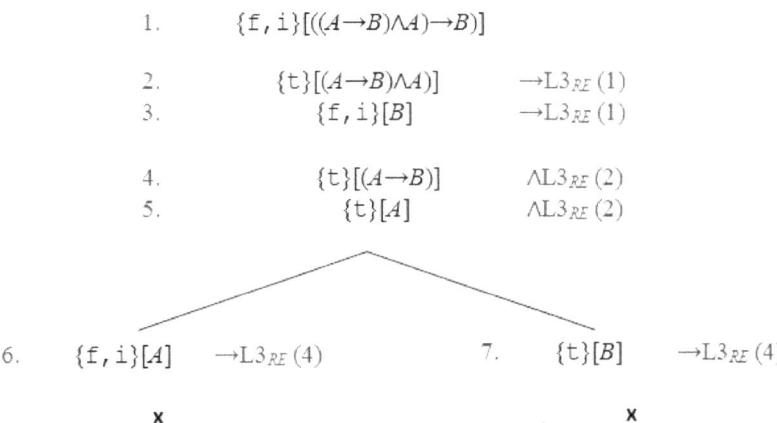

Figure 7.4.2.: A signed tableau proof in L₃ of the classical tautology *modus ponens*.

Now, in order to prove the soundness of the propositional signed tableaux calculus for finitely many-valued logics, we require some basic semantics. In particular, we need a specific notion of *signed tableaux satisfiability*. To this end, we need only elaborate further on Definition 6.1.5 with respect to the signed tableaux calculus. We repeat part of Definition 6.1.5, because the reader should now reread it bearing in mind the algebraic foundations above with respect to the sign $S$. Recall that the $n$-valued propositional logic L can be seen as the triple $L = (\mathfrak{L}, \mathfrak{M}_\mathfrak{U}, \mathscr{S}_L)$.

**7.4.17. (Def.)** Given some $n$-valued logic $L = (\mathfrak{L}, \mathfrak{M}_\mathfrak{U}, \mathscr{S}_L)$, let $\Phi^S$ be a set of signed formulae of L. We say that $\Phi^S$ is *satisfiable (in L)* iff there is an interpretation $\mathcal{I}$ such that for all $S[\phi] \in \Phi^S$ we have $val_\mathcal{I}(\phi) \in S$, and we write

$$\models_{val_\mathcal{I}} S[\phi] \quad \text{iff} \quad val_\mathcal{I}(\phi) \in S.$$

Then, $val_\mathcal{I}$ is a model for $\Phi^S$. A signed tableau branch $\mathcal{B}$ is satisfiable (open) if its set of node labels is. A signed tableau $\mathcal{T}^S_\Phi$ is satisfiable iff at least one of its branches is satisfiable (open).

**Theorem 7.4.2.** *(Soundness of the propositional signed tableaux calcu-*

## 7. Signed tableaux for the MV-SAT

lus) Let $S[\phi]$ be a propositional signed formula with sign $S \subseteq W_n, S \neq \emptyset$, in an n-valued logic L and let $D \subset W_n$ be a set of designated values. If there is a closed signed tableau with root $\overline{S}[\phi]$ for $\overline{S} = \overline{D}$, then $S[\phi]$ is a tautology of L.

*Proof:*[10] Let $\mathcal{T}_\Phi^S$ be a closed signed tableau for $S[\phi]$. Clearly, by Definitions 7.4.14-6 $\mathcal{T}_\Phi^S$ cannot be satisfiable, as for every branch $\mathcal{B} \in \mathcal{T}_\Phi^S$ either $\mathcal{B}$ contains an inconsistent signed pair of atoms, or $\mathcal{B}$ has a self-contradictory formula, i.e. for each branch $\mathcal{B}$, $\perp_S \in \mathcal{B}$, and each branch is closed. Now, let the root formula have sign $\overline{S} = \overline{D}$; then, $\overline{S}[\phi]$ is not satisfiable because the tree $\mathcal{T}_\Phi^S$ is not satisfiable, and $\mathcal{T}_\Phi^S$ was built from the root by applying sound tableau rules (cf. Def. 7.4.10).[11] Because a closed tableau for some formula $\overline{S}[\phi]$ is a proof of the validity of $S[\phi]$ (see Theorem 7.4.1), then if $S = D$, the formula $S[\phi]$ is a tautology.
**QED**

In order to prove the completeness of the propositional signed tableaux calculus, we need to make some changes with respect to the notion of Hintikka set (cf. Def. 4.5.16).

**7.4.18. (Def.)** Let $\Phi^S$ be a set of signed formulae of an n-valued logic L. We say that $\Phi^S$ is an *(atomic) Hintikka sign set*, denoted by $H^S$, iff the following conditions hold:

1. for all signed atomic formulae $S_i[p], p \in \mathcal{L}_0^S$, if $S_1[p], ..., S_r[p] \in \Phi^S$, then $\bigcap_{i=1}^r S_i \neq \emptyset$, i.e. the "mapping" $\Phi^S \longrightarrow \mathcal{B}$ produces an open branch (cf. Def. 7.4.15);

2. if $\phi = S[\heartsuit(\phi_1, ..., \phi_m)] \in \Phi^S$ and $\bigvee_{i=1}^r \mathcal{E}_r$ is a signed DNF representation of $\phi$, then $\dot{\mathcal{E}_r} \in \Phi^S$ for at least one $r$, $1 \leq r \leq M$.

**Theorem 7.4.3.** *Every Hintikka sign set $H^S$ has a model.*

*Proof:* (Hähnle, 1994) For each $p \in \Upsilon_{\mathcal{L}^*}$, let $H^S(p)$ be the set of signed literals with atom $p$ in $H^S$. Now, define a function $\mathcal{I}^S : \Upsilon_{\mathcal{L}^*} \longrightarrow \mathscr{S}$ as:

$$\mathcal{I}^S(p) = \begin{cases} \bigcap_{S[p] \in H^S(p)} S & \text{if } H^S(p) \neq \emptyset \\ W & \text{if } H^S(p) = \emptyset \end{cases}$$

---

[10] This is the more convoluted proof; repeated application of Theorem 6.2.1, with $\mathscr{S}$, $S$, and $\{i\}$ specified as above in this Section, provides a more straightforward proof.

[11] Here one could apply a notion of preservation of satisfiability by means of a tableau $\mathcal{T}'$ created from $\mathcal{T}$ by means of the application of rules of expansion, but this would require a further, redundant, lemma (see Hähnle, 1994).

## 7.4. Signed tableaux for finitely many-valued logics

$\mathcal{I}^S$ is well defined because $H^S$ is atomically consistent. From $\mathcal{I}^S(p)$ we directly obtain $\mathcal{I}^S(L)$ for a literal $L$. Assume now that $\mathcal{I}^S$ satisfies complex formulae $S[\phi]$ of the form $S[\heartsuit(\phi_1,...,\phi_m)], m \geq 1$. Then, by Definition 7.4.12 and by Theorem 6.2.1, a formula $\phi$ has a signed DNF representation $\bigvee_{i=1}^{r} \mathcal{E}_r$, and so $\overset{\bullet}{\mathcal{E}}_r \in H^S$ for some $r$, $1 \leq r \leq M$. By the induction hypothesis, $\mathcal{I}^S$ satisfies $\overset{\bullet}{\mathcal{E}}_r$ and again by Theorem 6.2.1 $\mathcal{I}^S$ satisfies $S[\phi]$. Thus, $\mathcal{I}^S$ satisfies $H^S$. **QED**

We are now ready to tackle the completeness theorem for the propositional signed tableaux calculus for many-valid logics.

**Theorem 7.4.4.** *(Completeness of the propositional signed tableaux calculus) Let $S[\phi]$ be a propositional signed formula with sign $S \subseteq W_n, S \neq \emptyset$, in an n-valued logic L and let $D \subset W_n$ be a set of designated values. Then, $S[\phi]$ is valid if there exists a closed signed tableau with root $\overline{S}[\phi]$ for $\overline{S} = \overline{D}$.*

Proof: (Sketch) Assume $\overline{S}[\phi]$ is valid for $\overline{S} = W_n - S, S = D \subset W_n$.[12] Now, given a many-valued signed tableau $\mathcal{T}_\phi^S$ for $\overline{S}[\phi]$, consider a complete branch $\mathcal{B} \in \mathcal{T}_\phi^S$. In fact, $\mathcal{B}$ is open, and so it has a Hintikka set $H^S$. Hence, it has a model, and this model contradicts the assumption that $\overline{S}[\phi]$ is valid. Therefore, $S[\phi], S = D$, is valid.[13] **QED**

**Example 7.4.6.** From the above, it is obvious that we are working with two sets only, namely $D$ and $\overline{D}$. We can make this "two-valuedness"–and "classicality"–more precise by, for instance, establishing the following assignments of sets of truth values to signs:

$$\{\texttt{t}\} = \texttt{T}, \{\texttt{f}, \texttt{i}\} = \texttt{NT}$$

Let us now establish the closure rules for this tableaux calculus:

$$\begin{array}{cc} \texttt{T}[\phi] & \texttt{T}[\phi] \\ \texttt{NT}[\phi] & \texttt{T}[\neg\phi] \\ \times & \times \end{array}$$

We have the basics of the signed tableaux calculus for the logical system MK, a hybrid logic of Kleene's and McCarthy's (1963) many-valued systems. The tableau rules are as shown in Figure 7.4.3. This is a signed tableaux calculus conceived by Konikowska, Tarlecki, & Blikle (1991).

---

[12] Or $\phi$ is $\overline{S}$-valid.
[13] Or $\phi$ is $S$-valid.

# 7. Signed tableaux for the MV-SAT

Figure 7.4.3.: Signed tableau rules for the logical system MK.

262

## 7.4.2. FO signed tableaux

We now approach the problematic topic of FO signed tableaux, and we do so via distribution quantifiers. More specifically, the problem is that of, given a signed formula $S\left[\blacklozenge x\phi\left(x\right)\right]$, finding a formula of the form

$$\bigvee_{i\in I}\bigwedge_{j\in J} S_{ij}\left[\phi\left(a_{ij}\right)\right]$$

where $S_{ij}\subseteq W_n$ and the $a_{ij}$ are ground terms of two kinds, to wit, a new Skolem constant $c$, or an arbitrary ground term $t$.

The distribution quantifiers, so coined by Carnielli (1987), are a very general class of quantifiers (the generalized quantifiers) of which $\forall$ and $\exists$ are special cases (see Section 5.7.1). They are important for the signed tableaux calculus, because we obtain tableau rules directly from them.

**Example 7.4.7.** Let $W_n = \left\{0, \frac{1}{n-1}, ..., \frac{n-2}{n-1}, 1\right\}$. We can define generalizations of the classical quantifiers with respect to the ordering in $W_n$ by $\widetilde{\forall} = min$ and $\widetilde{\exists} = max$ (cf. Example 5.7.1). For instance, given $W_2 = \{0, 1\}$, we have $\widetilde{\exists}(\{0\}) = 0$, $\widetilde{\exists}(\{1\}) = 1$, and $\widetilde{\exists}(\{0,1\}) = 1$.

**Example 7.4.8.** Consider the distribution quantifiers of Example 5.7.8. Note the relations *glb* and *lub* for $\forall$ and $\exists$, respectively, with respect to the truth values in the distribution $(\Delta \neq \emptyset) \subseteq W_3$ taken as a partially ordered set with $\mathtt{f} < \mathtt{i} < \mathtt{t}$, corresponding to $0 < \frac{1}{2} < 1$ (cf. Prop. 5.7.3 and Example 5.7.1). In fact, we say that $\forall$ is *based on the meet semi-lattice* $\mathcal{L}_\wedge = (W_3, \leq)$, i.e. for all non-empty $V \subseteq W$ we have $\widetilde{\forall}(V) = glb(V)$. In turn, $\exists$ is *based on the join semi-lattice* $\mathcal{L}_\vee = (W_3, \leq)$, i.e. for all non-empty $V \subseteq W$ we have $\widetilde{\exists}(V) = lub(V)$ (cf. Def. 5.7.2). We obtain the following signed tableau rules, where $t$ is any term and $c$ is a new Skolem constant:

- $$\frac{\{\mathtt{t}\}\left[\forall x A\left(x\right)\right]}{\{\mathtt{t}\}\left[A\left(t\right)\right]}$$

- $$\frac{\{\mathtt{i}\}\left[\forall x A\left(x\right)\right]}{\{\mathtt{i}\}\left[A\left(c\right)\right] \mid \begin{array}{l}\{\mathtt{t}\}\left[A\left(t\right)\right]\\ \mid \{\mathtt{i}\}\left[A\left(t\right)\right]\end{array}}$$

- $$\frac{\{\mathtt{f}\}\left[\forall x A\left(x\right)\right]}{\{\mathtt{f}\}\left[A\left(c\right)\right]}$$

## 7. Signed tableaux for the MV-SAT

- $$\frac{\{\mathtt{t}\}\,[\exists x A\,(x)]}{\{\mathtt{t}\}\,[A\,(c)]}$$

- $$\frac{\{\mathtt{i}\}\,[\exists x A\,(x)]}{\{\mathtt{i}\}\,[A\,(c)] \;\mid\; \begin{array}{l}\{\mathtt{i}\}\,[A\,(t)] \\ \{\mathtt{f}\}\,[A\,(t)]\end{array}}$$

- $$\frac{\{\mathtt{f}\}\,[\exists x A\,(x)]}{\{\mathtt{f}\}\,[A\,(t)]}$$

Note the satisfiability-preserving Skolemization (in a DNF): ∀-bound variables become term variables, and ∃-bound variables become Skolem constants. Contrast with Example 5.7.7, in which the reverse is true: ∀-bound variables become Skolem constants, and ∃-bound variables become term variables.

By comparing the tableau rules in Example 7.4.8 with Example 5.7.8, one will verify that the rules for $\{\mathtt{f}\}\,[\forall x A\,(x)]$ and $\{\mathtt{t}\}\,[\exists x A\,(x)]$ are already greatly simplified, but this simplification is not a trivial matter. Moreover, as stated above, working with sets of truth values (instead of with single truth values) has computational advantages. Thus, we now restrict our discussion to sets as signs for signed FO tableaux. In Example 7.4.6, we made brief reference to semi-lattices; keep this in the back of your mind until we ask you to bring it forth to your frontal cortex. We shall soon be occupied with lattices (see Section 9.4). We first work with Tarskian semantics, keeping in line with Section 5.7, and then approach distribution quantifiers from the viewpoint of Herbrand semantics.

In what respects the distribution quantifiers, we shall restrict our discussion to ∀ and ∃, as these are the two only quantifiers of interest in all the object languages approached in this text.

**Lemma 7.4.5.** *Let* $\widetilde{\blacklozenge}^{-1}(S) = \left\{(\Delta \neq \emptyset) \subseteq W \mid \widetilde{\blacklozenge}(\Delta) \in S\right\}$. *Then, a signed quantified formula* $S\,[\blacklozenge x \phi\,(x)]$ *is satisfiable iff there is a* $\Delta \in \widetilde{\blacklozenge}^{-1}(S)$ *such that the following conditions are both simultaneously satisfiable:*

1. *for each* $i \in \Delta$ *there is a ground term* $c_i$ *not occurring in* $\phi\,(x)$ *such that all* $\{i\}\,[\phi\,(c_i)]$ *are satisfiable;*

## 7.4. Signed tableaux for finitely many-valued logics

2. $\Delta [\phi(t)]$ is satisfiable for all ground terms t.

Intuitively, with respect to each distribution $\Delta \in \widetilde{\blacklozenge}^{-1}(S)$, condition 1 of Lemma 7.4.5 assures us by means of a "witness" $c_i$ that for every $i \in \Delta$ the truth value $i$ is reached by $\phi$, whereas condition 2 guarantees that no other truth values are taken by $\phi$. We write $W[\blacklozenge x \phi(x)]$ if $\widetilde{\blacklozenge}^{-1}(S) = (2^W - \emptyset)$ and $\emptyset [\blacklozenge x \phi(x)]$ if $\widetilde{\blacklozenge}^{-1}(S) = \emptyset$.

The following theorem, building up on Lemma 7.4.6, "licenses" us to use a SFE representation of a signed quantified formula in terms of certain signed instances by characterizing the distributions that are mapped to one of the truth values occurring in $S$, where $S$ is the sign of a quantified formula.

**Theorem 7.4.6.** *Given $\Delta \in \widetilde{\blacklozenge}^{-1}(S)$, a formula $S[\blacklozenge x \phi(x)]$ is satisfiable iff*

$$\bigvee_{\Delta \in \widetilde{\blacklozenge}^{-1}(S)} \left( \bigwedge_{i \in \Delta} \{i\} [\phi(c_i)] \wedge \bigwedge_{t \in T} \Delta [\phi(t)] \right)$$

*where the $c_i$ are new Skolem constants, is satisfiable.*

*Proof:* See, e.g., Hähnle (1999). **QED**

Informally, each disjunct in the representation above expresses the fact that the distribution of $\phi$ at $x$ is $\Delta$; as for the conjunctions, the first one guarantees that *at least* the elements of $\Delta$ occur in the distribution, whereas the second one states that *at most* such elements occur.

Given Theorem 7.4.6, we now have the following statement for signed first-order tableau rules:

**7.4.19. (Prop.)** For $\widetilde{\blacklozenge}^{-1}(S) = \{\Delta_1, ..., \Delta_m\}, \Delta_j = \{i_{j1}, ..., i_{jk_j}\}$, new Skolem constants $c_1, c_2, ...$ and $t_1, ..., t_m$ arbitrary ground terms, the following tableau rule is sound and complete by Theorem 7.4.6:

$$\frac{S[\blacklozenge x \phi(x)]}{\begin{array}{ccc} \{i_{11}\}[\phi(c_1)] \mid & \cdots \mid & \{i_{m1}\}[\phi(c_1)] \\ \vdots & \mid & \vdots \\ \{i_{1k_1}\}[\phi(c_{k_1})] \mid & \cdots \mid & \{i_{mk_m}\}[\phi(c_{k_m})] \\ \Delta_1[\phi(t_1)] \mid & \cdots \mid & \Delta_m[\phi(t_m)] \end{array}}$$

*Proof:* Left as an exercise.

**Example 7.4.9.** For $W_3 = \{0, \frac{1}{2}, 1\}$, consider now $\widetilde{\forall}^{-1}(\{\frac{1}{2}, 1\})$ (cf. Example 5.7.8). We have

## 7. Signed tableaux for the MV-SAT

$$\tilde{\forall}^{-1}\left(\left\{\tfrac{1}{2},1\right\}\right)=\left\{\left\{\tfrac{1}{2}\right\},\{1\},\left\{\tfrac{1}{2},1\right\}\right\}$$

i.e. all subsets of $W_3$ that contain *at least one* of $\tfrac{1}{2}$ and 1. We thus obtain the following tableau rule:

$$\frac{\{\tfrac{1}{2},1\}\,[\forall x A(x)]}{\{\tfrac{1}{2}\}\,[A(t_1)] \;\mid\; \begin{array}{c}\{\tfrac{1}{2}\}\,[A(c_1)]\\ \{1\}\,[A(c_2)]\end{array} \;\mid\; \{\tfrac{1}{2},1\}\,[A(t_2)] \;\mid\; \{1\}\,[A(t_3)]}$$

This shows that the tableaux can be too complex to be computationally efficient in terms of tree construction. (Solve Exercise 7.11 for even more dramatic examples.) In fact, the number of branches can be exponential with respect to $|W_n|$. Hähnle (1998) provided a way to obtain "slimmer" tableau rules from Theorem 7.4.6. by means of Herbrand semantics.

Recall now the main notions of Herbrand semantics, to wit, Herbrand interpretation (Def. 4.3.3) and Herbrand model (Def. 4.3.4). With respect to the latter, we make the following further specifications with respect to distribution quantifiers:

**7.4.20. (Def.)** Given a FO language $\mathsf{L}^*$, a Herbrand model $H\mathcal{M}$ over $\Upsilon_{\mathsf{L}^*}$ (containing at least one 0-ary function symbol) is an interpretation $H\mathcal{I}$ mapping predicate symbols $P \in \Upsilon_{\mathsf{L}^*}$ to functions $H\mathcal{I}(P) : T_0^{P(n)} \to W$, where $P(n)$ is the arity of the atom $P(t_1, ..., t_n)$ and $T_0$ is the set of ground terms in $T$. Now, let the distribution of $\phi(x)$ be given by

$$\Delta_{H\mathcal{M}}(\phi(x)) = \{val_{H\mathcal{M}}(\phi(t))\,|\,t \in T_0\}.$$

If $\psi = \blacklozenge x \phi(x)$, then

$$val_{H\mathcal{M}}(\psi) = \blacklozenge\left(\Delta_{H\mathcal{M}}(\phi(x))\right).$$

**7.4.21. (Def.)** Let $S[\phi]$ be a FO signed formula. We say that $S[\phi]$ is *satisfiable* iff there is a Herbrand model $H\mathcal{M}$ such that $val_{H\mathcal{M}} \in S \subseteq W$. Then, we say that $H\mathcal{M}$ is a model of $S[\phi]$ and write $\models_{H\mathcal{M}} \phi$. If every $H\mathcal{M}$ of some Herbrand interpretation $H\mathcal{I}$ satisfies $S[\phi]$, then we say that $S[\phi]$ is *valid*, and we write $\models S[\phi]$.

For what follows, consider the *Boolean set lattice* $2^W = (2^W, \emptyset, W, \cap, \cup)$ and let for $(Y \neq \emptyset) \subseteq W$ there be given the abbreviation

$$\mathcal{U}_Y = \{X \subseteq W \,|\, X \cap Y \neq \emptyset\}.$$

## 7.4. Signed tableaux for finitely many-valued logics

**Lemma 7.4.7.** *Assume* $\mathcal{U}_Y = \widetilde{\blacklozenge}^{-1}(S)$. *Then, any model $H\mathcal{M}$ over a FO signature $\Upsilon_{L^*}$ can be extended to a model $\underline{H\mathcal{M}}$ over a signature $\underline{\Upsilon} \supseteq \Upsilon$ such that $val_{H\mathcal{M}}(\blacklozenge x\phi(x)) \in S$ iff $val_{\underline{H\mathcal{M}}}(\phi(c)) \in Y$.*

**Example 7.4.10.** According to this lemma, the tableau rule in Example 7.4.9 can be simplified as:

$$\frac{\{\frac{1}{2},1\}\,[\forall x A(x)]}{\{\frac{1}{2},1\}\,[A(t)]}$$

We explain this simplification. Verify now Definition 9.2.8.3. The usefulness of Lemma 7.4.7 should become obvious in the light of this Definition and of the following lemma:

**Lemma 7.4.8.** *For any $Y \in 2^\mathcal{W}$, $\mathcal{U}_Y = \bigcup_{i\in Y} \uparrow \{i\}$. In particular, $\mathcal{U}_{\{i\}} = \uparrow\{i\}$.*

The informal explanation is as follows: A principal filter generated by an atom of $2^\mathcal{W}$ (cf. Def. 9.4.6) is a maximal filter of $2^\mathcal{W}$, or a maximal filter for $W$, for short. If finite, all maximal filters on $W$ are of this form. If $\widetilde{\blacklozenge}^{-1}(S)$ is a union of maximal filters, we can consider its upset representation, in order to obtain a single expansion rule. In particular, whenever we have it that $\widetilde{\blacklozenge}^{-1}(S)$ is a maximal filter on $W$, for example $\uparrow\{i\}$, then there is a *single-formula expansion* $\{i\}\,[\phi(c)]$. Generalizing:

**Lemma 7.4.9.** *Given finite $W$, the principal filter $\uparrow \Delta$ of $2^\mathcal{W}$ is equal to $\bigcap_{i\in\Delta} \uparrow \{i\}$.*

So, whenever $\widetilde{\blacklozenge}^{-1}(S)$ is a filter of $2^\mathcal{W}$, there is a single expansion rule containing the formulae $\{\{i\}\,[\phi(c_i)\,|\,i \in \Delta]\}$ for some $\Delta \subseteq W$. Thus, whenever a representation of $\widetilde{\blacklozenge}^{-1}(S)$ is of the form $\bigcup_{k\in K} Y_k$, the $Y_k$ are filters of $W$, by repeated application and disjunctive combination of Lemma 7.4.7 there is a tableau rule for a formula $S\,[\blacklozenge x\phi(x)]$ with $|K|$ expansions.

A similar process can be applied to signed formulae now of the form $\Delta(\phi(t))$ whose distributions are characterized by ideals of $2^\mathcal{W}$. The following lemma relies on defining $\mathcal{D}_\Delta = \{X \subseteq W\,|\,(X \neq \emptyset) \subseteq \Delta\}$.

**Lemma 7.4.10.** *Assume now that $\mathcal{D}_\Delta = \widetilde{\blacklozenge}^{-1}(S)$. Then, for all ground terms $t$ we have*

$$val_{H\mathcal{M}}(\psi) \in S \text{ iff } val_{H\mathcal{M}}(\phi(t) \in \Delta)$$

7. Signed tableaux for the MV-SAT

$$\frac{S\,[\blacklozenge x \phi(x)]}{\{i_1\}\,[\phi(c_1)]}$$

$$\vdots$$

$$\{i_r\}\,[\phi(c_r)]$$

1

$$\frac{S\,[\blacklozenge x \phi(x)]}{\Delta\,[\phi(t)]}$$

2

Figure 7.4.4.: FO tableau rules for filters (1) and ideals (2).

for $\psi = \blacklozenge x \phi(x)$.

In fact, $\mathcal{D}_\Delta \cup \{\emptyset\} = \downarrow \Delta$, the principal ideal of $2^W$ generated by $\Delta$. As in a finite lattice every ideal is principal (cf. Prop. 9.4.13), if $\widetilde{\blacklozenge}^{-1}(S) \cup \{\emptyset\}$ is an ideal $\downarrow \Delta$ of $2^W$, then there is a *single-formula expansion* $\Delta(\phi(t))$.

We finish with an important theorem of this FO signed tableaux calculus:

**Theorem 7.4.11.** *(Hähnle, 1998) Let $W_n$, $n$ is finite, be a set of truth values, and let $S \subseteq W_n$. If $\widetilde{\blacklozenge}^{-1}(S)$ is a filter of the Boolean set lattice $2^W$ generated by $Y = \{i_1, ..., i_r\} \subseteq W_n$, then the rule in Figure 7.4.4.1 is sound and complete. If $\widetilde{\blacklozenge}^{-1}(S) \cup \{\emptyset\}$ is an ideal of $2^W$ generated by $\Delta \subseteq W_n$, then the rule in Figure 7.4.4.2 is sound and complete.*

The mathematical explanation for the above is given by the following theorem:

**Theorem 7.4.12.** *(Hähnle, 1998) Let $W$ be partially ordered so that it constitutes a finite distributive lattice $\mathcal{L} = (W, \cap, \cup)$. Define the quantifiers $\blacktriangle = \cap$ and $\blacktriangledown = \cup$. Then the following tableau rules are sound and complete:*

1. *If $i$ is meet-irreducible:*

## 7.4. Signed tableaux for finitely many-valued logics

a)
$$\frac{\{i\}\,[\blacktriangle x\phi(x)]}{\{i\}\,[\phi(c)] \\ \uparrow \{i\}\,[\phi(t)]}$$

b)
$$\frac{\downarrow \{i\}\,[\blacktriangle x\phi(x)]}{\downarrow \{i\}\,[\phi(c)]}$$

2. *If $i$ is join-irreducible:*

   a)
   $$\frac{\{i\}\,[\blacktriangledown x\phi(x)]}{\{i\}\,[\phi(c)] \\ \downarrow \{i\}\,[\phi(t)]}$$

   b)
   $$\frac{\uparrow \{i\}\,[\blacktriangledown x\phi(x)]}{\uparrow \{i\}\,[\phi(c)]}$$

3. *For all $i$:*

   a)
   $$\frac{\uparrow \{i\}\,[\blacktriangle x\phi(x)]}{\uparrow \{i\}\,[\phi(t)]}$$

   b)
   $$\frac{\downarrow \{i\}\,[\blacktriangledown x\phi(x)]}{\downarrow \{i\}\,[\phi(t)]}$$

**Example 7.4.11.** We provide now the tableau rules for the Kleene quantifiers in the hybrid logic MK of Example 7.4.5:

7. Signed tableaux for the MV-SAT

- $\forall_{RE}$    $\dfrac{\text{T}[\forall x A(x)]}{\text{T}[A(t)]}$    $\dfrac{\text{NT}[\forall x A(x)]}{\text{NT}[A(c)]}$    $\dfrac{\text{T}[\neg \forall x A(x)]}{\text{T}[\neg A(c)]}$    $\dfrac{\text{NT}[\neg \forall x A(x)]}{\text{T}[\neg A(t)]}$

- $\forall_{RE}$    $\dfrac{\text{T}[\exists x A(x)]}{\text{T}[A(c)]}$    $\dfrac{\text{NT}[\exists x A(x)]}{\text{NT}[A(t)]}$    $\dfrac{\text{T}[\neg \exists x A(x)]}{\text{T}[\neg A(t)]}$    $\dfrac{\text{NT}[\neg \exists x A(x)]}{\text{T}[\neg A(c)]}$

Figure 7.4.5.: Signed tableau rules for quantified formulae in MK.

## 7.5. Signed tableaux for infinitely many-valued logics

By infinitely many-valued logics we mean here the (t-norm) fuzzy logics. Given the importance of these logics, it is no wonder that much effort has been dedicated to find efficient proof calculi for them. In this context, several tableaux calculi have been proposed (e.g., Di Lascio, 2001; Olivetti, 2003; Straccia, 2001), but none can actually accommodate all FLs. In particular, semantical innovations appear to be required in order to create tableaux calculi for FLs.

One such example is Gerla's (2001) signed tableaux calculus. This calculus is restricted to a special class of FLs that possess a so-called *tableaux-manageable semantics* $\mathfrak{S}^{\mathcal{T}}$ defined via a geometrical characterization. We merely sketch the main aspects of this approach. We begin by expanding on lattices (cf. Section 9.4) and by making slight adaptations both to our and Gerla's notation and definitions. We elaborate on Gerla's fuzzy semantics to some extent, because we think it interesting to provide a kind of non-matrix semantics for many-valued logics.

**7.5.1. (Def.)** Let $\mathcal{L}$ be a complete lattice. A class $\mathscr{C}$ of elements of $\mathcal{L}$ is called a *closure system* if the meet of any family of elements of $\mathscr{C}$ is an element of $\mathscr{C}$.

Every closure system contains 1, because $\emptyset$ is an element of any class and $inf(\emptyset) = 1$.

**7.5.2. (Def.)** Let $W$ be a set of truth values. A *constraint* is any subset $S \subseteq W$. Let $W$ be a lattice and $\mathbb{C}$ a closure system on $W$. Then we say that $\mathbb{C}$ is a *constraint frame*. Given a constraint frame $\mathbb{C}$, a *signed formula* $S\phi$ is a pair $(\phi, S)$ where $\phi$ is a formula and $S \in \mathbb{C}$ is

## 7.5. Signed tableaux for infinitely many-valued logics

a constraint. The information carried by a signed formula $S\phi$ is: the truth value v of $\phi$ satisfies $S$, i.e. $v \in S$.

**7.5.3. (Def.)** Let $\mathbb{C}$ be a constraint frame. Given a subset $S$ of $W$, we denote by $\langle S \rangle$ the constraint generated by $S$:

$$\langle S \rangle = \bigcap \{X \in \mathbb{C} | X \supseteq S\}$$

**7.5.4. (Def.)** Let $W$ be a lattice and $\mathbb{C}$ a constraint frame in $W$. Given a logical language $\mathsf{L}$, we define a $\mathbb{C}$-*valuation* to be any mapping $val_\mathbb{C} : F_\mathsf{L} \longrightarrow \mathbb{C}$. The information carried by $val$ (henceforth abbreviating $val_\mathbb{C}$) is that given any formula $\phi \in F_\mathsf{L}$, the actual truth value of $\phi$ satisfies $val(\phi)$. Let now $\mathfrak{S}$ be a $W$-semantics. Then, given a valuation $val$, we say that $\mathcal{M} \in \mathfrak{S}$ is a model of $val$ if $\mathcal{M}$ satisfies $val(\phi)$ for every formula $\phi$, and we write $\models_{val_\mathcal{M}} \phi$. Then, a $\mathbb{C}$-*consequence* $Cn_\mathbb{C}$ is defined with respect to a formula $\phi$ as

$$Cn_\mathbb{C}(\phi) = \langle \{\mathcal{M}_\phi | \mathcal{M} \in \mathfrak{S}, \models_{val_\mathcal{M}} \phi\} \rangle.$$

and the set of $\mathbb{C}$-tautologies $\mathsf{T}_\mathbb{C}$ is established by the mapping $\mathsf{T} : F_\mathsf{L} \longrightarrow \mathbb{C}$ such that for every formula $\phi \in F_\mathsf{L}$

$$\mathsf{T}_\phi = \langle \{\mathcal{M}_\phi | \mathcal{M} \in \mathfrak{S}\} \rangle.$$

When $S$ is infinite, then the class of constraints $2^S$ is too big, being thus more convenient to refer only to particular constraints, namely interval-constraints. We can thus assume that $\mathbb{C}$ is the set of closed intervals. In fact, $\mathbb{C}$ can be made to be equal to the set of closed constraints by setting

$$\mathbb{C} = \{[a,b] | a, b \in W \text{ and } a \leq b\} \cup \{\emptyset\}.$$

$\mathbb{C}$ is indeed a closure system: If $([a_i, b_i])_{i \in I}$ is a family of intervals such that $a_i \leq b_i$, then $\bigcap_{i \in I} [a_i, b_i]$ is the interval $[sup_{i \in I} a_i, inf_{i \in I} b_i]$. This provides a fuzzy semantics in the following way: a valuation $val$ is defined by a pair $(val_1, val_2)$ of $W$-subsets of formulae such that $val(\phi) = [val_1(\phi), val_2(\phi)]$ where $val$ carries the information that $\phi$ is at least true with degree $val_1$ and at most true with degree $val_2$.

**7.5.5. (Def.)** A truth-functional semantics $\mathfrak{S}$ is said to be *tableaux-manageable* provided that, given any $n$-ary logical connective $\heartsuit$, a formula $S[\heartsuit(\phi_1, ..., \phi_n)]$ is equivalent to $\dot{\mathcal{E}}_1 \vee ... \vee \dot{\mathcal{E}}_m$ where $\dot{\mathcal{E}}_i$ is a conjunction $S_1^j \phi_1 \wedge ... \wedge S_n^j \phi_n$ of signed formulae in which all the formulae $\phi_1, ..., \phi_n$ occur.[14]

---

[14]If $\phi_j$ does not occur in $\dot{\mathcal{E}}_i$, we can add the tautology $W\phi_j$.

271

# 7. Signed tableaux for the MV-SAT

**7.5.6.** This implies that there is a *rewriting rule*

$$S[\heartsuit(\phi_1, ..., \phi_n)] \implies \bigvee_i \left(\bigwedge_j S_i^j \phi_i\right)$$

such that the left and the right sides of $\implies$ are equivalent (cf. Def. 6.2.5.2). Given a system of such rewriting rules, we can transform any signed formula $S\phi$ into an equivalent DNF of signed atoms of the form $Sp$.

A DNF is seen as a *positive Boolean combination*, i.e. a combination of disjunctions and conjunctions in which the connective $\neg$ does not occur. Two Boolean combinations are equivalent if they assume the same truth value for any valuation. Examples are the following equivalences:

$$X\phi \wedge Y\phi \equiv (X \cap Y)\phi$$

$$X\phi \vee Y\phi \equiv (X \cup Y)\phi$$

**Example 7.5.1.** The following are correct rewriting rules for the Zadeh logic $Z_\aleph$, a fuzzy logic characterized by a semantics of closed intervals as constraints:

$[a, b] (\phi \wedge \psi) \implies ([a, b]\phi \wedge [a, 1]\psi) \vee ([a, 1]\phi \wedge [a, b]\psi)$
$[a, b] (\phi \vee \psi) \implies ([a, b]\phi \wedge [0, b]\psi) \vee ([0, b]\phi \wedge [a, b]\psi)$
$[a, b] (\neg\phi) \implies [1 - b, 1 - a]\phi$

And these are the derived rewriting rules:

$[a, 1] (\phi \wedge \psi) \implies [a, 1]\phi \wedge [a, 1]\psi$
$[a, 1] (\phi \vee \psi) \implies [a, 1]\phi \vee [a, 1]\psi$
$[0, b] (\phi \wedge \psi) \implies [0, b]\phi \wedge [0, b]\psi$
$[0, b] (\phi \vee \psi) \implies [0, b]\phi \vee [0, b]\psi$

**Example 7.5.2.** From the formula $[0.7, 1] (p \wedge \neg p)$ we obtain the sequence:

$[0.7, 1] (p \wedge \neg p); [0.7, 1] p \wedge [0.7, 1] (\neg p); [0.7, 1] p \wedge [0, 0.3] p; [0.7, 1] \cap [0, 0.3] p; \emptyset p$

**Example 7.5.3.** We start now from the formula $[0.7, 1] (\neg p \wedge (\neg q \vee p))$:

$$[0.7, 1] (\neg p \wedge (\neg q \vee p));$$

$$[0.7, 1] (\neg p) \wedge [0.7, 1] (\neg q \vee p);$$

## 7.5. Signed tableaux for infinitely many-valued logics

$$[0.7, 1] (\neg p) \wedge ([0.7, 1] (\neg q) \vee [0.7, 1] p) ;$$

$$[0.7, 1] (\neg p) \wedge ([0, 0.3] q \vee [0.7, 1] p) ;$$

$$[0, 0.3] p \wedge ([0, 0.3] q \vee [0.7, 1] p) ;$$

$$([0, 0.3] p \wedge [0, 0.3] q) \vee ([0, 0.3] p \wedge [0.7, 1] p) ;$$

$$([0, 0.3] p \wedge [0, 0.3] q) \vee ([0, 0.3] \cap [0.7, 1]) p;$$

$$([0, 0.3] p \wedge [0, 0.3] q) \vee \emptyset p;$$

$$[0, 0.3] p \wedge [0, 0.3] q$$

**7.5.7. (Def.)** Now, given a constraint frame $\mathbb{C}$, we say that any Cartesian product of $n$ elements in $\mathbb{C}$ is an *n-rectangle*. If $\heartsuit$ is a logical connective, then we denote its interpretation in $W$ by $\widetilde{\heartsuit}$.

**Theorem 7.5.1.** *A truth-functional semantics $\mathfrak{S}$ is tableaux-manageable, denoted by $\mathfrak{S}^{\mathcal{T}}$, iff, given any n-ary logical connective $\heartsuit$ and $S \in \mathbb{C}$, $\widetilde{\heartsuit}^{-1}(S)$ is a finite union of n-rectangles.*

*Proof:* (Gerla, 2001): Assume that $\mathfrak{S}$ is tableaux-manageable. Then,

$$\widetilde{\heartsuit}^{-1}(S) = \left(S_1^1 \times ... \times S_n^1\right) \cup ... \cup \left(S_1^m \times ... \times S_n^m\right).$$

Conversely, let $S[\heartsuit(\phi_1, ..., \phi_n)]$ be any signed formula and assume that $\widetilde{\heartsuit}^{-1}(S)$ is equal to $\left(S_1^1 \times ... \times S_n^1\right) \cup ... \cup \left(S_1^m \times ... \times S_n^m\right)$ where the $S_i^j$ are suitable elements in $\mathbb{C}$. Then, for every model $\mathcal{M}$ of $\mathfrak{S}^{\mathcal{T}}$ we have

$$\models_{\mathcal{M}} S[\heartsuit(\phi_1, ..., \phi_n)] \text{ iff } val_{\mathcal{M}}(\heartsuit(\phi_1, ..., \phi_n)) \in S$$

iff

$$\widetilde{\heartsuit}(val_{\mathcal{M}}(\phi_1), ..., val_{\mathcal{M}}(\phi_n)) \in S$$

iff

$$(val_{\mathcal{M}}(\phi_1), ..., val_{\mathcal{M}}(\phi_n)) \in \widetilde{\heartsuit}^{-1}(S).$$

The last condition is equivalent to saying that $\mathcal{M}$ satisfies the positive Boolean combination:

$$\left(S_1^1 \phi_1 \wedge ... \wedge S_n^1 \phi_n\right) \vee ... \vee \left(S_1^m \phi_1 \wedge ... \wedge S_n^m \phi_n\right)$$

**QED**

**Example 7.5.4.** The formula $[0.7, 1] (\neg p \wedge (\neg q \vee p))$ (cf. Example 7.5.3) is satisfiable in the rectangle $[0, 0.3] \times [0, 0.3]$. Because no model satisfies $\emptyset p$, we conclude that $[0.7, 1] (p \wedge \neg p)$ (cf. Example 7.5.2) is unsatisfiable.

## 7. Signed tableaux for the MV-SAT

**7.5.8. (Def.)** Let $\mathfrak{S}^{\mathcal{T}}$ be a tableaux-manageable fuzzy semantics. Then, a *tableau proof* is a finite labeled tree whose nodes are signed formulae constructed as follows: Let the signed formula $S\left[\overset{\bullet}{\heartsuit}(\phi_1,...,\phi_n)\right]$ be equivalent to $\overset{\bullet}{\mathcal{E}}_1 \vee ... \vee \overset{\bullet}{\mathcal{E}}_m$ where $\overset{\bullet}{\mathcal{E}}_i$ is a conjunction $S_1^j \phi_1 \wedge ... \wedge S_n^j \phi_n$ of signed formulae. Then, an *expansion rule* is a tree with the form:

$$\frac{S\left[\heartsuit(\phi_1,...,\phi_n)\right]}{\begin{array}{c|c|c|c} S_1^1\phi_1 & S_1^2\phi_1 & \cdots & S_1^m\phi_1 \\ \vdots & \vdots & \vdots & \vdots \\ S_n^1\phi_n & S_n^2\phi_n & \cdots & S_n^m\phi_n \end{array}}$$

**7.5.9. (Def.)** The set of all tableaux for a given signed formula $S\left[\heartsuit(\phi_1,...,\phi_n)\right]$ is defined as the set of labeled trees that can be produced by a finite number of applications of the following rules:

1. For $\phi = S\left[\heartsuit(\phi_1,...,\phi_n)\right]$ a signed formula, $S\left[\heartsuit(\phi_1,...,\phi_n)\right]$ is a signed tableau for $\phi$, denoted by $\mathcal{T}_\phi^S$;

2. If $\mathcal{T}_\phi^S$ is a signed tableau for $\phi$ and $X\left[\heartsuit(\psi_1,...,\psi_l)\right]$ is a node label in $\mathcal{T}_\phi^S$, then a new tableau $\left(\mathcal{T}_\phi^S\right)'$ is constructed by expanding all branches of $\mathcal{T}_\phi^S$ containing $X\left[\heartsuit(\psi_1,...,\psi_l)\right]$ in accordance with the related expansion rule;

3. If $\mathcal{T}_\phi^S$ is a signed tableau for $\phi$ and $X\psi, Y\psi$ are two signed formulae in $\mathcal{T}_\phi^S$, then we can expand any branch containing both $X\psi$ and $Y\psi$ by the signed formula $(X \cap Y)\psi$.

**7.5.10. (Prop.)** Let $S\phi$ be a signed formula and $\mathcal{T}_\phi^S$ a tableau for it. Let $\mathcal{B}_i \in \mathcal{T}_\phi^S$, $1 < i \leq m$ be the branches. Then, $\mathcal{M}$ is a model of $S\phi$ iff $\mathcal{M}$ is a model of the formulae in $\mathcal{B}_i$.

*Proof:* Left as an exercise.

**7.5.11. (Def.)** A branch in a signed tableau is *closed* if it contains a *trivially inconsistent* signed formula $\emptyset\phi$. A branch is *complete* if either it is closed or no rule application produces a formula not already to be found in that branch. A tableau is *closed* (*complete*) if all its branches are closed (complete, respectively).

**7.5.12. (Def.)** A signed formula $S\phi$ is *tableau-refutable* if a closed tree $\mathcal{T}_\phi^S$ for it exists.

**Theorem 7.5.2.** Let $\mathfrak{S}^{\mathcal{T}}$ be a tableaux-manageable fuzzy semantics. Then, a signed formula is tableau-refutable iff it is a contradiction.

## 7.5. Signed tableaux for infinitely many-valued logics

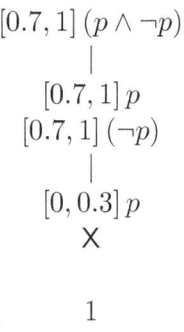

1

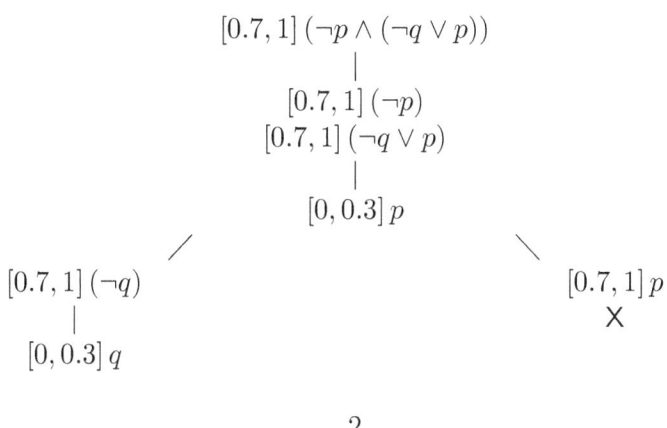

2

Figure 7.5.1.: Signed fuzzy tableaux.

*Proof:* The proof is left as an exercise.

**Example 7.5.5.** Figures 7.5.1.1 and 7.5.1.2 show the signed tableaux of the formulae in Examples 5.7.2-3.

Above, we remarked that Gerla's method is not applicable to all the FLs. In effect, by Theorem 7.5.1, a FL has a tableaux-manageable semantics iff, given an $n$-ary connective $\heartsuit$ and some $S \in \mathbb{C}$, $\widetilde{\heartsuit}^{-1}(S)$ is a finite union of $n$-rectangles.

**Example 7.5.6.** In Łukasiewicz semantics, we interpret strong disjunction in the interval $[0, 1]$ as $x \nabla y = min \{x + y, 1\}$. Furthermore,

## 7. Signed tableaux for the MV-SAT

given $a, b \in [0, 1]$, for $a \leq b$ we have

$$U = \{(x, y) \in [0, 1] \times [0, 1] \,|\, a \leq x \nabla y \leq b\}$$

is the set of points in the square $[0, 1] \times [0, 1]$ between the lines $x + y = a$ and $x + y = b$. The set $U$ is not a finite union of rectangles. Therefore, Łukasiewicz semantics is not tableaux-manageable.

A problem with this tableaux calculus is the exponential character of the tableaux. This can nevertheless be mitigated in the following way: if for every connective $\heartsuit$ and every sign $S \in \mathbb{C}$ we have it that $\widetilde{\heartsuit}^{-1}(S)$ is a union of two rectangles, then we are left only with binary trees.

# Exercises

**Exercise 7.1.** Single out a binary connective shared by all the finitely many-valued logics of Chapter 5 (except $B_4$) and produce its decomposition rules according to Surma's algorithm.

**Exercise 7.2.** Verify by means of Surma's algorithm whether the formulae of Exercise 5.8 are theorems of the respective 3-valued logics.

**Exercise 7.3.** Write down the expansion rules of the many-valued signed tableaux calculus for the logical system $Rn_3$.

**Exercise 7.4.** Verify whether the following signed formulae are valid in the indicated many-valued systems by means of the propositional signed tableaux calculus elaborated on in Section 7.4.1:

1. $(P \vee \neg P) \to (Q \wedge P)$ in $Ł_3$, $B_3^E$, $F_3$, $Hn_3$, $Rn_3$, and $G_3$.
2. $\phi \to (\chi \circ \psi)$ in $Rn_3$
3. $P \to (Q \to P)$ in $Ł_3$, $B_3^E$, $F_3$, $Hn_3$, $Rn_3$, and $G_3$.

**Exercise 7.5.** Consider Example 7.4.1. Obtain the tableau rules for the remaining connectives of the logic $Ł_3$ using individual truth values as signs.

**Exercise 7.6.** Consider the logical system $Ł_3$ of Example 7.4.4.

1. Determine the signed tableau rules for the connectives $\odot$ and $\sim$.
2. Verify if $\mathscr{S}_{Ł3} = \{\{f\}, \{i\}, \{t\}, \{f,i\}\}$ is (not) a complete set of signs with respect to $Ł_3$.
3. Verify by means of the signed tableaux calculus $\mathcal{L}_3$ if the classical tautologies of Proposition 3.5.3 (except MP) are tautologies in this many-valued logical system.

**Exercise 7.7.** Give the signed tableau rules for the following signed propositional formulae of $Ł_3$ given the signs $S_1, S_2 \in \mathscr{S}_{Ł3}^{>}$, $S_1 = \{f, t\}$ and $S_2 = \{i, t\}$:

1. $S_1 [A \wedge B]$
2. $S_2 [A \wedge B]$

7. Signed tableaux for the MV-SAT

3. $S_1 [A \lor B]$

4. $S_2 [A \lor B]$

5. $S_1 [A \to B]$

6. $S_2 [A \to B]$

7. $S_1 [\neg A]$

8. $S_2 [\neg A]$

**Exercise 7.8.** Write the tableau closure rules for the signed tableau calculus for the logic Ł3 with $\mathscr{S}^>_{Ł3}$.

**Exercise 7.9.** Consider the propositional signed tableaux calculus elaborated on in Section 7.4.1.

1. Verify whether the classical tautologies in Proposition 3.5.3 are theorems in the different finitely many-valued logics studied in Sections 5.4-5.

2. Verify the validity of the formulae of Exercise 5.8.

**Exercise 7.10.** Verify whether the classical tautologies in Proposition 3.5.3 are theorems in the logical system MK by means of the propositional signed tableau calculus of Example 7.4.5.

**Exercise 7.11.** For $W_3 = \{0, \frac{1}{2}, 1\}$, determine the *non-simplified* tableau rules for the following Ł3 distribution quantifiers:

1. $\widetilde{\forall}^{-1}(\{0, \frac{1}{2}\})$

2. $\widetilde{\forall}^{-1}(\{0, 1\})$

3. $\widetilde{\exists}^{-1}(\{0, \frac{1}{2}\})$

4. $\widetilde{\exists}^{-1}(\{\frac{1}{2}, 1\})$

5. $\widetilde{\exists}^{-1}(\{0, 1\})$

**Exercise 7.12.** Consider the FO formulae of Exercise 5.23. Verify their validity:

1. By means of the signed tableaux calculus for FO many-valued formulae in Section 7.4.2.

2. With the first-order MK calculus.

**Exercise 7.13.** Build tableau proofs for the following fuzzy signed formulae of $Z_\aleph$ by applying Gerla's (2001) method:

1. $[0.2, 1]\, p \to [0, 0.8]\, (p \vee \neg p)$
2. $[0.4, 1]\, (p \wedge \neg p) \to [0, 1]\, q$
3. $[0.3, 1]\, ((p \wedge q) \to \neg p)$
4. $[0, 0.1]\, (p \wedge (p \to q)) \to [0.9, 1]\, (\neg q)$

**Exercise 7.14.** Prove (Complete the proof of) the propositions and theorems in this Chapter that were left without a proof (with a sketchy proof, respectively).

**Exercise 7.15.** Formulate and prove the theorems of soundness and completeness of the signed tableaux calculus for FO many-valued formulae in Section 7.4.2.

# 8. Signed resolution for the MV-SAT

It would probably be not exaggerated to say that *signed resolution* is one of the most efficient calculi to tackle the MV-SAT. Although research into it has been somehow stagnant lately, we believe that its past successes will further future advances. In the meantime, we give here the essential aspects of this calculus, including important *refinements* such as *signed hyper-resolution* and *signed ($<_A$-ordering) macro-resolution*. After some *preliminary remarks* in Section 8.1, we discuss at length the application of signed resolution to both *the finitely* and the *infinitely many-valued logics*. Because these pose different challenges, we discuss them in two distinct Sections (8.2-3).

## 8.1. Introductory remarks

As seen above, the application of the resolution principle as an inference rule requires that the formulae be transformed into clausal form, namely into CNF. As there are no normal forms in many-valued logics, strategies had to be developed for equivalent results. Attempts to obtain CNFs directly in many-valued logics were only partially satisfactory, though refutational equivalence can be achieved. In effect, these attempts are commonly highly inefficient and remain theoretical rather than applicable constructs (e.g., Morgan, 1976), or only particular logical systems are considered (e.g., Schmitt, 1986). Stachniak (e.g., 1996) came up with a more general resolution-based proof system for many-valued logics based on the fact that the Tarskian classical consequence relation (see above; see Augusto, 2017, for details and developments) can be carried over to great extent to these logics. However, this proof system is only efficiently applicable to functionally complete logics or to Łukasiewicz logics, namely because the number of verifiers (a kind of truth values) can be arbitrarily large for certain other classes of logics. For this and other reasons, Stachniak's remains a largely non-automatable resolution procedure. Interesting, but now rather historical (as most of the above-mentioned) further attempts at implementing the

## 8. Signed resolution for the MV-SAT

resolution principle in many-valued logics, are, for instance, Di Zenzo (1988), Lee (1972), and Orłowska (1978).

The first aspect to bear in mind when attempting to implement the resolution principle in many-valued logics is that these are *generalizations* of CL. This is particularly so in that the proof procedures implementable in these logics can apply precisely the same semantics as CL, contrarily to other non-classical logics, which often have a largely different semantics (e.g., possible world semantics; see Augusto, 2017). We saw above (Sections 5.2-3) that the generalization of CL entails structural properties (normality, uniformity, K-regularity, and decisiveness) that, in turn, are related to properties such as quasi-validity, quasi-entailment, quasi-tautologousness, and quasi-contradictoriness; all these allow us to "translate" many-valued logics into CL.

In particular, matrix semantics for many-valued logics proves itself useful in that the choice of sets of (anti-)designated values is at the root of this "natural translation." This is particularly important in the context of resolution applied to many-valued logics, as it is possible, by means of signed (clause) logic, to implement the resolution principle in a highly automated way. In effect, signed (clause) logic permits the generalization of the important notions of (in)validity and (un)satisfiability to the various many-valued logics approached above. By allowing also the building of CNFs, it allows the direct application of resolution proof procedures to these logics.

Signed resolution techniques date basically from the early 1990s (e.g., Hähnle, 1994; Murray & Rosenthal, 1993) and they are largely based on matrix semantics. As seen above (Section 5.2.3), we can select the sets of designated and anti-designated truth values of any finitely many-valued logical system in such a way that we end up with as few sets as possible–namely, we can end up with only two sets of truth values, $D^+$ and $D^-$, which can obviously be associated with truth and falsity, respectively (cf. Example 5.2.1). This allows us to apply the resolution techniques of CL to finitely many-valued logics directly. Indeed, given two clauses $D^+[P] \vee A$ and $D^-[P] \vee B$, they resolve unproblematically into the resolvent $A \vee B$ if $D^+ \cap D^- = \emptyset$ or into the resolvent $(D^+ \cap D^-)[P] \vee A \vee B$ if $D^+ \cap D^- \neq \emptyset$, by applying rule R1 below.

As remarked above (cf. Section 6.2.1), signed logical systems differ in important points from the many-valued logical systems whose formulae we test for validity within them, but they are equivalent in terms of satisfiability. Although the most important advantage of signed logic from the viewpoint of automated deduction is that it allows us to apply the resolution techniques of CL (which, in fact, is just the simplest case of a resolution calculus) to many-valued logics (which, as seen, just generalize CL), this does not mean that we are facing a complete reduction to CL. In particular, as Caleiro & Marcos (2009) point out, for *genuinely n-valued* logical systems (i.e. systems for which $n$ is the cardinality of the smallest collection of truth values by virtue of which they are given a truth-functional semantics), this bivalent characterization entails a loss of a truth-functional characterization. If we opt for a minimization of the trade-off between expressivity and complexity, we then are conditioned to approaches that in a larger or smaller measure safeguard many-valuedness.

We next provide the basic definitions and results that shall allow us to explore the implementation of the resolution principle in finitely many-valued logics. Resolution in infinitely many-valued logics poses many challenges and requires different techniques, and we approach these in a separate section.

## 8.2. Signed resolution for finitely many-valued logics

### 8.2.1. Signed resolution proof procedures

#### 8.2.1.1. Main rules

In what follows, $S \subseteq W_n$ is a sign, $S[L]$ is a signed literal (cf. Def.s 2.5.1-2), and $\mathcal{C}_i$ are clauses (cf. Def. 2.4.2). $\square$ denotes the empty clause.

**8.2.1. (Prop.)** There are two rules that constitute a straightforward, refutation complete[1] resolution calculus for signed clauses/CNF formulae:

$$(R1) \quad \frac{S_1[P] \vee \mathcal{C}_1 \quad S_2[P] \vee \mathcal{C}_2}{(S_1 \cap S_2)[P] \vee \mathcal{C}_1 \vee \mathcal{C}_2}$$

and

$$(R2) \quad \frac{\{\}[P] \vee \mathcal{C}}{\mathcal{C}}$$

*Proof:* Left as an exercise.

**8.2.2. (Def.)** R1 is known as *signed binary resolution* and the conclusion $(S_1 \cap S_2)[P] \vee \mathcal{C}_1 \vee \mathcal{C}_2$ is called a *signed binary resolvent*. R2 is the *simplification* rule: a clause $\mathcal{C}$ is said to be a simplification of $\mathcal{C}'$ if it results from $\mathcal{C}'$ by deleting all literals with the empty sign.

Note that in R1, if $\{\} \in (S_1 \cap S_2)$, then $(S_1 \cap S_2)[P]$ is unsatisfiable and can be removed from the resolvent.

---
[1] For a proof: Ansótegui et al. (2002).

## 8. Signed resolution for the MV-SAT

**Example 8.2.1.** Let $W = \{0, 1, 2\}$. We apply R1 in the following examples:

1.
$$\frac{\{2\}[P] \quad \overline{\{2\}}[P]}{\square}$$

2.
$$\frac{\{0,1\}[P] \vee \{2\}[Q] \quad \overline{\{1\}}[P]}{\{0\}[P] \vee \{2\}[Q]}$$

**8.2.3. (Def.)** *Signed factoring* is expressed by the rule:

(R3) $\quad \dfrac{S_1[P] \vee C \vee S_2[P]}{S[P] \vee C} \quad$ if $S_1 = S_2$

**Example 8.2.2.** Let $W = \{0, \frac{1}{2}, 1\}$. We apply R3:

$$\frac{\{0\}[P] \vee \{\frac{1}{2}\}[Q] \vee \{0\}[P] \vee \{1\}[R]}{\{0\}[P] \vee \{\frac{1}{2}\}[Q] \vee \{1\}[R]}$$

With regard to R1, it is obvious that the literal resolved upon may remain (as a so-called *residue*) as a result of its application in signed resolution, unlike the application of binary resolution in classical resolution (cf. Theorem 4.6.2). This is easily avoided by applying the next rule.

**8.2.4. (Def.)** Given a totally ordered truth-value set $W$, completeness of signed binary resolution is preserved for mono-signed and regular CNF formulae if R1 is simplified as

(R4) $\quad \dfrac{S_1[P] \vee C_1 \quad S_2[P] \vee C_2}{C_1 \vee C_2} \quad$ if $S_1 \cap S_2 = \emptyset$

called *mono-signed/regular binary resolution*.

Rule R4 obviously avoids the generation of a residue allowed by R1. (See also Example 8.2.5.)

**Example 8.2.3.** Let $W = \{0, \frac{1}{2}, 1\}$. We apply R4 in the following example:

$$\frac{\{0\}[P] \vee \overline{\{0,1\}}[R] \quad \{1\}[P] \vee \{0\}[Z]}{\overline{\{0,1\}}[R] \vee \{0\}[Z]}$$

## 8.2. Signed resolution for finitely many-valued logics

Although the next rule (R5) is not required for completeness, it is useful for the reduction of the search space (as well as to simplify the completeness proof). Basically, R5 unites literals that have the same atom:

$$\text{(R5)} \quad \frac{S_1\,[P] \vee \ldots \vee S_m\,[P] \vee C}{(S_1 \cup \ldots \cup S_m)\,[P] \vee C}$$

**8.2.5. (Def.)** R5 is called the *(signed) merging rule*. In R5, $C$ is called a *normal form* or *normalized* version of $C'$ if it results from $C'$ by repeated (i.e. as often as possible) application of R2 and R5 to $C'$.

The result of this strategy is the removal of all literals with empty signs; moreover, atoms are different in all different literals in $C$.

Extension of the above rules to the first-order case is straightforward by introduction of unification techniques via the extension of Herbrand semantics to FO signed logic.

**8.2.6. (Prop.)** The following versions of R1 and R3-5 are all valid inference rules in the FO signed resolution calculus:

1.

$$\text{(R1*)} \quad \frac{S_1\,[P_1] \vee C_1 \quad S_2\,[P_2] \vee C_2}{((S_1 \cap S_2)\,[P] \vee C_1 \vee C_2)\,\sigma}, \quad \sigma = mgu\,(P_1, P_2)$$

2.

$$\text{(R3*)} \quad \frac{S_1\,[P_1] \vee C \vee S_2\,[P_2]}{(S\,[P] \vee C)\,\sigma}, \quad \sigma = mgu\,(P_1, P_2),\, S_1 = S_2$$

3.

$$\text{(R4*)} \quad \frac{S_1\,[P_1] \quad S_2\,[P_2] \vee C}{C\sigma} \quad \text{if } S_1 \cap S_2 = \emptyset, \sigma = mgu\,(P_1, P_2)$$

4.

$$\text{(R5*)} \quad \frac{S_1\,[P_1] \vee \ldots \vee S_m\,[P_m] \vee C}{((S_1 \cup \ldots \cup S_m)\,[P] \vee C)\,\sigma}, \quad \sigma = mgu\,(P_1, \ldots, P_m)$$

*Proof:* Left as an exercise.

The generalization of Herbrand's theorem (see Section 4.3) to sets of signed clauses entails the following important result:

## 8. Signed resolution for the MV-SAT

**8.2.7. (Prop.)** A set $C$ of signed clauses is unsatisfiable iff it is H-unsatisfiable.

*Proof:* ($\Rightarrow$) By Definition 4.3.3 and by the definition of the set of designated values $D$, it is obvious that any set $C$ of signed clauses is H-unsatisfiable if there is no H-interpretation $H\mathcal{I}_C$ such that $val_{H\mathcal{I}}(C_i) \in D$ for any clause $C_i \in C$. Then, and by the definition of unsatisfiability, if $C$ is H-unsatisfiable, then it is unsatisfiable.

($\Leftarrow$) For an arbitrary set of clauses $C$ every interpretation $\mathcal{I}_C$ induces a H-interpretation $H\mathcal{I}_C = \{\{v_i\}[P] \,|\, val_{\mathcal{I}}(P) = v_i, P \in H(C)\}$. Obviously, if $\mathcal{I}_C$ does not satisfy $C$, then $H\mathcal{I}_C$ does not satisfy $C$ either.
**QED**

**Example 8.2.4.** Let $W = \{\mathtt{f}, \mathtt{i}, \mathtt{t}\}$, $D = \{\mathtt{t}\}$, and

$$C = \{\|\{\mathtt{t}\}[Q(a)]\|, \|\{\mathtt{f}\}[Q(x)], \{\mathtt{i}\}[Q(x)], \{\mathtt{t}\}[Q(f(x))]\|\}.$$

Then:

$$H_C = \{a, f(a), f(f(a)), ...\}$$

and

$$H(C) = \{Q(a), Q(f(a)), Q(f(f(a))), ...\}.$$

The interpretation

$$H\mathcal{I}_C = \{\{\mathtt{t}\}[Q(a)], \{\mathtt{t}\}[Q(f(a))], \{\mathtt{t}\}[Q(f(f(a)))], ...\}$$

is satisfiable, whereas the interpretation

$$H\mathcal{I}_C = \{\{\{\mathtt{t}\}[Q(a)], \{\mathtt{f}\}[Q(f(a))], \{\mathtt{t}\}[Q(f(f(a)))], ...\}\}$$

is unsatisfiable.

### 8.2.1.2. Refinements of signed resolution

It is possible to combine the rules above; for example, the following rule R6*, a combination of a series of steps of R1-2, is a refinement of signed resolution (cf. Section 4.6.4).[2]

---

[2] Actually, given that rules R1-2 suffice for a complete and sound signed resolution calculus, rules R4 and R5, in particular, can already be said to be refinements.

## 8.2. Signed resolution for finitely many-valued logics

**8.2.8. (Def.)** The rule

(R6)  $\dfrac{S_1\,[P_1] \vee \mathcal{C}_1 ... S_k\,[P_k] \vee \mathcal{C}_k}{(\mathcal{C}_1 \cup ... \cup \mathcal{C}_k)\,\sigma}$, $\displaystyle\bigcap_{1\leq i\leq k} S_i = \emptyset,\ \sigma = mgu\,(P_i),\, 1 \leq i \leq k$

is called *signed macro-* or *hyper-resolution* and $(\mathcal{C}_1 \cup ... \cup \mathcal{C}_k)\,\sigma$ is called the *signed macro-* or *hyper-resolvent*.

**Example 8.2.5.** Let $W = \{a, b, c, d\}$. We apply R6–the propositional "version" of R6*, sometimes called *signed parallel resolution*–in the following example:

$$\dfrac{\{a\}\,[P] \vee \{a\}\,[Q] \quad \{b,c\}\,[P] \vee \{c,d\}\,[R] \quad \{d\}\,[P]}{\{a\}\,[Q] \vee \{c,d\}\,[R]}$$

See Example 8.2.6 for an application of hyper-resolution with quantified formulae.

The double terminology of Definition 8.2.8 is accounted for by the fact that hyper-resolution is a version of macro-resolution (see Section 4.6.4.2).

The importance of the generalization of Herbrand's theorem to the many-valued logics can be verified in that the Herbrand base provides a means for yet another refinement of signed resolution. We speak here of A-ordering (cf. Section 4.6.4.1): because this refers to atoms (vs. literals), it can be applied directly to many clauses.

**8.2.9. (Def.)** The rule

(R7)  $\dfrac{S_1\,[P_1] \vee \mathcal{C}_1 ... S_k\,[P_k] \vee \mathcal{C}_k}{(\mathcal{C}_1 \cup ... \cup \mathcal{C}_k)\,\sigma}$

where $\bigcap_{1\leq i\leq k} S_i = \emptyset,\ \sigma = mgu\,(P_i),\, 1 \leq i \leq k$, and $P_i\sigma \not<_A Q$ for all $R\,[Q] \in \bar{\mathcal{C}}_i\sigma$, is called *signed $<_A$-ordering macro-resolution*, and the conclusion is the *signed $<_A$-ordering macro-resolvent*.

Just as in the case of CL, we need to specify the ordering of the atoms, which typically falls on the depth $\vartheta$ of the terms and of the variable atoms, with the stipulation that the resolved atom is not lesser than any literal in the macro-resolvent.

### 8.2.2. The main theorem of signed resolution

As seen above, and particularly with automation in mind, the generalization of deduction to the many-valued logics by means of the resolution calculus should be as conservative as possible. Comparing Definition

## 8. Signed resolution for the MV-SAT

4.6.4 with the definition that follows below, it is easy to see that this conservativeness is a straightforward matter.

**8.2.10. (Def.)** Given a set of clauses $C$ and a clause $\mathcal{C}$, we say that $\mathcal{C}$ is derivable from $C$ by many-valued resolution, denoted by $C \vdash_{mvres} \mathcal{C}$, if there is a finite number of clauses $\mathcal{C}_1, ..., \mathcal{C}_n = \mathcal{C}$ such that for each $\mathcal{C}_i$, $1 \leq i \leq n$, one of the following holds true:

1. $\mathcal{C}_i$ is a variant of a clause in $C$;

2. $\mathcal{C}_i$ is a resolvent of the clauses $\mathcal{C}_j$ and $\mathcal{C}_k$, $j, k < i$.

Henceforth, we identify *many-valued resolution* with *signed (many-valued) resolution*, i.e. with the signed resolution calculus above constituted by the rules R1-7.[3] Our approach to signed resolution gives a prominent role to the selection of a set of designated values, namely because this places this calculus in the context of the SAT and of the decision problem. Indeed, it is often the case in practical applications of signed resolution that our interest falls on proving or refuting assertions such as "$val_\mathcal{I}(\phi) \in U$ for every $\mathcal{I}$" given some $U \subset W$. Thus, we now turn our attention to this aspect; we shall see the impact this aspect has on proofs of (quasi-)validity of many-valued formulae when this is formulated as "$val_\mathcal{I}(\phi) \in D$ for every $\mathcal{I}$" given some $D \subset W$, $D$ is a set of designated values.

**Example 8.2.6.** Let there be given the formula

$$A = \forall x P(x) \to \exists y P(y)$$

and the corresponding signed formula

$$\{\mathtt{i}\}[A] = \{\mathtt{i}\}[\forall x P(x) \to \exists y P(y)]$$

in $Ł_3^*$. The formula $A$ is closed, and we want to know whether $A$ can be assigned the truth value i in $Ł_3^*$. We first need to obtain the clauses of $\{\mathtt{i}\}[A]$, i.e. the set $C_{\{\mathtt{i}\}[A]}$; we already did this above in Example 6.2.9. If we show, by applying rules R1-7 above, that $C_{\{\mathtt{i}\}[A]} \vdash_{mvres} \square$, then we show that $A$ cannot take the value i in $Ł_3^*$, i.e. $\{\mathtt{i}\}[A]$ is not valid in $Ł_3^*$. We show the resolution procedure in Figure 8.2.1. Note that $C_{\{\mathtt{i}\}[A]} = \{\mathcal{C}_1, \mathcal{C}_2, \mathcal{C}_3\}$, $\mathcal{C}_1$ and $\mathcal{C}_2$ are taken as parent clauses. It is obvious that $\{\mathtt{i}\}[A]$ was refuted: $\{\mathtt{i}\}[A]$ is unsatisfiable in $Ł_3^*$.

---

[3] We henceforth omit the * in signed resolution rules R1-6.

## 8.2. Signed resolution for finitely many-valued logics

Note in Example 8.2.6 that in the resolution procedure clause $C_3$ was removed according to Proposition 6.2.2, items 1. and 3., once condensation of $C_3$ was carried out. Note also that we could have obtained $C_8$ directly from $C_1\sigma$ and $C_2\sigma$ by hyper-resolution given the MGU $\sigma = \{x \mapsto a, y \mapsto a\}$.

Above, the notion of semantic tree (cf. Def. 4.3.5) proved to be very relevant for proving completeness of the resolution calculus for the classical logic K. It so happens that we can naturally generalize this notion to many-valued systems, namely via signed clause logic. Recall that, given a set of clauses $C$, in a semantic tree $\mathcal{T}_C$ every branch $I(N)$ is a truth-value assignment to the ground atoms of $H(C)$. Let $At(C)$ be the set of ground atoms of the ground clauses in $C$, i.e. $At(C) \subseteq H(C)$; then, each branch $I(N)$ in $\mathcal{T}_C$ has at least a truth-value assignment to $At(C)$, and every complete truth-value assignment to $At(C)$ (i.e. a H-interpretation) is a subset of the union of the set of literals in the edges of $I(N)$. Recall also from Definition 4.3.6 that $N$ is a failure node if $I(N)$ falsifies some ground instance of some clause $\mathcal{C}$ in $C$, but $I(N')$ does not falsify any ground instance of some some clause $\mathcal{C}$ in $C$. We have it thus that:

**8.2.11. (Prop.)** A semantic tree of $C$ is closed if for every $I(N)$ the intersection of the CNFs obtained from $C$ with the corresponding DNFs that constitute the edges of $\mathcal{T}_C$ (cf. Def. 4.3.5.1) is empty, that is, if for every $I(N)$ and every atom $A_i$ of some clause $\mathcal{C}$ we have:

$$\left(\bigvee_{i=1}^{n} A_i\right) \cap \left(\bigwedge_{i=1}^{n} \neg A_i\right) = \emptyset$$

**8.2.12. (Prop.)** In effect, $\bigwedge_{i=1}^{n} \neg A_i$ is a refutation of $\bigvee_{i=1}^{n} A_i$ (cf. Prop. 2.4.15).

**8.2.13. (Prop.)** Let now $S$ be the sign of some atom $A_i \in At(C)$. It is evident that for every $I(N) \in \mathcal{T}_C$ we have:

$$\left(\bigvee_{i=1}^{n} S[A_i]\right) \cap \left(\bigwedge_{i=1}^{n} \overline{S}[A_i]\right) = \emptyset$$

Thus, $I(N)$ omits at least one assignment of a truth value $v_i \in S$ to a set of literals of some clause $\mathcal{C}$ and similarly with respect to the set of atoms of the ground clause corresponding to $\mathcal{C}$, which entails that every $I(N)$ omits exactly one H-interpretation of $C$.

Proofs of the above Propositions are left as exercises.

## 8. Signed resolution for the MV-SAT

$C_1$    $\{\mathbf{t}\}[P(x)] \vee \{\mathbf{f}\}[P(y)]$
$C_2$    $\{\mathbf{i}\}[P(a)]$
$C_3$    $\{\mathbf{i}, \mathbf{t}\}[P(x)] \vee \{\mathbf{f}, \mathbf{i}\}[P(y)]$    Resolvent of $C_1\theta$ and $C_2\theta$, $\theta = \{x \mapsto a\}$
$C_4$    $\{\}[P(a)] \vee \{\mathbf{f}\}[P(y)]$    Resolvent of $C_1\lambda$ and $C_2\lambda$, $\lambda = \{y \mapsto a\}$
$C_5$    $\{\mathbf{t}\}[P(x)] \vee \{\}[P(a)]$    $C_4$, R2 and renaming
$C_6$    $\{\mathbf{f}\}[P(y_1)]$    $C_5$, R2 and renaming
$C_7$    $\{\mathbf{t}\}[P(x_1)]$
$C_8$    $\square$    Resolvent of $C_6\sigma$ and $C_7\sigma$, $\sigma = \{x_1 \mapsto c, y_1 \mapsto c\}$, by R1

Figure 8.2.1.: A signed resolution procedure.

## 8.2. Signed resolution for finitely many-valued logics

The following theorem makes this generalization of H-unsatisfiability precise:

**Theorem 8.2.1.** *A set of clauses $C$ is H-unsatisfiable iff there is a finite set $At(C) \subseteq H(C)$ such that every semantic tree for $At(C)$ is closed for $C$.*

*Proof:* ($\Rightarrow$) Let $C$ be an unsatisfiable set of clauses. By Theorem 4.3.4, there is a finite closed semantic tree for $C$. Let $\mathcal{T}_C$ be a closed semantic tree for $C$. Being closed, every branch of $\mathcal{T}_C$ is closed, i.e. they end in a failure node (cf. Def. 4.3.6). By the definition of semantic tree, the number of nodes in $\mathcal{T}_C$ above a failure node of a finite semantic tree is also finite. This means that for every unsatisfiable finite set of clauses $C$ there is a finite set $C'$ of ground clauses built from $H(C)$ that is H-unsatisfiable. Hence, every closed semantic tree $\mathcal{T}'_C$ for $C$ entails that $\mathcal{T}_C$ is also closed. Thus, in order to build a closed semantic tree for $C$ it suffices a finite subset $At(C) \subseteq H(C)$, where $At(C)$ is the set of ground atoms in $C'$.

($\Leftarrow$) Let $\mathcal{T}_C$ be a closed semantic tree for a finite subset $At(C) \subseteq H(C)$. Let $C'$ be the set of ground clauses of $C$, and let $At(C)$ be the set of ground atoms of $C'$. Assume now that $C'$ is H-satisfiable. There is an open semantic tree $\mathcal{T}'_C$ for $C'$. This means that there is at least one ground instance in $C'$ of $C$ that is not falsified by an H-interpretation in a failure node of $\mathcal{T}'_C$. Thus, there is at least one branch of $\mathcal{T}'_C$ that does not end in a failure node. But this goes against the assumption that $\mathcal{T}_C$, a semantic tree for $At(C) \subseteq H(C)$, is closed. Hence, $C$ must be H-unsatisfiable. **QED**

We are now ready to present the central results for the many-valued resolution calculus.

**Theorem 8.2.2.** *(Soundness of the many-valued resolution calculus) For some set of clauses $C$, if $C \vdash_{mvres} \square$, then $C$ is H-unsatisfiable.*

*Proof:* We have already seen for CL that no interpretation satisfies the empty clause $\square$. Hence, a set of clauses $C$ is unsatisfiable every time one can derive $\square$ from $C$. Besides, given that $\square$ has no atom belonging to $At(C) \subseteq H(C)$ that can be satisfied by an H-interpretation, we have it that if $\square$ can be derived from $C$, then $C$ is H-unsatisfiable, namely via a subset $C' \subseteq C$ of ground clauses of $C$. **QED**

**Theorem 8.2.3.** *(Completeness of the many-valued resolution calculus) For some set of clauses $C$, if $C$ is H-unsatisfiable, then $C \vdash_{mvres} \square$.*

## 8. Signed resolution for the MV-SAT

*Proof:* We know that if $C$ is an unsatisfiable set of clauses, then every semantic tree for $At(C)$ is closed. In particular, every semantic tree $\mathcal{T}_C$ is closed for $C$. Let $C'$ be the set of signed ground clauses of $C$. We select a tree $\mathcal{T}'_C$ whose edges are mono-signed literals. We start with the failure node $N$ of $\mathcal{T}'_C$; because this is a failure node, given two clauses $C_1, C_2 \in C$ and a ground substitution $\sigma$, $(C_1 - L_1)\sigma \cup (C_2 - L_2)\sigma$ is a subset of the refutation of $N$ for $L_1\sigma = \{v_i\}[A]$ and $L_2\sigma = \{v_j\}[A]$, $v_i \neq v_j$, $v_i, v_j \in W$. By the definition of closed semantic tree, there is a resolvent $C_3$ of (factors of) the clauses $C_1$ and $C_2$ that fails in $N$, and a failure node for $C \cup \{C_3\}$ can only be found $I(N) - 2N'$ above $N$. For $C_n$, $n$ is the number of resolvents in $C$, it is then evident that the failure node for $C \cup \{C_3\}$ is at the root of $\mathcal{T}'_C$, which corresponds to the empty clause $\square$. **QED**

**Theorem 8.2.4.** *Let $\phi$ be any closed formula and let $C_{\overline{U}\Phi}$ be the set of clauses of the clausal translation $\overline{U}\Phi$ of $\{v\}[\phi]$ for any truth value $v \in \overline{U}$, $U \subset W$. Then, all interpretations give a truth value $u \in U$ to $\phi$ iff $C_{\overline{U}\Phi} \vdash_{mvres} \square$, where mvres designates any of the rules R1-7.*

*Proof:* ($\Rightarrow$) Given a formula $\phi$ and a truth-value set $U \subset W$, if all the interpretations assign a truth value $u \in U$ to $\phi$, then $\{v\}[\phi]$ is unsatisfiable for any truth value $v \in \overline{U}$ by Definition 3.2.6; by Propositions 6.2.10 and 6.2.13, the clausal translation $\overline{U}\Phi$ of $\{v\}[\phi]$ is unsatisfiable. Let $C_{\overline{U}\Phi}$ be the set of clauses corresponding to $\overline{U}\Phi$; then, by Proposition 8.2.7 $C_{\overline{U}\Phi}$ is H-unsatisfiable. Hence, by Theorem 8.2.3 we have $C_{\overline{U}\Phi} \vdash_{mvres} \square$.

($\Leftarrow$) By theorem 8.2.2, if $C_{\overline{U}\Phi} \vdash_{mvres} \square$, then $C_{\overline{U}\Phi}$ is H-unsatisfiable. By Proposition 8.2.7, then $C_{\overline{U}\Phi}$ is unsatisfiable. Hence, by Theorem 3.2.4 we have $\models \{u\}[\phi]$, i.e. all the interpretations assign to $\phi$ the truth value $u$. **QED**

From the $\Leftarrow$-direction of this proof, it is often concluded that rather than *completeness* for the signed resolution calculus for many-valued logics one has to speak solely of *refutation completeness*, a feature inherited from classical resolution.

**8.2.14. (Prop.)** *Refutation completeness* – For a set of clauses $C$ and a formula $\phi$,

$$C \models \phi \quad \text{iff} \quad C \cup \{\neg\phi\} \vdash_{(mv)res} \square.$$

*Proof:* Left as an exercise.

In any case, following the main theorem above we now have the fundamental results, whose proofs we leave as exercises:

## 8.2. Signed resolution for finitely many-valued logics

**Algorithm 8.1** Signed resolution procedure for MV-SAT (a.k.a. MVRES)

**Input:** A formula $\phi$ in a many-valued logical system $S_n$ with a set of truth values $W_n$ for finite $n > 2$

**Output:** A proof of validity of $\phi$

---

1. Obtain the clausal form $\overline{D}\Phi$ of the signed formula $\{v\}[\phi]$, $v \in \overline{D}$, where $D \subset W_n$ is the set of designated values.

2. Obtain the set of clauses $C_{\overline{D}\Phi}$ from $\overline{D}\Phi$.

3. Apply the many-valued resolution calculus (rules R1-7) to $C_{\overline{D}\Phi}$ to test for unsatisfiability: if $C_{\overline{D}\Phi}$ is unsatisfiable, then $\{u\}[\phi]$, $u \in D$, is a valid formula in S.

---

**8.2.15. (Prop.)** MVRES constitutes a refutation proof.

**8.2.16. (Prop.)** Algorithm 8.1 constitutes an automation of theorem proving in the signed resolution calculus for many-valued logics.

**Example 8.2.7.** In Example 8.2.6, we obtained the result that the signed formula $\{i\}[\forall x P(x) \to \exists y P(y)]$ is not valid in $L_3^*$. Let $D_{L_3^*} \subset W_{L_3^*} = \{t\}$ and let $\forall x P(x) \to \exists y P(y) = A$. We can now conclude that $\models_{L_3^*} \{t\}[A]$, given that $\overline{\{D_{L_3^*}\}} = \{f, i\}$.

We now conclude our discussion of the signed resolution calculus for many-valued logics with what can be considered its main theorem (alternatively to Theorem 8.2.4):

**Theorem 8.2.5.** *Given a formula $\phi$ in a many-valued logic in which $D \subset W$ is the set of designated values, the translation of the signed formula $\{v\}[\phi]$, $v \in \overline{D}$, into the SFE $\overline{D}\Phi$ and the corresponding transformation into the set of clauses $C_{\overline{D}\Phi}$ combined with its respective refutation is a proof of the validity of $\{u\}[\phi]$ for $u \in D$.*

*Proof:* The proof is trivial given the above. **QED**

**8.2.17. (Prop.)** The signed resolution calculus for many-valued logics is an adequate (automatable) proof system for finitely many-valued logics.

8. Signed resolution for the MV-SAT

*Proof:* Left as an exercise.

## 8.3. Signed resolution for infinitely many-valued logics

Rules R1-7 of the many-valued signed resolution calculus are not straightforwardly applicable to the infinitely many-valued logics, i.e. to the *(signed) MV-SAT$_\infty$*. We shall mostly concentrate on the fuzzy logics (FLs) approached above in Section 5.6. These logics have fundamental applications in many areas, and as such their axiomatization and automation in terms of deduction are important topics of research. However, as seen above, their axiomatization poses problems that, in turn, impact on the automation of deduction in these logics. By and large, research in this matter is not very developed, and this includes the topic of signed resolution for FLs. For instance, Baaz, Fermüller, & Salzer (2001), an article that somehow crystallizes the state of the art in signed resolution for many-valued logics, provides no more than a few remarks on this subject, and none of these remarks actually points to an application of this calculus to the FLs. We offer some suggestions on the application of the signed resolution principle to these logics.

To begin with, it is important to recall that classical logic just is the *simplest* non-trivial case of FLs, given that a trivial (fuzzy) logical system would be one with a single truth value. As a matter of fact, FL connectives are normal, so the reduction to a truth-value set $W = \{\mathtt{f}, \mathtt{t}\}$ via sets of designated and anti-designated truth values simulates classical logic semantics.

We can, for instance, based on Definition 5.6.2 and on Proposition 5.6.3, start by setting $D_\varepsilon = [0.5, 1]$. This allows us a reduction to classical semantics by defining $\varepsilon[\phi] = \{\mathtt{t}\}[\phi]$ and $\eta[\phi] = \{\mathtt{f}\}[\phi]$ for $\eta \in \overline{D_\varepsilon}$, i.e. $\eta \in [0, 0.5)$, where $\phi$ may or may not be a quantified formula. This would allow us to implement directly signed resolution in a prover such as Prover9-Mace4, as we have the equalities $\{\mathtt{t}\}[\phi] = \phi$ and $\{\mathtt{f}\}[\phi] = \neg\phi$ (or vice-versa). One such implementation is that of Example 4.6.13. Indeed, if we want to be mathematically rigorous, then the property *is much smaller than* is a fuzzy property (cf. Example 5.7.10), which makes of Schubert's steamroller a fuzzy theory. In Example 4.6.13, we denoted this property by means of the binary predicate $Smaller\,(x, y)$, but it would have been more correct if we had applied the binary fuzzy predicate $Ms\,(x, y)$ of Example 5.7.10, given that *much* is a fuzzy modifier of the property *smaller than*. Between the valuations $val\,(Ms\,(x,y)) = 0$ and $val\,(Ms\,(x,y)) = 1$, i.e. between the valuation that denies absolutely that $x$ has the property of being much smaller than $y$ and the valuation that asserts without a doubt that $x$ is much smaller than $y$, there is an infinity of degrees of *vagueness*.

However, a reduction to the bivalent semantics of classical logic entails the inability to tackle vagueness. And, what is more: any reduction or restriction defined by $|W_{FL}| = n$, $n$ is finite, entails the same consequence. A highly undesirable consequence, as this goes against the original mathematical motivation of creating fuzzy sets (cf. Zadeh, 1965, 1975), which then motivated the FLs. Moreover, the transformation and translation of FL formulae to clausal form is generally not possible, with only a few fragments allowing to undergo such operations. For instance, it has been proven (Mundici & Olivetti, 1998) that resolution and model building for at least the infinitely many-valued Łukasiewicz logic are polynomially tractable in the fragments given by Horn clauses (cf. Def. 2.4.5), as well as by Krom clauses (clauses with at most two literals), built from the set of connectives $\{\neg, \wedge, \vee, \&, \nabla\}$. We discuss Mundici & Olivetti's (1998) in a more or less loose way, given the very restricted character of this work. Their starting point is a notion of $\varepsilon$-*support*.[4]

**8.3.1. (Def.)** The $\varepsilon$-support of a non-negative literal $L$ is the set $supp_\varepsilon = \{val^* | val\,(L) \geq \varepsilon\} \neq \emptyset$ where $\varepsilon \in (0,1]$.

This motivates a definition of $\varepsilon$-satisfiability for which we require an additional definition.

**8.3.2. (Def.)** Each formula $\phi = \phi\,(x_1, ..., x_n)$ is associated with a mapping $\Phi = \Phi\,(x_1, ..., x_n) : [0,1]^n \longrightarrow [0,1]$ that is a McNaughton function in the following way (cf. 5.7.15):[5]

1. $X_i =$ the $i$-th projection $(x_1, ..., x_n) \longmapsto x_i$

2. $\neg \Phi = 1 - \Phi$

3. $\Phi \nabla \Psi = min\,(1, \Phi + \Psi)$

4. $\Phi \& \Psi = max\,(0, \Phi + \Psi - 1)$

5. $\Phi \wedge \Psi = min\,(\Phi, \Psi)$

---

[4]Originally $\theta$-support.
[5]A mapping $f : [0,1]^n \longrightarrow [0,1]$ is called a McNaughton function iff it is continuous and there are linear polynomials $p_1, ..., p_t$ with integer coefficients,

$$p_j\,(\mathbf{x}) = p_j\,(x_1, ..., x_n) = a_{j1}x_1 + ... + a_{jn}x_n + b_j$$

such that for each point $\mathbf{x} \in [0,1]^n$ there is an index $j = 1, ..., t$ with $f\,(\mathbf{x}) = p_j\,(\mathbf{x})$.

## 8. Signed resolution for the MV-SAT

6. $\Phi \vee \Psi = max(\Phi, \Psi)$

**8.3.3. (Def.)** *$\varepsilon$-Satisfiability* – Given a set $supp_\varepsilon$, a formula $\phi = \phi(x_1, ..., x_n)$ is said to be *$\varepsilon$-satisfiable* if there is some valuation $val$ such that $val(\phi) \geq \varepsilon$. A clause $\mathcal{C} = \|L_1, ..., L_k\|$ is $\varepsilon$-satisfiable if so is the formula $L_1 \vee ... \vee L_k$. A set of formulae $F$ (clauses $C$) is $\varepsilon$-satisfiable iff there is some assignment $val$ such that for all formulae $\phi \in F$ (all $\phi \in C$) we have $\models_\varepsilon \phi$. Then, we say that $val$ is a $\varepsilon$-model of $F$ ($C$, respectively). The empty clause is never $\varepsilon$-satisfiable and the empty set of clauses $\emptyset$ is always $\varepsilon$-satisfiable.

**8.3.4. (Prop.)** Let $\varepsilon \in (0, 1]$ and let $F$ be a set of formulae. Then, if every finite subset $F' \subseteq F$ is $\varepsilon$-satisfiable, so is $F$.

*Proof:* Left as an exercise.

This proposition is obviously a statement of *compactness* (cf. Def. 3.4.3), a property that distinguishes between $\varepsilon$-satisfiability and fuzzy satisfiability. In effect, compactness does not hold with respect to the latter.

**8.3.5. (Def.)** *Fuzzy satisfiability* – A formula $\phi = \phi(x_1, ..., x_n)$ is *fuzzy-satisfiable* iff there is some $val \in [0, 1]^n$ such that $val(\phi) > 0$. A clause $\mathcal{C} = \|L_1, ..., L_k\|$ is fuzzy-satisfiable iff the formula $L_1 \vee ... \vee L_k$ is fuzzy-satisfiable. A set $F$ of formulae is fuzzy-satisfiable iff for some $val$ it is the case that $val(\phi) > 0$ for every $\phi \in F$.

Then, we can generalize the SAT as the *generalized satisfiability problem* GEN-SAT$_\infty$ in the infinitely many-valued calculus, defined as:

**8.3.6. (Def.)** GEN-SAT$_\infty$ – Given a formula $\phi$ and a rational number $\varepsilon = c/d$, $1 \leq c \leq d$, $c$ and $d$ are integers, we ask: Is $\phi$ $\varepsilon$-satisfiable?

**Theorem 8.3.1.** *GEN-SAT$_\infty$ is in NP. Moreover, there exists a reduction of MV-SAT$_\infty$ to GEN-SAT$_\infty$ which is computable in polynomial time. Thus GEN-SAT$_\infty$ is also NP-hard.*

*Proof:* See Mundici & Olivetti (1998). **QED**

The GEN-SAT$_\infty$ is then reduced to the KROM-SAT$_\infty$ and to the HORN-SAT$_\infty$ for Krom clauses and Horn clauses, respectively, both of which are equally computable in polynomial time.

**8.3.7.** From the above, a notion of *$\varepsilon$-complementarity* for literals $L$ and $M$ is obvious, and Mundici & Olivetti produce a binary resolution rule for $Ł_\aleph$ given two clauses $\mathcal{C}_1 = L_1 \vee ... \vee L_n$ and $\mathcal{C}_2 = M_1 \vee ... \vee M_q$ and a propositional variable $p$ such that

1. $p$ occurs solely in the $i$-th positive literal $L_i \in \mathcal{C}_1$,

2. $p$ occurs solely in the $j$-th negative literal $M_j \in \mathcal{C}_2$, and

3. $L_i$ and $M_j$ are $\varepsilon$-complementary.

**8.3.8. (Def.)** If conditions 8.3.7.1-3 are satisfied, then $\mathcal{C}_1$ and $\mathcal{C}_2$ are said to be *p-resolvable with respect to* $\varepsilon$ in the implementation of the rule:
$$\frac{\mathcal{C}_1 \quad \mathcal{C}_2}{(\mathcal{C}_1 - \{L_i\}) \cup (\mathcal{C}_2 - \{M_j\})}$$

It is evident that the notion of $\varepsilon$-satisfiability allows a characterization of $p$ as a regular signed literal (cf. Def. 2.5.3), i.e. as $\uparrow i\,[p]$ or $\downarrow i\,[p]$. However, the translation into clausal logic is based on the connective & (instead of $\wedge$; cf. Def. 5.6.5) and in one-variable McNaughton functions (see above), which actually accounts for an internal proof procedure (cf. Baaz, Fermüller, & Salzer, 2001). The main disadvantage is that the translation of formulae from $L_\aleph$ into clausal logic by this means is far from trivial (see Metcalfe, Olivetti, & Gabbay, 2009, p. 159ff).

Beckert, Hähnle, & Manyà (1999) propose a method based on a partial ordering $\leq$ on $W$, i.e. the lattice $\mathcal{W} = (W, \leq)$, from which we can obtain resolution procedures for signed regular Horn formulae.

**8.3.9. (Def.)** Let $\varsigma$ be a regular Horn formula on a (possibly infinite) lattice $\mathcal{W} = (W, \leq)$ with infimum $\bot$ and supremum $\top$. We define $\mathcal{W}^\varsigma = (W_\varsigma, \leq)$ to be the sub-lattice of $\mathcal{W}$ generated by the set of elements $\{i \in W | i \text{ occurs in } \varsigma\}$.

**Theorem 8.3.2.** $\varsigma$ *is satisfiable by an interpretation for the lattice* $\mathcal{W}$ *iff it is satisfiable in the lattice* $\mathcal{W}^\varsigma$.

*Proof:* (Sketch) The $\Rightarrow$-direction is trivial (every interpretation for $\mathcal{W}^\varsigma$ is an interpretation for $\mathcal{W}$). In order to prove the $\Leftarrow$-direction, it suffices to show that for all truth values in $\varsigma$ we have

$$\models_\mathcal{I} \uparrow i\,[p] \quad \text{iff} \quad \models_{\mathcal{I}_\varsigma} \uparrow i\,[p]$$

for which we need a supremum operator $\cup$ shared by the lattices $\mathcal{W}$ and $\mathcal{W}_\varsigma$. **QED**

**8.3.10. (Def.)** We then have the inference rules

(1) $$\frac{\rightarrow \uparrow i\,[p] \qquad \uparrow i_1\,[p_1],...,\uparrow i_l\,[p_l],...,\uparrow i_k\,[p_k] \rightarrow \uparrow j\,[q]}{\uparrow i_1\,[p_1],...,\uparrow i_{l-1}\,[p_{l-1}],\uparrow i_{l+1}\,[p_{l+1}]...,\uparrow i_k\,[p_k] \rightarrow \uparrow j\,[q]}$$

## 8. Signed resolution for the MV-SAT

and

$$(2) \quad \frac{\rightarrow i\,[p] \quad \rightarrow j\,[q]}{\rightarrow \uparrow (i \cup j)\,[p]}$$

called *regular positive unitary resolution* (rule 1) and *regular reduction* (rule 2).

Rule 1 requires the condition $i \geq i_l$, while rule 2 requires that neither $i \geq j$ nor $j \geq i$. This procedure appears to be applicable in principle to the FLs (or fragments of some of the FLs) when we make $\bot$ ($\top$) coincide with 0 (1, respectively).

The suggestions above work only, as seen, for very restricted instances, and they are not generalizable to the FO case. We next present a method of signed resolution for FLs that considers FO formulae (Lu, Murray, & Rosenthal, 1994; 1998). Besides this aspect, this signed resolution method is also interesting because some of its specific features show the flexibility and wide-range applicability of signed resolution. We elaborate on the propositional case and leave most of the generalization to FO as exercises.

**8.3.11. (Def.)** *Fuzzy operator logic* – Let there be given fuzzy formulae of the forms $\lambda A = \phi$, $\lambda (\phi \wedge \psi)$, and $\lambda (\phi \vee \psi)$, where $\lambda \in [0,1]$ is a *fuzzy operator* (a generalized negation), $A$ is an atom, and $\lambda A$ is a fuzzy atom. Let us call this logic a *fuzzy operator logic*. We denote this logic by $L_\aleph$ and its language as just defined by L$\lambda$. In L$\lambda$, the formulae $S\phi$, $S(\phi \wedge \psi)$, and $S(\phi \vee \psi)$, in which $S \subseteq [0,1]$ is a sign, are signed formulae.[6] If $A$ is an atom, a signed formula $SA$ is a signed base atom in L$\lambda$. Regardless of the size or complexity of $\phi$, a formula $S\phi$ is always a signed atom and as such has no subformulae in L$\lambda$. If whenever $SA$ is an atom in a formula, then $A$ is an atom in L$\lambda$, i.e. if the only atoms in the formula are signed base atoms, then we call this a L$\lambda$-*atomic* formula.

**Example 8.3.1.** An example of a fuzzy formula is

$$\phi = A \wedge 0.3\,(0.9B \vee 0.2C)$$

where real numbers denote both truth values[7] and fuzzy operators. Note that $A = 1A$, i.e. any formula without an explicit fuzzy operator is understood to have 1 as a fuzzy operator.

**8.3.12. (Def.)** Consider now a L$\lambda$-atomic formula $\phi \in F_{L\lambda}$ in CNF.

---

[6] In $L_\aleph$, instead of $S\,[\phi]$ we write $S\phi$ or $(S)\,\phi$.
[7] This is what makes of this FL a many-valued logic. Indeed, a FL need not be considered a many-valued logic.

## 8.3. Signed resolution for infinitely many-valued logics

Let $C_j$, $1 \leq j \leq r$, be clauses in $\phi$ containing signed base atoms $S_j A$. Thus, each $C_j$ has the form $K_j \vee S_j A$. Then the *resolvent* of the $C_j$ is defined as the clause

$$\left( \bigvee_{j=1}^{r} K_j \right) \vee \left( \left( \bigcap_{j=1}^{r} S_j \right) A \right)$$

where signed binary resolution falls on the rightmost disjunct by applying the rules above for signed resolution. In particular, Proposition 6.2.2 holds for L$\lambda$-atomic formulae; merging (Def. 8.2.5) and subsumption (Def.s 2.5.4 and 2.6.5) also hold.

**8.3.13. (Prop.)** Most of the syntactical and semantical properties of FLs hold for L$_{\aleph}$. In particular, $W_{L_{\aleph}}$ is a complete lattice, with the connectives $\wedge$ and $\vee$ naturally generalized as the infimum and supremum operators, respectively.

*Proof:* Left as an exercise.

However, there are some important specifications.

**8.3.14. (Def.)** An interpretation $\mathcal{I}$ over L assigns to each atom and thus to each formula $\phi$ a truth value in $W_2$, and the corresponding L-*consistent interpretation* $\mathcal{I}_\lambda$ is defined as:

$$\mathcal{I}_\lambda (S\phi) = \begin{cases} \mathtt{t} & \text{if } \mathcal{I}(\phi) \in S \\ \mathtt{f} & \text{if } \mathcal{I}(\phi) \notin S \end{cases}$$

This shows that L$\lambda$ is actually the range of a mapping whose domain is L (i.e. formulae of L are operated upon so as to be converted to signed formulae in L$\lambda$). Another way to put this is to say that L is the base language of L$\lambda$. Thus, the above is a one-to-one correspondence (a bijection) between the set of all interpretations over L and the set of L-consistent interpretations over L$\lambda$, according to which specific notions of equivalence and consequence can be defined as follows:

**8.3.15. (Def.)** Let $\phi_1, \phi_2 \in F_{L\lambda}$. Whenever there is an interpretation $\mathcal{I}_\lambda$ such that if $\mathcal{I}_\lambda (\phi_1) = \mathtt{t}$, then $\mathcal{I}_\lambda (\phi_2) = \mathtt{t}$, we write

$$\phi_1 \models_\lambda \phi_2.$$

**8.3.16. (Def.)** Two formulae $\phi_1, \phi_2 \in F_{L\lambda}$ are said to be L-*equivalent*, written $\phi_1 \equiv_\lambda \phi_2$, if $\mathcal{I}_\lambda (\phi_1) = \mathcal{I}_\lambda (\phi_2)$ for any interpretation $\mathcal{I}_\lambda$.

**8.3.17. (Prop.)** Let $\mathcal{I}_\lambda$ be a L-equivalent interpretation. Let further $A$ be an atom, $\phi \in F_{L\lambda}$, $S_i$ be a sign. Then,

8. Signed resolution for the MV-SAT

1. $\mathcal{I}_\lambda(\emptyset\phi) = \mathbf{f}$;

2. $\mathcal{I}_\lambda(W\phi) = \mathbf{t}$;

3. $S_1 \subseteq S_2$ iff $S_1\phi \models_\lambda S_2\phi$ for all formulae $\phi$;

4. there is exactly one $\nu \in W$ such that $\mathcal{I}_\lambda(\nu A) = \mathbf{t}$.

*Proof:* Left as an exercise.

**8.3.18. (Def.)** The semantics of fuzzy operators is given by means of a kind of *fuzzy product* $\otimes$ that is commutative and associative:

If $\lambda, \nu \in [0,1]$, then $\lambda \otimes \nu = \lambda \cdot \nu + (1-\lambda) \cdot (1-\nu) = (2\lambda - 1) \cdot \nu - \lambda + 1$

Intuitively, if $\lambda$ were the likelihood that $A_1$ is **true** and $\nu$ the likelihood that $A_2$ is **true**, then $\lambda \otimes \nu$ would be the likelihood that $A_1$ and $A_2$ are both **true** or both **false**. This actually means that there is a confidence function mapping fuzzy formulae to the set $[0,1]$.

**8.3.19. (Def.)** *Confidence* – Let $\mathcal{C}_\mathcal{I}$ denote the *confidence function* $\mathcal{C}$ corresponding to the interpretation $\mathcal{I}$. Let further $A$ be an atom, $\phi, \psi$ be fuzzy formulae, and $\lambda$ be an operator. Then,

1. $\mathcal{C}_\mathcal{I}(A) = \mathcal{I}(A)$, and $\mathcal{C}_\mathcal{I}(\lambda\phi) = \mathcal{I}(\lambda\phi)$;

2. $\mathcal{C}_\mathcal{I}(\lambda\phi) = \lambda \otimes \mathcal{C}_\mathcal{I}(\phi) = \Lambda_\lambda(\mathcal{C}_\mathcal{I}(\phi))$, where $\Lambda_\lambda := \Lambda_\lambda(\nu) = \lambda \otimes \nu$;

3. $\mathcal{C}_\mathcal{I}(\phi \wedge \psi) = min\{\mathcal{C}_\mathcal{I}(\phi), \mathcal{C}_\mathcal{I}(\psi)\}$;

4. $\mathcal{C}_\mathcal{I}(\phi \vee \psi) = max\{\mathcal{C}_\mathcal{I}(\phi), \mathcal{C}_\mathcal{I}(\psi)\}$.

Intuitively, we have it that $1\phi$ asserts the certainty of the truth value assigned to $\phi$, $0.8\phi$ asserts that the truth value assigned to $\psi$ is likely to be right, $0.2\phi$ asserts that the truth value assigned to $\phi$ is likely to be wrong, and $0\phi$ asserts that the truth value assigned to $\phi$ is certainly wrong. Note, however, that it is important to distinguish between the fuzzy operator as the *likelihood of the accuracy* of the truth value of a formula and its *actual* truth value, which is given by $\mathcal{C}_\mathcal{I}$.

**Example 8.3.2.** Let $0\phi$ be given. If $\mathcal{I}(\phi) = \mathbf{f}$, then the truth value of $0\phi$ is $\mathbf{t}$. Note how this corresponds to standard negation.

**8.3.20. (Prop.)** Let $\phi, \psi$ be fuzzy formulae and let $\lambda$ be a fuzzy operator. Then, the De Morgan's laws $DM_\wedge$ and $DM_\vee$ (cf. 2.4.6.2) are generalized to $L_\aleph$ as follows:[8]

1. If $\lambda > 0.5$, then:
$$\lambda(\phi \wedge \psi) \equiv_\lambda \lambda\phi \wedge \lambda\psi; \lambda(\phi \vee \psi) \equiv_\lambda \lambda\phi \vee \lambda\psi$$

2. If $\lambda < 0.5$, then:
$$\lambda(\phi \wedge \psi) \equiv_\lambda \lambda\phi \vee \lambda\psi; \lambda(\phi \vee \psi) \equiv_\lambda \lambda\phi \wedge \lambda\psi$$

*Proof:* Left as an exercise.

**8.3.21. (Prop.)** Every fuzzy formula is equivalent to a formula in which 1 is the only fuzzy operator applied to non-atomic arguments.
*Proof:* Left as an exercise.

The semantics of $L_\aleph$ requires yet some further notions, in particular with respect to a specific notion of satisfiability in $L_\aleph$.

**8.3.22. (Def.)** *T-Satisfiability* – A real number $T \in [0.5, 1]$ is called a *threshold of acceptability*. Given some $T$, we say of an interpretation $\mathcal{I}$ that it *T-satisfies* a formula $\phi$ iff $\mathcal{I}(\phi) \geq T$. In particular, an interpretation $\mathcal{I}$ is said to $T$-satisfy a formula $\phi$ iff $\mathscr{C}_\mathcal{I}(\phi) \geq T$.

In practical terms, $T$ redefines the designated set $D_{FL}$ as $[T, 1]$ where $T$ provides a variable. Compare with Definition 5.6.1 and Propositions 5.6.2-3 above. Intuitively, we require the confidence of a formula under a specific interpretation to be close to 1.

**Example 8.3.3.** (Lu, Murray, & Rosenthal, 1994) Let $T = 0.7$. Then, the formula $0.8A$ is $T$-satisfiable. We show why: Suppose $A$ is

---

[8] Note that if $\lambda > 0, 5$, then
$$\lambda \otimes min\{\nu_1, \nu_2\} = min\{\lambda \otimes \nu_1, \lambda \otimes \nu_2\}$$
and
$$\lambda \otimes max\{\nu_1, \nu_2\} = max\{\lambda \otimes \nu_1, \lambda \otimes \nu_2\}.$$
If, however, $\lambda < 0.5$, then
$$\lambda \otimes min\{\nu_1, \nu_2\} = max\{\lambda \otimes \nu_1, \lambda \otimes \nu_2\}$$
and
$$\lambda \otimes max\{\nu_1, \nu_2\} = min\{\lambda \otimes \nu_1, \lambda \otimes \nu_2\}.$$

## 8. Signed resolution for the MV-SAT

1, i.e. $\mathcal{I}(A) = 1$ for some interpretation $\mathcal{I}$. Then, $\mathcal{I}(0.8A) = 0.8 \geq T$. For the same reasons, the formula $0.6A$ is not $T$-satisfiable. Now let $\mathcal{I}(A) = 0$. Then the first formula evaluates to $0.8 \otimes 0 = 0.2$, and the second evaluates to $0.2 \otimes 0 = 0.8$, so that now it is the second formula that is $T$-satisfiable. Because 0.2 is in fact a negative sign (i.e. $0.2 < 1 - T$), by assigning the truth value 0 to $A$ we obtain a true formula. A formula like $0.6A$ is always $T$-unsatisfiable.

**8.3.23. (Def.)** A clause is said to be $T$-*empty* if for every fuzzy operator $\lambda$ of every literal in the clause we have $1 - T \leq \lambda \leq T$.

**8.3.24. (Prop.)** Every $T$-empty clause is $T$-unsatisfiable.
*Proof:* Left as an exercise.

We now need a notion of complementarity that allows us to resolve upon clauses.

**8.3.25. (Def.)** Two literals $\lambda_1 A$ and $\lambda_2 A$ are said to be $T$-*complementary* if $\lambda_1 \leq 1 - T$ and $\lambda_2 \geq T$.

We can now define resolution for the fuzzy operator logic $L_\aleph$ with respect to the threshold $T$.

**8.3.26. (Def.)** Let two literals $\lambda_1 A = L_1$ and $\lambda_2 A = L_2$ be $T$-complementary and let $\mathcal{C}_1 = L_1 \vee \mathcal{C}_1'$ and $\mathcal{C}_2 = L_2 \vee \mathcal{C}_2'$ be fuzzy clauses. Then:

$$\frac{L_1 \vee \mathcal{C}_1' \quad L_2 \vee \mathcal{C}_2'}{\mathcal{C}_1' \vee \mathcal{C}_2'}$$

$\mathcal{C}_1' \vee \mathcal{C}_2'$ is called the $T$-*resolvent* of $\mathcal{C}_1$ and $\mathcal{C}_2$ on the literals $L_1$ and $L_2$.

We now require a precise conception of the sign $S \subseteq [0,1]$. In particular, we wish to analyze formulae of the form $S\phi$. Let us first consider the formula $\{0.5\}\phi$. For any interpretation $\mathcal{I}_\lambda$ we have $\mathcal{I}_\lambda(\{0.5\}\phi) = \mathfrak{t}$ iff $\mathcal{I}(\phi) = 0.5$. The latter can only be the case if some atom in $\phi$ is assigned 0.5 by $\mathcal{I}$ or if some fuzzy operator in $\phi$ is 0.5, it being so that either case is trivial, as 0.5 indicates complete uncertainty with respect to the truth value of $\phi$ (see above Def. 8.3.19), reason why we shall henceforth assume that $0.5 \notin S$. The reader should start to realize that we are going in the direction of *regular signs* (cf. Def. 2.5.2), which are specified in $L_\aleph$ as follows:

**8.3.27. (Def.)** A regular sign $S$ is an interval containing 0 or 1. If $S \subset (0.5, 1]$, then $S$ is called a *positive regular sign*; if $S \subset [0, 0.5)$, then $S$ is a *negative regular sign*.

In order to apply signed resolution to signed formulae in $L_\aleph$, we need to move all signs inward to the atomic level. To this end, Proposition 8.3.15 then generalizes to signs $S$ in the following way for formulae of the form $S(\phi \heartsuit \psi)$:

**8.3.28. (Prop.)** Let $S$ be a regular sign, and let $S(\phi \heartsuit \psi)$, $\heartsuit = \wedge, \vee$, be a signed formula in $L_{\aleph}$. Then,

1. if $S$ is positive, we have:
$$S(\phi \wedge \psi) \equiv_\lambda S\phi \wedge S\psi; S(\phi \vee \psi) \equiv_\lambda S\phi \vee S\psi$$

2. if $S$ is negative, we have:
$$S(\phi \wedge \psi) \equiv_\lambda S\phi \vee S\psi; S(\phi \vee \psi) \equiv_\lambda S\phi \wedge S\psi$$

*Proof:* Left as an exercise.

We now consider a formula of the form $(S)\lambda A$. Let $\lambda = 0.5 + \lambda'$. Then, $\lambda' < 0$ if $\lambda \in [0, 0.5)$ and $\lambda' > 0$ if $\lambda \in (0.5, 1]$. If for $\nu$ any truth value we have $\nu \in [0, 1]$, then manipulating $(\lambda' + 0.5) \otimes \nu$ gives $\lambda \otimes \nu = 0.5 - \lambda' + 2\lambda'\nu$. If $\lambda > 0.5$ (i.e. if $\lambda' > 0$), then $\Lambda_\lambda$ is monotonically increasing in $\nu$. Let now $S$ be the regular positive sign $[\tau, 1]$ and let $T = \Lambda_\lambda(\tau)$. Because $\Lambda_\lambda(0.5) = 0.5$, $T = 0.5$, we have it that $\Lambda_\lambda(S) = [T, \lambda]$ is a positive sign. Unless $\lambda = 1$, in general $\Lambda_\lambda(S)$ will not be regular. Nonetheless, we have

$$\Lambda_\lambda^{-1}([T, 1]) = \Lambda_\lambda^{-1}([T, \lambda] = [\tau, 1])$$

which is regular and positive.

**8.3.29. (Prop.)** Let $S$ be a regular sign and $\lambda$ a positive fuzzy operator. Then, $\Lambda_\lambda^{-1}(S)$ is regular and positive or negative as $S$ is positive or negative. If $\lambda$ is negative, then $\Lambda_\lambda^{-1}(S)$ is regular and positive or negative as $S$ is negative or positive.

**8.3.30. (Prop.)** Let $A$ be an atom in $L\lambda$ and let $S$ be a (positive or negative) regular sign. Then:

$$(S)\lambda A \equiv_\lambda \Lambda_\lambda^{-1}(S) A$$

$\Lambda_\lambda^{-1}(S)$ will be empty unless there is at least one $\tau$ such that $\tau < \lambda$ or $\lambda < 1 - \tau$ if $S$ is positive (the reverse is the case for negative $S$). When $S$ is empty, we have $\Lambda_\lambda^{-1}(S) A = \mathtt{f}$.

**Theorem 8.3.3.** *Let $\phi$ be a fuzzy formula, $S$ is a positive or negative regular sign, and let $\phi'$ be the signed formula that is produced by repeated applications of Propositions (i.e. Lemmas) 8.3.20 and 8.3.28 to $S\phi$ until all fuzzy operators other than 1 and all signs are at the literal level. Then,*

## 8. Signed resolution for the MV-SAT

$S\phi \equiv_\lambda \varphi$ where $\varphi$ is the formula obtained from $\phi'$ by replacing each literal $(S)\lambda A$ in $\phi'$ with $\Lambda_\lambda^{-1}(S)A$.

*Proof:* Straightforward. **QED**

**Example 8.3.4.** (Lu, Murray, & Rosenthal, 1994) Let $T = 0.7$, so that we have $S = [0.7, 1]$. Consider now the following signed atoms: $(S)\,0.8A$, $(S)\,0.25A$, and $(S)\,0.6A$. The corresponding L$\lambda$-atomic literals are $[0.33, 1]\,A$, $[0, 1]\,A$, and $\emptyset A$.

**8.3.31. (Prop.)** Given a threshold of acceptability $T$, let $S = [T, 1]$. Let further $\lambda A$, $\lambda_1 A$, and $\lambda_2 A$ be fuzzy literals with $\lambda_1 < \lambda_2$, and let also $S'A$, $S'_1 A$, and $S'_2 A$ be the L$\lambda$-atomic literals corresponding to $(S)\lambda A$, $(S)\lambda_1 A$, and $(S)\lambda_2 A$, respectively. Then, $1 - T < \lambda < T$ iff $S' = \emptyset$. Moreover, $\lambda_1 A$ and $\lambda_2 A$ are complementary iff $S'_1$ and $S'_2$ are non-empty and $S'_1 \cap S'_2 = \emptyset$.

*Proof:* Left as an exercise.

**Example 8.3.5.** In Example 8.3.4, note that $0.8A$ and $0.25A$ are $T$-complementary and the corresponding signed atoms have an empty intersection. The fuzzy literal $0.6A$ is $T$-unsatisfiable and the corresponding signed atom has an empty sign.

Extension to the FO case is straightforward, by augmenting L$\lambda$ with a signature $\Upsilon$ so as to obtain $\overset{*}{L\lambda}$.

**8.3.32. (Def.)** Let the language $\overset{*}{L\lambda}$ be given and let $\phi \in F_{\overset{*}{L\lambda}}$. Just as in the propositional case, $S\phi$ is an atom in $\overset{*}{L\lambda}$. Its *signed predicate* is $S\hat{\phi}$, where $\hat{\phi}$ is $\phi$ without arguments. That is, if $\phi$ is the atom $P(t_1, ..., t_n)$, then $\hat{\phi} = P$; if $\phi$ is the formula $\heartsuit(\phi_1, ..., \phi_n)$, then $\hat{\phi} = \heartsuit\left(\hat{\phi}_1, ..., \hat{\phi}_n\right)$. In $\overset{*}{L\lambda}$, the formulae $\widetilde{\forall} x\phi$ and $\widetilde{\exists} x\phi$, where $\blacklozenge$ denotes a quantifier of L*, are fuzzy formulae.

**Example 8.3.6.** Let $\phi = P(x, f(x, y)) \otimes Q(a)$ be a formula. Then, the signed formula $S\phi$ is the atom $S(P \otimes Q(x, f(x, y), a))$. The predicate $S(P \otimes Q)$ in $\overset{*}{L\lambda}$ denotes a function from a domain $\mathscr{D}^3$ to $W_2$, and $S\phi$ denotes a function from $\mathscr{D}^2$ to $W_2$.

The definitions of interpretation and valuation (Def.s 2.2.9-11) extend straightforwardly to $\overset{*}{L\lambda}$, with the additional *val'* defined to be a *x-variant* of *val* if *val* and *val'* assign the same value to all variables except possibly to $x$.

## 8.3. Signed resolution for infinitely many-valued logics

**8.3.33. (Prop.)** The quantifiers in $\overset{*}{\mathsf{L}\lambda}$ behave classically. Thus,

1. $\blacklozenge y\,(\phi\,[x/y]) = \blacklozenge x\phi$, where $y$ does not occur in $\phi$;
2. $\forall x\,(\phi \wedge \psi) = \forall x\phi \wedge \forall x\psi$;
3. $\exists x\,(\phi \vee \psi) = \exists x\phi \vee \exists x\psi$.

*Proof:* Left as an exercise.

**Theorem 8.3.4.** *Let $\forall x\phi$ be a fuzzy formula. Let further $\mathcal{I}$ be any interpretation and let $\mathcal{I}_\lambda$ be the correspondent L-consistent interpretation. Then, $\mathcal{I}_\lambda$ maps $\forall x\,[T,1]\,\phi$ to* true *iff $\mathcal{I}$ T-satisfies $\forall x\phi$, and $\mathcal{I}_\lambda$ maps $\exists x\,[T,1]\,\phi$ to* true *iff $\mathcal{I}$ T-satisfies $\exists x\phi$.*

Distribution quantifiers are defined in the standard way. Prenex normal forms and Skolemization are just as in the classical case. The Herbrand universe of a formula $\phi$ is defined in the standard way. Substitutions are carried out in the same way as for classical logic.

**Theorem 8.3.5.** *Let $\phi$ be a (ground) formula in $\mathsf{L}\lambda$ and let $S$ be any subset of $W$ such that no interpretation maps $\phi$ into $S$. Let further $\varphi$ be a $\mathsf{L}\lambda$-atomic formula such that $\varphi \equiv_\lambda S\phi$ and let $C_\varphi$ be a CNF equivalent of $\varphi$. Then $C_\varphi \vdash_{res} \square$.*

*Proof:* See Lu, Murray, & Rosenthal (1998). **QED**

Theorem 8.3.5 corresponds to a soundness theorem for signed resolution in $\overset{*}{\mathsf{L}\lambda}$. In terms of completeness of $\overset{*}{\mathsf{L}\lambda}$, note however the following passage:

> We are dealing with a much wider class of first-order many-valued logics and are allowing more general queries about the formulae in those many-valued logics. The resolution rule and the signed formulae on which it operates, however, are defined in a classical logic. In this setting, completeness is not a statement about what formulae in L* can be proved inconsistent; rather, completeness means that we can answer certain queries in L* by deriving the empty clause in $\overset{*}{\mathsf{L}\lambda}$. (Lu, Murray, & Rosenthal, 1998, p. 49; slightly modified)

**Theorem 8.3.6.** *A deduction by fuzzy resolution is a special case of a deduction by signed resolution.*

## 8. Signed resolution for the MV-SAT

**Example 8.3.7.** (Lu, Murray, & Rosenthal, 1994) Let $C_1 = 0.1A \vee 0.2A$, $C_2 = 0.8A$. Let $T = 0.7$. Then:

$$[T, 1]\, C_1 \equiv_\lambda ([0, 0.25]\, A) \vee ([0, 0.167]\, A)\,, [T, 1]\, C_2 \equiv_\lambda ([0.833, 1]\, A)$$

Applying signed resolution produces the empty clause.

# Exercises

**Exercise 8.1.** Given the set of truth values $W = \{0, 1, 2\}$, apply signed binary resolution to the following sets of clauses:

1. $C = \left\{\|\{0\}\,[P]\|, \left\|\overline{\{0\}}\,[P], \{1\}\,[Q]\right\|\right\}$

2. $C = \left\{\|\{0\}\,[P], \{2\}\,[R]\|, \|\{1\}\,[P]\|, \left\|\overline{\{2\}}\,[R]\right\|\right\}$

3. $C = \left\{\left\|\overline{\{2\}}\,[P], \{1\}\,[Q]\right\|, \|\{1\}\,[P], \{0\}\,[Q]\|\right\}$

**Exercise 8.2.** Given the set of truth values $W = \{a, b, c\}$, apply signed resolution to the following sets of clauses:

1. $C = \left\{\left\|\overline{\{a\}}\,[Q]\right\|, \left\|\overline{\{c\}}\,[Q]\right\|, \left\|\overline{\{b\}}\,[Q]\right\|\right\}$

2. $C = \{\|\{a\}\,[P], \{b\}\,[Q]\|, \|\{b\}\,[P], \{b\}\,[Q]\|, \|\{b\}\,[Q], \{c\}\,[P]\|\}$

3. $C = \left\{\left\|\overline{\{a\}}\,[P], \{b\}\,[Q]\right\|, \left\|\overline{\{b\}}\,[P], \overline{\{b\}}\,[Q]\right\|, \left\|\overline{\{c\}}\,[P], \{c\}\,[Q]\right\|\right\}$

**Exercise 8.3.** Given a truth-value set $W = \{1, 2, 3, 4\}$, apply the signed resolution calculus to the following sets of clauses:

1. $C = \{\|\!\uparrow 2\,[P], \downarrow 2\,[Q]\|, \|\!\downarrow 2\,[P], \downarrow 4\,[Q]\|\}$

2. $C = \{\|\!\uparrow 3\,[R]\|, \|\!\downarrow 2\,[R], \uparrow 1\,[T]\|, \|\!\downarrow 1\,[T], \downarrow 3\,[R]\|\}$

3. $C = \left\{\begin{array}{c}\|\{1,3\}\,[P], \{3,4\}\,[Q]\|, \|\{2,4\}\,[P], \{2\}\,Q, \{3\}\,[P]\|, \\ \|\{1\}\,[Q], W\,[P]\|\end{array}\right\}$

4. $C = \left\{\begin{array}{c}\|\{3\}\,P, \{1,2\}\,[Q]\|, \|\{2,3\}\,[P], \{1\}\,[Q], \{2,4\}\,[R]\|, \\ \|\{1,3\}\,[R], \{4\}\,[Q], \{1,4\}\,[P]\|, \|\{\}\,[R]\|\end{array}\right\}$

**Exercise 8.4.** Prove by means of signed resolution the following formulae of Łukasiewicz's 3-valued logic $\mathrm{L}_3^{(*)}$:

1. $\neg P \to (P \to Q)$ (Example 5.4.1)

2. $\neg(P \to Q) \to P$ (Example 5.4.2.7)

3. $\forall x P(x) \to \exists x P(x)$ (Example 5.7.5)

## 8. Signed resolution for the MV-SAT

**Exercise 8.5.** Provide informal accounts for the following additional signed resolution rules in Beckert, Hähnle, & Manyà (2000):

1. *Regular resolution* – If $(max_{1 \leq k \leq m} i_k) > j$:

$$\frac{\uparrow i_1 [P] \vee C_1 \quad \cdots \quad \uparrow i_m [P] \vee C_m \quad \downarrow j [P] \vee C}{C_1 \vee \ldots \vee C_m \vee C}$$

2. *Regular negative hyper-resolution* – Provided that $m \geq 1$, $i_l < j_l$ for all $1 \leq l \leq m$, and $C_1, \ldots, C_m, \mathcal{D}$ contain only negative literals:

$$\frac{\downarrow i_1 [P_1] \vee C_1 \quad \cdots \quad \downarrow i_m [P_m] \vee C_m \quad \uparrow j_1 [P_1] \vee \ldots \vee \uparrow j_m [P_m] \vee \mathcal{D}}{C_1 \vee \ldots \vee C_m \vee \mathcal{D}}$$

3. In case $W$ forms an upper semi-lattice:

   a) *Lattice-regular positive unit resolution:* If $i \geq j$

$$\frac{\uparrow i [P] \quad \overline{\uparrow j} [P] \vee C}{C}$$

   b) *Lattice-regular reduction:* If neither $i \geq j$ nor $j \geq i$

$$\frac{\uparrow i [P] \quad \uparrow j [P]}{\uparrow (i \cup j) [P]}$$

**Exercise 8.6.** Prove the completeness of signed macro-resolution (R6) and $<_A$-ordering macro-resolution (R7).

**Exercise 8.7.** Define $T$-consequence for $L_\aleph$.

**Exercise 8.8.** Let the language $L\overset{*}{\lambda}$ be given.

1. Extend Definition 8.3.12 to formulae of $L\overset{\wedge}{\lambda}$.

2. Extend the notion of $L\lambda$-consistency (cf. Def. 8.3.14) to formulae of $L\overset{*}{\lambda}$, specifying the cases for formulae of the form $\forall x S \phi$ and $\exists x S \phi$.

3. Extend the notion of confidence (Def. 8.3.19) to the formulae $\widetilde{\forall} x \phi$ and $\widetilde{\exists} x \phi$.

4. Extend Proposition 8.3.20 to the formulae $\widetilde{\forall} x \phi$ and $\widetilde{\exists} x \phi$.

5. Extend Proposition 8.3.28 to the formulae $\widetilde{\forall} x \phi$ and $\widetilde{\exists} x \phi$.

6. Extend Herbrand's theorem to $L\overset{*}{\lambda}$. (Hint: L-consistent interpretations must feature in it.)

7. Define a semantic tree for signed fuzzy formulae.

**Exercise 8.9.** Resolve the following clauses of the fuzzy operator logic $L_\aleph$ by applying signed resolution:

1. $C_1 = 0.2A \vee 0.9B$, $C_2 = 0.98A \vee 0.87A$; $T = 0.75$
2. $C_1 = 0.11A \vee 0.43A \vee 0.19A$, $C_2 = 0.63B \vee 0.66A$; $T = 0.65$

**Exercise 8.10.** Prove (Complete the proofs of) the propositions in this Chapter that were left without a proof (with sketchy proofs, respectively).

# Part IV.
# APPENDIX

# 9. Mathematical notions

## 9.1. Sets

Set theory is the foundational discipline for mathematics and hence for mathematical logic. The central notion of set theory is that of a set. In this Section we discuss sets as they are typically–or "classically"–conceived in these disciplines. Below, in Section 9.2.1, we approach a generalized notion of set, to wit, fuzzy set, that is relevant for infinitely many-valued logics, namely for the fuzzy logics.

**9.1.1. (Def.)** A *set* is a well-defined collection of objects; these are its *members* or *elements*.

1. The elements of a set can be specified *intensionally*, i.e. in terms of one or more properties that all its elements share, or *extensionally*, i.e. by naming all its elements. For example:

$$A = \{x | x \text{ is an even positive integer}, x \leq 10\}$$

$$B = \{2, 4, 6, 8, 10\}$$

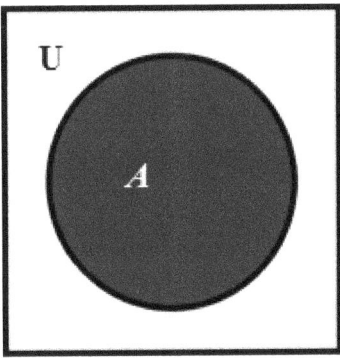

Figure 9.1.1.: Venn diagram of the set $A$.

## 9. Mathematical notions

and $A = B$, as two sets are equal if they have exactly the same elements. (Order and repetition of the elements of a set are irrelevant.) This specification entails a *universe*, denoted by **U**, the collection of all the (classes of) objects one wishes to consider. For instance, the universe of the set $A$ can be taken as the set of all the numbers. This relation between the set $A$ and its universe **U** is intuitively represented in a Venn diagram as in Figure 9.1.1: The area within the rectangle represents the universe and the circle represents $A$.[1]

2. If $x$ is an element of a set $A$, we write $x \in A$; otherwise, we write $x \notin A$. The *complement of* $A$, denoted by $\overline{A}$, is the set $\{x | x \notin A\}$. In Figure 9.1.1, $\overline{A}$ is the portion of the rectangle outside the region representing $A$.

3. A set with one single element is a *singleton*. The *empty* set, i.e. the set with no elements, is denoted by $\emptyset$.

4. A set is *finite* if it is either empty or else it can be put in a one-to-one correspondence with the set $\{1, 2, ..., n\}$, for $n \in \mathbb{Z}^+$, for $\mathbb{Z}^+$ the set of positive integers. A set is *infinite* if it is not finite. A set is *denumerable*, or *countably infinite*, if it can be put in a one-to-one correspondence with the set $\mathbb{N}$ of natural numbers $\{0, 1, 2, ...\}$. We say that a set is *countable* if it is either finite or denumerable; otherwise, it is *uncountable*.

5. The *cardinality* of the set $A$, designated by $|A|$, is the number of elements of $A$.

6. A set $B$ is said to be a *subset* of a set $A$ if $B \subseteq A$, where $\subseteq$ is a *containment* relation (read "the set $B$ is contained or equal to the set $A$"). In turn, a set $A$ is called a *superset* of $B$ if $A \supseteq B$ (read "the set $A$ contains or is equal to the set $B$"). Given a set $C$, the *power set* of $C$, denoted by $2^C$, is the set of all the subsets of $C$. Let $C = \{x, y, z\}$; then

$$2^C = \{\emptyset, \{x\}, \{y\}, \{z\}, \{x, y\}, \{x, z\}, \{y, z\}, \{x, y, z\}\}.$$

---

[1] Sometimes, it is relevant to reduce the universe to a more restricted set that nevertheless is still large enough to allow for generalization over some set $A$; this restriction of the universe is called the *domain (of discourse)*, and we denote it by $\mathscr{D}$. In our example, we can specify the domain as $\mathscr{D}_A = \mathbb{Z}^+$ (or $\mathbb{N}^+$, if zero is not considered a natural number). We remark that it is often the case that *universe* and *domain* are taken as synonyms. Furthermore, the specific logical concept of *domain (of discourse)* is tied up with quantification (cf. Def.s 2.2.9-11 and in particular Def. 2.2.12).

9.1. Sets

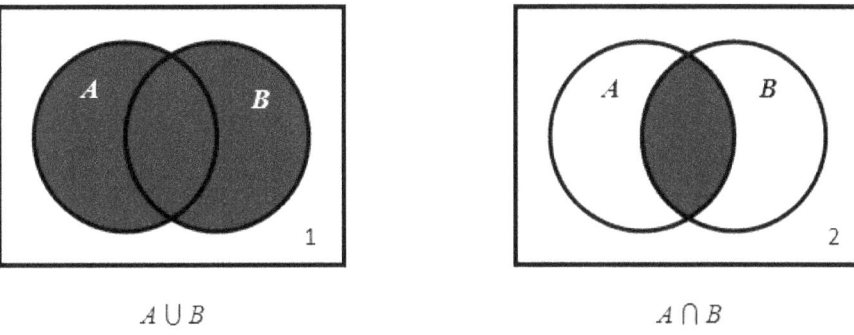

$A \cup B$ $\qquad\qquad\qquad$ $A \cap B$

Figure 9.1.2.: Venn diagrams of the union (1) and the intersection (2) of two sets.

7. For two sets $A$ and $B$, the *union* of $A$ and $B$ is the set $A \cup B = \{x | x \in A \text{ or } x \in B\}$ (cf. Fig. 1.1.2.1). Now let $\{A_i | i \in I\}$ be the family of sets $A_i$ indexed by the set $I$; then the union of the sets $A_i$, written

$$\bigcup_{i \in I} A_i = \{x | \exists i \in I, x \in A_i\}$$

is the set of objects that are members of at least one set $A_i$ in the family.

8. Let $A$ and $B$ be sets. The *intersection* of $A$ and $B$ is the set $A \cap B = \{x | x \in A \text{ and } x \in B\}$ (cf. Fig. 1.1.2.2). Now let $\{A_i | i \in I\}$ be the family of sets $A_i$ indexed by the set $I$; then the intersection of the sets $A_i$, written

$$\bigcap_{i \in I} A_i = \{x | \forall i \in I, x \in A_i\}$$

is the set of objects that are members of every set $A_i$. $A$ and $B$ are *disjoint* sets if $A \cap B = \emptyset$ and a collection of sets $\{A_i | i \in I\}$ is disjoint if $\bigcap_{i \in I} A_i = \emptyset$.

9. Given two sets $A$ and $B$, the *set difference* is the set $A - B = A \backslash B = A \cap \overline{B} = \{x | x \in A \text{ and } x \notin B\}$. $A - B$ and $A \backslash B$ are alternative ways to write the same set operation.

## 9.2. Functions, operations, and relations

**9.2.1. (Def.)** A *function* (or *mapping*) $f$ defined on two sets $A$ (the *domain* of $f$) and $B$ (the *range* of $f$), denoted by $f : A \longrightarrow B$, is a subset $C \subseteq A \times B$ such that for each $x \in A$ there is one and only one pair in $C$ of the form $(x, y)$, $y \in B$, such that $f(x) = y$.

1. Let $f : A \longrightarrow B$ be a function, $D \subseteq A$ and $E \subseteq B$. The *image of $D$ under $f$* is the subset $f(D) = \{f(x) \,|\, x \in D\}$. The *inverse image of $E$ under $f$* is the subset $f^{-1}(E) = \{x \,|\, f(x) \in E\}$.

2. For $n \geq 1$, we say that $f$ is an *$n$-ary function* over a set $A$ if $f : A^n \longrightarrow A$, i.e. if $f(x_1, ..., x_n) \in A$ whenever $x_1, ..., x_n \in A$. If $n = 0$, we may speak of a *degenerate* function. A 0-ary function over $A$ denotes a fixed element of $A$, i.e. a *constant*.

3. We say that a function $f : A \longrightarrow B$ is *one-to-one*, or *injective*, when for all $x, y \in A$ if $x \neq y$, then $f(x) \neq f(y)$. An *injection* is also called a *monomorphism* (cf. Def. 9.3.2.4).

4. A function $f : A \longrightarrow B$ is said to be *onto*, or *surjective*, when for every $y \in B$ there is a $x \in A$ such that $f(x) = y$. A *surjection* is also called an *epimorphism* (cf. Def. 9.3.2.4).

5. A function is said to be *bijective*, or a *bijection*, if it is both one-to-one and onto. A bijection is also called a *one-to-one correspondence*, or an *isomorphism* (cf. Def. 9.3.2.4).

6. Let $f : A \longrightarrow B$ and $g : B \longrightarrow C$. Then, the *composition of $f$ and $g$* is the function $g \circ f : A \longrightarrow C$ defined as $(g \circ f)(x) = g(f(x))$ for all $x \in A$.

**9.2.2. (Def.)** Let $A$ be a set. An *$n$-ary operation* $O$ on $A$, for $n \geq 1$, is a function $f$ that assigns an element of $A$ to every $n$-tuple of elements of $A$, i.e. $f : A^n \longrightarrow A$. When $n = 0$, we may speak of a *degenerate* operation. In a logical language, a 0-ary operation is a *constant*.

**9.2.3. (Def.)** Let $A$ be a set. For $n \geq 1$, $R$ is a *$n$-ary relation* over $A$ if $R \subseteq A^n$, i.e. $R$ is a subset of $\underbrace{A \times ... \times A}_{n}$. Thus, a *unary* relation $R$ on $A$ is a relation $R \subseteq A$, and a *binary* relation $R$ on $A$ is a relation $R \subseteq A \times A$. Let $A$ and $B$ be sets. A relation $R$ from $A$ to $B$ is a binary relation from $A$ to $B$, i.e. $R \subseteq A \times B$. When $n = 0$, we may speak of a *degenerate* relation.

## 9.2. Functions, operations, and relations

**9.2.4. (Def.)** A binary relation $R$ over a set $A$ is a relation $R \subseteq A \times A$. We say that a binary relation is

1. *reflexive* if for every $x \in A$ we have it that $xRx$.

2. *irreflexive* if for every $x \in A$ it is not the case that $xRx$.

3. *symmetric* if for every $x, y \in A$ we have it that if $xRy$, then $yRx$.

4. *antisymmetric* if for every $x, y \in A$ we have it that if $xRy$ and $yRx$, then $x = y$.

5. *transitive* if for every $x, y, z \in A$ we have it that if $xRy$ and $yRz$, then $xRz$.

6. an *equivalence relation* if it is reflexive, symmetric, and transitive.

**9.2.5. (Def.)** A binary relation $\leq$ on a set $A$ that is reflexive and transitive is called a *preorder*, and $\mathcal{R} = (A, \leq)$ is correspondingly called a preorder.

- $\mathcal{R}$ is a *partial order*, or a *poset*, if $\leq$ is reflexive, antisymmetric, and transitive.

- If $\leq$ is irreflexive and transitive, then $\mathcal{R}$ is a *strict partial order* (often denoted by $<$ or $\lneq$). Let $\mathcal{R}$ be a strict partial order; then for $x, y, z \in A$ we say that $y$ *covers* $x$ in $A$ if $x < y$ and for $x \leq z \leq y$ it is the case that $z = x$ or $z = y$.

- A poset $\mathcal{R}$ is *totally ordered* if every $x, y \in A$ are comparable, i.e. we have $x \leq y$ or $y \leq x$. We then speak of a *total* or *linear order*.

**Example 9.2.1.** Let the set $A$ be given whose members and partial order are those of Figure 9.2.1. Clearly, the pairs ($\square$, ♠) and ($\triangle$, @) are incomparable. Comparable are, for instance, $\square < $ ♦ and $\triangle > $ ♠.

**9.2.6. (Def.)** Given a poset $\mathcal{R} = (A, \leq)$ and a subset $B \subseteq A$, we have the following order relations for $x, y \in A$ and $z \in B$:

1. $x$ is a *maximal element* of $A$, written $max\,(A) = x$, if there is no $y \in A$ such that $x \leq y$.

## 9. Mathematical notions

Figure 9.2.1.: A partially ordered set.

2. $x$ is a *minimal element* of $A$, written $min(A) = x$, if there is no $y \in A$ such that $y \leq x$.

3. $x$ is the *least element* of $A$ if for every $y \in A$ we have $x \leq y$.

4. $x$ is the *greatest element* of $A$ if for every $y \in A$ we have $y \leq x$.

5. $y$ is an *upper bound* of $B$ if for every $z \in B$ we have $z \leq y$.

6. $x$ is a *lower bound* of $B$ if for every $z \in B$ we have $x \leq z$.

7. $y$ is a *least upper bound (lub)*, or *supremum*, of $B$, written $lub(B) = y$ or $sup(B) = y$, if $y$ is the upper bound of $B$ and $y \leq y'$ for any other upper bound $y'$ of $B$.

8. $x$ is a *greatest lower bound (glb)*, or *infimum*, of $B$, written $glb(B) = x$ or $inf(B) = x$, if $x$ is the lower bound of $B$ and $x' \leq x$ for any other lower bound $x'$ of $B$.

**9.2.7. (Def.)** Let $R$ be a binary relation on a set $A$. The *transitive closure* of $R$ is the connectivity relation $R'$ associated with $R$ such that $R' = \bigcup_{i=1}^{\infty} R^i$ of all the positive powers of $R$. In other words, for a relation $R$, the transitive closure of $R$ is the smallest transitive relation containing $R$.

**9.2.8. (Def.)** Let $\mathcal{R} = (A, \leq)$ be a poset.

## 9.2. Functions, operations, and relations

1. The *interval* between $x, y \in A$ is defined as

$$[x, y] = \{z \in A | x \leq z \leq y\}$$
$$(x, y) = \{z \in A | x < z < y\}$$
$$[x, y) = \{z \in A | x \leq z < y\}$$
$$(x, y] = \{z \in A | x < z \leq y\}$$

2. An *ideal*, or *downset*, in $\mathcal{R}$ is a subposet $\mathcal{S}$ such that if $x \in \mathcal{S}$ and $y < x$, then $y \in \mathcal{S}$. The *ideal generated by an element* $x$ in a poset $\mathcal{R}$ is the downset $\downarrow \{x\} = \{y \in \mathcal{R} | y \leq x\}$. The *ideal generated by a subset* $A$ in $\mathcal{R}$ is the downset $\mathcal{D}_A = \bigcup_{x \in A} \downarrow \{x\}$.

3. A *filter*, or *upset*, in $\mathcal{R}$ is a subposet $\mathcal{S}$ such that if $x \in \mathcal{S}$ and $y > x$, then $y \in \mathcal{S}$. The *filter generated by an element* $x$ in a poset $\mathcal{R}$ is the upset $\uparrow \{x\} = \{y \in \mathcal{R} | y \geq x\}$. The *filter generated by a subset* $A$ in $\mathcal{R}$ is the upset $\mathcal{U}_A = \bigcap_{x \in A} \uparrow \{x\}$.

### 9.2.1. Functions, ordered intervals, and sets: Crisp sets and fuzzy sets

Taken in a broader set-theoretical context, Definition 9.1.1.2 defines a *crisp set*, i.e. a set membership to which is a "Yes/No" matter. Given the example of Definition 9.1.1.1, for $x \in \mathbb{Z}^+$ we have, for instance, $2 \in A$ and $8 \in A$, but $3 \notin A$ and $12 \notin A$. This allows for a clear intensional definition of the complement of $A$ as

$$\overline{A} = \{x | x \text{ is a positive odd integer}, x \leq 9\} \cup \{x | x > 11\}$$

corresponding to the extensional specification

$$\overline{A} = \{1, 3, 5, 7, 9\} \cup \{11, 12, 13, ...\} = \{1, 3, 5, 7, 9, 11, 12, ...\}.$$

Formally, this "crispness" is expressed by a mapping of the elements of some universe $\mathbf{U}$ to the set $\{0, 1\}$ such that every element of $\mathbf{U}$ either belongs to some set $A$, in which case the output is 1, or it does not belong to $A$, and the output is 0. We formalize this as follows:

**9.2.9. (Def.)** A *crisp set* $A$ is defined by a function $\chi : \mathbf{U} \longrightarrow \{0, 1\}$ that declares which elements of $\mathbf{U}$ belong or do not belong to $A$.[2] This

---
[2] We emphasize that the term "crisp set" is only used in the context of fuzzy set theory or related contexts, such as rough set theory.

# 9. Mathematical notions

function is called the *characteristic function*. Let $x$ be an arbitrary element of $\mathbf{U}$; then, the characteristic function of $A$, denoted by $\chi_A$, is the function

$$\chi_A(x) = \begin{cases} 1 & \text{if } x \in A \\ 0 & \text{otherwise} \end{cases}.$$

However, this can be generalized in the sense that membership to a set is a matter of degree, a generalization that appears essential when the elements in consideration are vaguely characterizable, such as "old" (How old is "45 years old"?; Is a 90-year-old person just *old, rather old,* or *very old?*), "tall," "bald," "hot," "nice," etc. This vagueness rests typically on natural-language usage, but scientific concepts are also often vague (e.g., Is a virus a living organism, and if so, how living is it?), which sanctions this generalization to a scientific context. Put briefly, this generalization is called *fuzzy set theory*, and its central concept is that of *fuzzy set*, where *fuzziness* is formalizable by means of a membership function mapping elements of some universe $\mathbf{U}$–always a crisp set–to the interval $[0,1]$, which is uncountably infinite.

**9.2.10. (Def.)** Let a *membership function* be defined as

$$\mu : \mathbf{U} \longrightarrow [0,1]$$

for a universe $\mathbf{U}$. Then, the set $A$ whose elements are determined by the membership function

$$\mu_A : \mathbf{U} \longrightarrow [0,1]$$

is said to be a *fuzzy set*.

**Example 9.2.2.** Let $\mathbf{U} = \{0, 1, 2, ..., 90\}$ and let the elements of $\mathbf{U}$ (the integers in the interval $[0, 90]$) denote years of life. With respect to $\mathbf{U}$, we consider the three sets $Y, M, O$ of respectively young, middle-aged, and old persons. A 1-year-old is clearly young, i.e.

$$\mu_Y(1) = 1$$

and

$$\mu_O(1) = 0.$$

However, whether a 30-year-old is a young person depends on cultural and even personal perceptions. For instance,

$$\mu_Y(30) = 0.70,$$

$$\mu_M(30) = 0.50,$$

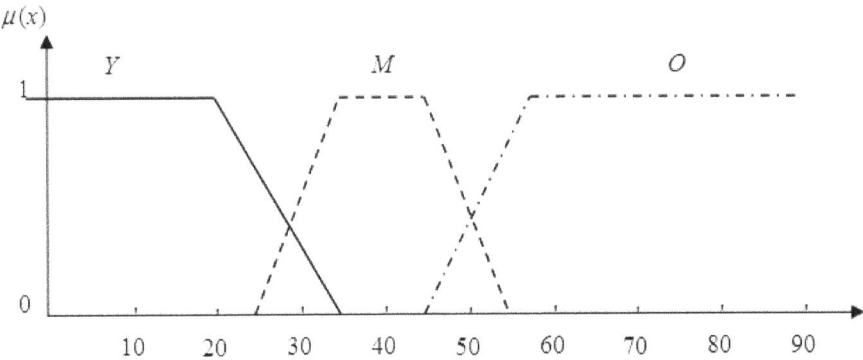

Figure 9.2.2.: Membership functions for young, middle, and old age in humans.

and
$$\mu_O(30) = 0,05.$$

Finally, we may safely specify
$$\mu_Y(90) = 0$$

and
$$\mu_O(90) = 1.$$

Figure 9.2.2 shows a likely graph of the membership functions for the elements of **U** with respect to the sets $Y$, $M$, and $O$.

## 9.3. Algebras and algebraic structures

**9.3.1. (Def.)** An *algebraic structure* (or *system*) is a triple $\mathfrak{Z} = (A, O, R)$ consisting of a non-empty set $A$, a set $O$ of operations on $A$, and a set $R$ of relations on $A$. If $O = \emptyset$, $\mathfrak{Z}$ is called a *model*, or *relational structure*. If $R = \emptyset$, $\mathfrak{Z}$ is called an *algebra*.

**Example 9.3.1.** For $\mathbb{Z}$ the set of integers, the structure $(\mathbb{Z}, +)$ is an algebra, the structure $(\mathbb{Z}, \leq)$ is a model.

**9.3.2. (Def.)** An *(abstract) algebra* is a pair
$$\mathfrak{A} = (U, O) = (U, \{o_i | i \in I\})$$

## 9. Mathematical notions

where $U$ is a non-empty set, called the *universe* or *carrier* of $\mathfrak{A}$, and for each $i \in I$, $o_i$ is a *basic operation* on $U$. If $U$ is a singleton, then $\mathfrak{A}$ is said to be *degenerate*. $\mathfrak{A}$ is a *finite* algebra if $I$, the *index set* of $\mathfrak{A}$, is finite; otherwise, it is *infinite*. The *type* of $\mathfrak{A}$ is a function $t : I \longrightarrow \mathbb{N}$ where $t(i)$ equals the arity of the operation $o_i$.

Let $\mathfrak{A} = (U, \{o_i | i \in I\})$ be an algebra.

1. An algebra $\mathfrak{B} = (Z, \{o_i | i \in I\})$ is a *subalgebra* of $\mathfrak{A}$ iff $Z \subseteq U$ and, for $i \in I$, $Z$ is closed with respect to each $o_i$.

2. An algebra $\mathfrak{C} = (U, \{o_i | i \in K\})$ is a *reduct* of $\mathfrak{A}$ iff $K \subsetneq I$.

3. An algebra $\mathfrak{D} = (U, \{p_k | k \in K\})$ is *similar* to the algebra $\mathfrak{A}$ iff $I = K$ and $o_i^m = p_i^m$ for each $i \in I$ and $m$ the arity of $o_i, p_i$. In other words, two algebras are similar iff they have the same type.

4. A *homomorphism* of $\mathfrak{A}$ into a similar algebra $\mathfrak{D} = (Y, \{p_i | i \in I\})$, denoted by $Hom\,(\mathfrak{A}, \mathfrak{D})$, is a mapping

$$h : U \longrightarrow Y$$

such that for each $i \in I$ and for arbitrary $u_1, ..., u_{m(i)} \in U$ we have

$$h\left(o_i\left(u_1, ..., u_{m(i)}\right)\right) = p_i\left(h\left(u_1\right), ..., h\left(u_{m(i)}\right)\right).$$

A one-to-one homomorphism is a *monomorphism* or *embedding* of $\mathfrak{A}$ into $\mathfrak{D}$. A homomorphism of $\mathfrak{A}$ into itself is an *endomorphism*. If $h : \mathfrak{A} \longrightarrow \mathfrak{D}$ is a homomorphism such that $h(U) = Y$, then $h$ is an *epimorphism*. An epimorphism that is also a monomorphism is an *isomorphism* of $\mathfrak{A}$ and $\mathfrak{D}$.

5. If $Z \subseteq U$ is non-empty, then there exists the least subalgebra $\mathfrak{E} = (X, \{o_i | i \in I\})$ of $\mathfrak{A}$ such that $Z \subseteq X$, i.e. the intersection of all subalgebras of $\mathfrak{A}$ with universes that contain $Z$.

6. $Z \subseteq U$, $Z \neq \emptyset$, is a set of *generators* of $\mathfrak{A}$ if the least subalgebra of $\mathfrak{A}$ containing $Z$ in its universe is $\mathfrak{A}$ itself.

7. $\mathfrak{A}$ is an *absolutely free* algebra if there is a set $Z \subseteq U$ of generators of $\mathfrak{A}$ such that every mapping $s : Z \longrightarrow Y$ of a similar algebra $\mathfrak{D} = (Y, \{p_k | k \in K\})$ can be extended to a homomorphism from $\mathfrak{A}$ into $\mathfrak{D}$. Then, the elements of $Z$ are said to be *free generators* of $\mathfrak{A}$ and $\mathfrak{A}$ is said to be *freely generated* by $Z$.

8. Let $\mathscr{C}$ be a class of similar algebras. The algebra $\mathfrak{A} \in \mathscr{C}$ is said to be a *$\mathscr{C}$-free algebra* iff it has a set of generators $G$ such that any mapping $f : G \longrightarrow Z$, for $Z$ the universe of an arbitrary algebra $\mathfrak{B} \in \mathscr{C}$, can be extended to a homomorphism $h : \mathfrak{A} \longrightarrow \mathfrak{B}$. If $\mathfrak{A}$ is $\mathscr{C}$-free, then $G$ is said to be a set of *$\mathscr{C}$-free generators* for the algebra $\mathfrak{A}$.

**9.3.3. (Def.)** A *Boolean algebra* is an algebraic structure $\mathfrak{B} = (A, ', \wedge, \vee, 0, 1)$ of type $(1, 2, 2, 0, 0)$ where the following properties are satisfied for all $x, y, z \in A$:

$$
\begin{aligned}
x \wedge y &= y \wedge x & &\text{Commutativity of } \wedge \\
x \vee y &= y \vee x & &\text{Commutativity of } \vee \\
x \wedge (y \wedge z) &= (x \wedge y) \wedge z & &\text{Associativity of } \wedge \\
x \vee (y \vee z) &= (x \vee y) \vee z & &\text{Associativity of } \vee \\
x \wedge (y \vee z) &= (x \wedge y) \vee (x \wedge z) & &\text{Distributivity of } \wedge \\
x \vee (y \wedge z) &= (x \vee y) \wedge (x \vee z) & &\text{Distributivity of } \vee \\
x \wedge x' &= 0 & &\text{Complementation} \\
x \vee x' &= 1 & &\text{Complementation} \\
x \wedge 1 &= x & &\text{Identity element of } \wedge \\
x \vee 0 &= x & &\text{Identity element of } \vee
\end{aligned}
$$

1. From the properties of distributivity, complementation, and identity it follows that in a Boolean algebra we have

$$x = x \wedge y \text{ iff } x \vee y = y$$

2. Then, the relation $x \leq y$ is a partial order with least element 0 and greatest element 1, and with respect to $\leq$ we have for the operations of *meet* ($\wedge$) and *join* ($\vee$) respectively

$$x \wedge y = \inf(x, y)$$

$$x \vee y = \sup(x, y)$$

**9.3.4. (Prop.)** The power set of any non-empty set forms a Boolean algebra with the two operations meet and join defined as *intersection* ($\cap$) and *union* ($\cup$), respectively.

## 9. Mathematical notions

**9.3.5. (Prop.)** Let $\mathfrak{B} = (A, ', \wedge, \vee, 0, 1)$ be a finite Boolean algebra. Then, every element $x \in A$ can uniquely be represented in two forms, to wit,

$$x = \bigvee_{\substack{a \in A \\ a \leq x}} \left( a \wedge \bigwedge_{b \in (A - \{a\})} b' \right)$$

$$x = \bigwedge_{\substack{a \in A \\ a \leq x'}} \left( a' \vee \bigvee_{b \in (A - \{a\})} b \right)$$

called the *perfect disjunctive normal form* and the *perfect conjunctive normal form*, respectively.

**9.3.6. (Def.)** *Semigroups and monoids* – An algebraic structure $\mathfrak{X} = (X, \star)$ such that $X$ is a non-empty set and $\star$ is an associative binary operation is a semigroup. If $X$ is a poset, then $\mathfrak{X}$ is called a partially-ordered semigroup. A (partially-ordered) semigroup with an identity element (usually: 1) is a monoid.

**9.3.7. (Def.)** A *BL-algebra* is a structure

$$\mathfrak{BL} = (A, \wedge, \vee, \star, \rightarrow, 0, 1)$$

where $\wedge, \vee, \star, \rightarrow$ are binary operations and 0 and 1 are constants, i.e. an algebra of type $(2, 2, 2, 2, 0, 0)$ such that[3]

(BL1) $(A, \wedge, \vee, 0, 1)$ is a bounded lattice
(BL2) $(A, \star, 1)$ is a commutative monoid
(BL3) $c \leq a \rightarrow b$ iff $a \star c \leq b$ for all $a, b, c \in A$
(BL4) $a \wedge b = a \star (a \rightarrow b)$
(BL5) $(a \rightarrow b) \vee (b \rightarrow a) = 1$

1. A BL-algebra is called an *MV-algebra* (i.e. many-valued algebra) if, for all $x \in A$ and where $x' = x \rightarrow 0$,

$$x'' = x.$$

---
[3] Cf. Definition 9.4.6 for a bounded lattice.

2. A BL-algebra is a *G-algebra* (i.e. a Gödel algebra) if for all $x \in A$,
$$x^2 = x \star x = x.$$

3. A BL-algebra is a $\Pi$-*algebra* (i.e. a product algebra) if for all $x, y, z \in A$,
$$x'' = (((x \star y) \to (z \star y)) \to (x \to y))$$
and
$$x \wedge x' = 0.$$

## 9.4. Lattices

**9.4.1. (Def.)** A poset $\mathcal{R} = (A, \leq)$ in which every two elements $x, y \in A$ have a $lub(x, y) := x \vee y = y$, called the *join* of $x$ and $y$, or a $glb(x, y) := x \wedge y = x$, called the *meet* of $x$ and $y$, is called a *semi-lattice*. If $\mathcal{R}$ has only a meet (a join), then it is a *lower* (respectively, *upper*) *semi-lattice* or a *meet semi-lattice* (a *join semi-lattice*, respectively). A poset $\mathcal{R}$ that is both an upper and a lower semi-lattice is a *lattice*; in turn, every lattice is a poset $\mathcal{R}$. A lattice $\mathcal{L}' = ((A', \leq), \vee, \wedge)$ is a *sublattice* of the lattice $\mathcal{L} = (\mathcal{R}, \vee, \wedge)$ if $A' \subseteq A$ and $\mathcal{L}'$ has the same operations as $\mathcal{L}$. A lattice is *complete* if every subset of $A$ has both a join (a supremum, or $\bigvee A$) and a meet (an infimum, or $\bigwedge A$).

**Theorem 9.4.1.** *Any non-empty finite lattice $\mathcal{L} = (\mathcal{R}, \vee, \wedge)$ is complete.*

**Theorem 9.4.2.** *The lattices of all transitive relations and of all equivalence relations on a set are complete lattices.*

**9.4.2. (Prop.)** Let for any $x, y \in A$, $lub(x, y) := x \cup y$ and $glb(x, y) := x \cap y$, called the *union* and *intersection* of $x$ and $y$, respectively. Let further the pair $\mathcal{S} = (2^A, \subseteq)$ be given, where $2^A$ denotes the power set of $A$, and $\subseteq$ denotes *set inclusion*. Then the structure $(\mathcal{S}, \cup, \cap)$ is a *complete (distributive) lattice*.

**9.4.3. (Prop.)** The lattice $\mathcal{L} = (\mathcal{S}, \cup, \cap)$ satisfies the following properties for every $x, y, z \in A$:

1. *Commutativity*: $x \cup y = y \cup x$, $x \cap y = y \cap x$.

2. *Associativity:* $(x \cup y) \cup z = x \cup (y \cup z)$, $(x \cap y) \cap z = x \cap (y \cap z)$.

3. *Idempotency:* $x \cup x = x$, $x \cap x = x$.

4. *Absorption:* $x \cup (x \cap y) = x$, $x \cap (x \cup y) = x$.

**Example 9.4.1.** Let $A = \{a, b, c\}$. Then,
$$2^A = \{\emptyset, \{a\}, \{b\}, \{c\}, \{a,b\}, \{a,c\}, \{b,c\}, \{a,b,c\}\}.$$

We show the join and meet tables of $(\mathcal{S}, \cup, \cap)$ (Fig.s 9.4.1-2). Figure 9.4.3 shows the Hasse diagram (also: poset diagram or lattice diagram) of the same structure. Obviously, $\mathcal{L} = (\mathcal{S}, \cup, \cap)$ is a complete lattice with $\bigwedge A = \emptyset$ and $\bigvee A = \{a, b, c\}$.

**9.4.4. (Def.)** A lattice $\mathcal{L} = (\mathcal{R}, \vee, \wedge)$ is *distributive* iff the following are true for all $x, y, z \in \mathcal{R}$:

1. $x \wedge (y \vee z) = (x \wedge y) \vee (x \wedge z)$

2. $x \vee (y \wedge z) = (x \vee y) \wedge (x \vee z)$

**9.4.5. (Prop.)** If a lattice is non-distributive, it must contain a sublattice isomorphic to one of the lattices in Figure 9.4.4.

**9.4.6. (Def.)** A lower bound in a lattice $\mathcal{L}$ is an element $0 \in \mathcal{R}$ such that $0 \wedge x = 0$ (equivalently, $0 \leq x$) for all $x \in \mathcal{R}$. An upper bound in $\mathcal{L}$ is an element $1 \in \mathcal{R}$ such that $1 \vee x = 1$ (equivalently, $x \leq 1$) for all $x \in \mathcal{R}$. A lattice $\mathcal{L}$ is *bounded* if it contains a lower bound and an upper bound. An element $x$ in a bounded lattice $\mathcal{L}$ such that $0 < x$ and there is no $y \in \mathcal{L}$ such that $0 < y < x$ is called an *atom* of $\mathcal{L}$.

**9.4.7. (Def.)** A lattice $\mathcal{L}$ is *complemented* if

1. it is bounded;

2. for each $x \in \mathcal{R}$ there is a $y \in \mathcal{R}$, called the complement of $x$, such that $x \vee y = 1$ and $x \wedge y = 0$.

**9.4.8. (Prop.)** Every finite lattice is bounded. If $\mathcal{R} = \{x_1, ..., x_n\}$, then we have the following equalities:

$$1 = x_1 \vee ... \vee x_n = \bigvee_{i=1}^{n} x_i$$

| ∪ | ∅ | {a} | {b} | {c} | {a,b} | {a,c} | {b,c} | {a,b,c} |
|---|---|---|---|---|---|---|---|---|
| ∅ | ∅ | {a} | {b} | {c} | {a,b} | {a,c} | {b,c} | {a,b,c} |
| {a} | {a} | {a} | {a,b} | {a,c} | {a,b} | {a,c} | {a,b,c} | {a,b,c} |
| {b} | {b} | {a,b} | {b} | {b,c} | {a,b} | {a,b,c} | {b,c} | {a,b,c} |
| {c} | {c} | {a,c} | {b,c} | {c} | {a,b,c} | {a,c} | {b,c} | {a,b,c} |
| {a,b} | {a,b} | {a,b} | {a,b} | {a,b,c} | {a,b} | {a,b,c} | {a,b,c} | {a,b,c} |
| {a,c} | {a,c} | {a,c} | {a,b,c} | {a,c} | {a,b,c} | {a,c} | {a,b,c} | {a,b,c} |
| {b,c} | {b,c} | {a,b,c} | {b,c} | {b,c} | {a,b,c} | {a,b,c} | {b,c} | {a,b,c} |
| {a,b,c} | {a,b,c} | {a,b,c} | {a,b,c} | {a,b,c} | {a,b,c} | {a,b,c} | {a,b,c} | {a,b,c} |

Figure 9.4.1.: Join table of $2^A$.

## 9. Mathematical notions

| ∩ | ∅ | {a} | {b} | {c} | {a,b} | {a,c} | {b,c} | {a,b,c} |
|---|---|---|---|---|---|---|---|---|
| ∅ | ∅ | ∅ | ∅ | ∅ | ∅ | ∅ | ∅ | ∅ |
| {a} | ∅ | {a} | ∅ | ∅ | {a} | {a} | ∅ | {a} |
| {b} | ∅ | ∅ | {b} | ∅ | {b} | ∅ | {b} | {b} |
| {c} | ∅ | ∅ | ∅ | {c} | ∅ | {c} | {c} | {c} |
| {a,b} | ∅ | {a} | {b} | ∅ | {a,b} | {a} | {b} | {a,b} |
| {a,c} | ∅ | {a} | ∅ | {c} | {a} | {a,c} | {c} | {a,c} |
| {b,c} | ∅ | ∅ | {b} | {c} | {b} | {c} | {b,c} | {b,c} |
| {a,b,c} | ∅ | {a} | {b} | {c} | {a,b} | {a,c} | {b,c} | {a,b,c} |

Figure 9.4.2.: Meet table of $2^A$.

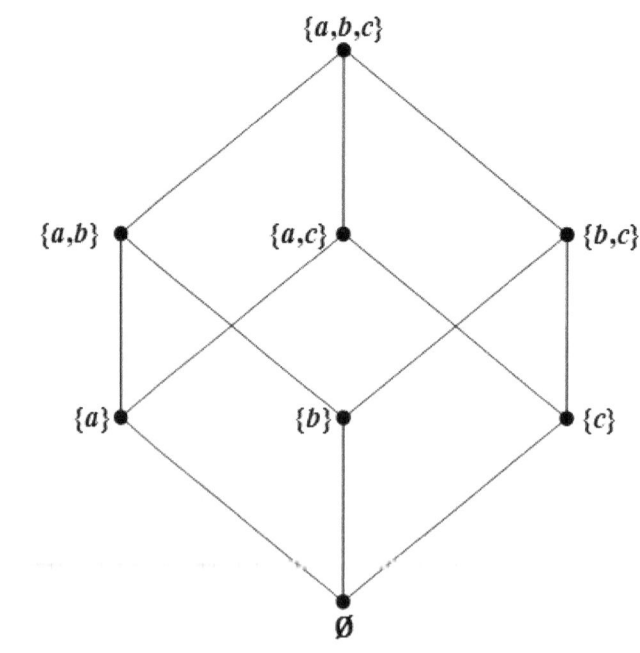

Figure 9.4.3.: The lattice $(\mathcal{S}, \cup, \cap)$.

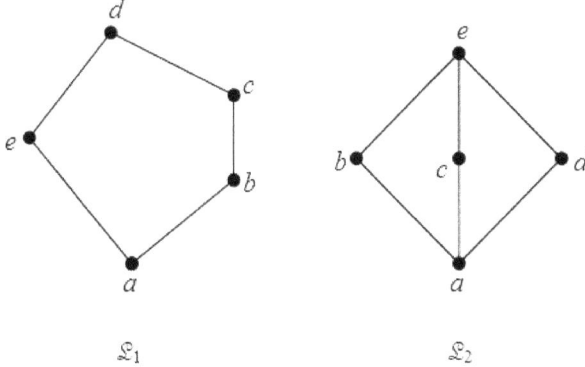

Figure 9.4.4.: The non-distributive lattices $\mathcal{L}_1$ and $\mathcal{L}_2$.

$$0 = x_1 \wedge ... \wedge x_n = \bigwedge_{i=1}^{n} x_i$$

**9.4.9. (Def.)** Let the algebra $\mathfrak{A} = (\mathcal{R}, \circ, /, \backslash)$ of type $(2,2,2)$ be given and for all $x, y, z \in \mathcal{R}$ let the following condition be satisfied by $\star, /, \backslash$:

(RES) $\quad x \star y \leq z \quad$ iff $\quad x \leq z/y \quad$ iff $\quad y \leq x \backslash z$

RES is known as the *law of residuation*, and $/$ and $\backslash$ are known as the left and right *residuals of* $\star$, respectively.

**9.4.10. (Def.)** An algebra $\mathfrak{A} = (\mathcal{R}, \vee, \wedge, \circ, /, \backslash, 1)$ is a *residuated lattice* iff

1. $(\mathcal{R}, \vee, \wedge)$ is a lattice,

2. $(\mathcal{R}, \star, 1)$ is a monoid such that $/$ and $\backslash$ are the left and right residuals of $\star$, respectively.

**9.4.11. (Def.)** Let $\mathcal{L}$ be a lattice and $A, B$ sets of elements of $\mathcal{L}$.

1. $A$ is called an *ideal* of $\mathcal{L}$ if

   a) $x, y \in A$ implies that $x \cup y \in A$, and
   
   b) $x \in \mathcal{L}, y \in A$ and $x \leq y$ imply that $x \in A$.

## 9. Mathematical notions

2. $B$ is called a *filter* of $\mathcal{L}$ if

   a) $x, y \in B$ implies that $x \cap y \in B$, and
   b) $x \in \mathcal{L}$, $y \in B$ and $x \geq y$ imply that $x \in B$.

**9.4.12. (Def.)** Let $\mathcal{L}$ be a lattice.

1. For each $x \in \mathcal{L}$, $\downarrow x$ (cf. Def. 9.2.8.2) is an ideal of $\mathcal{L}$. This particular ideal is called the *principal ideal generated by $x$*.

2. For each $x \in \mathcal{L}$, $\uparrow x$ (cf. Def. 9.2.8.3) is a filter of $\mathcal{L}$. This particular filter is called the *principal filter generated by $x$*.

**9.4.13. (Prop.)** In a finite lattice, every filter and every ideal is principal.

**9.4.14. (Prop.)** Let $\mathcal{L}$ be a lattice. Then, the following holds for all $x, y \in \mathcal{L}$:

$$[x, y] = (\uparrow x) \cap (\downarrow y)$$

**9.4.15. (Def.)** Let $\mathcal{L}$ be a lattice and let $x \in \mathcal{L}$, $x$ is a non-zero element of $\mathcal{L}$.

1. $x$ is said to be *meet-irreducible* if $x = a \wedge b$ implies that $x = a$ or $x = b$ for all $a, b \in \mathcal{L}$. $\underline{M_\mathcal{L}}$ denotes the set of meet-irrreducible elements of $\mathcal{L}$.

2. $x$ is said to be *join-irreducible* if $x = a \vee b$ implies that $x = a$ or $x = b$ for all $a, b \in \mathcal{L}$. $\underline{J_\mathcal{L}}$ denotes the set of join-irrreducible elements of $\mathcal{L}$.

## 9.5. Graphs and trees

**9.5.1. (Def.)** *Graphs* – Let $V = \{v_0, v_1, ..., v_n\}$ be a set of vertices and $E = \{e_0, e_1, ..., e_n\}$, $e_i = \{v_j, v_k\}$, be a set of edges. A *graph* $\mathfrak{G}$ is a pair $(V, E)$ such that all the endpoints $v_j, v_k$ of $E$ (i.e. nodes) are contained in $V$.

1. A *loop* is an edge or arc joining a vertex to itself.

2. A *multi-edge* is a set of $\geq 2$ edges with the same endpoints.

3. In a *directed graph* (or *digraph*) $\vec{\mathfrak{G}} = \left(V, E, \vec{f}\right)$, $V$ is as for $\mathfrak{G}$, $E$ is a set of arcs, i.e. directed edges ("arrows"), and $\vec{f}$ is an incidence function assigning to each member of $E$ a tail and a head, indicating the origin and the end of the edges, respectively.

4. A *simple* graph (digraph) is a graph $\mathfrak{G}$ with no loops or multi-edges (a digraph $\vec{\mathfrak{G}}$ with no self-loops and no pair of arcs with the same head and tail, respectively).

5. The edge $e_i = \{v_j, v_k\}$ *connects* the vertices $v_j, v_k$.

6. A *path* in a graph $\mathfrak{G}$ is an alternating sequence of vertices and edges of the form $v_0, e_1, v_1, e_2, ..., e_{n-1}, v_{n-1}, e_n, v_n$ in which each $e_i$ contains both $v_{i-1}$ and $v_i$.

7. A graph $\mathfrak{G}$ is *connected* if each pair of vertices is joined by a path. The number of edges is the *length* of the path.

8. If $v_0 = v_n$ in a path, then the path is said to be *closed*.

9. A closed path is *directed* if $v_0, v_1, ..., v_n$ are pairwise distinct vertices of a digraph $\vec{\mathfrak{G}}$.

10. A *cycle* is a closed path of length $\geq 3$.

11. A cycle is *directed* if it is a closed directed path.

**9.5.2. (Def.)** *Trees* – A *tree* is a connected graph containing no cycles.

1. A *subtree* is a subgraph of a graph that is also a tree.

2. A tree is said to be *finite* if it has a finite number of vertices (*nodes*) and edges (*branches*). Otherwise, it is said to be *infinite*.

3. A tree is *labeled* if it has labels attached to its vertices.

4. A tree is *rooted* if it has a least node $s_0$.

5. In a rooted tree, the *ancestor* (the *descendant*) of a vertex $v$ is either $v$ itself or any vertex that is the *predecessor* (a *successor*, respectively) of $v$ on a path from the root. A *proper ancestor* (*proper descendant*) of a vertex $v$ is any ancestor (descendant, respectively) except $v$ itself.

9. Mathematical notions

6. In a rooted tree, the *child* of a vertex $v$ is a vertex such that $v$ is its immediate ancestor. A vertex that is the immediate predecessor of $v$ on the unique path from the root to $v$ is called the *parent* of vertex $v$.

7. In a rooted tree, a *leaf* is a vertex that has no children.

8. A tree is *ordered* if it is rooted and the children of each internal vertex are linearly ordered.

9. A tree is said to be *binary* if it is an ordered rooted tree and each of its vertices has at most two children, a right and a left child.

10. The *depth* of a vertex $v$ in a rooted tree is the number of edges in the unique path from the root to $v$.

11. A binary tree is said to be *complete* if every parent thereof has two children and all leaves are at the same depth.

# Bibliography

- Ansótegui, C., Béjar, R., Cabiscol, A., Li, C. M., & Manyà, F. (2002). Resolution methods for many-valued CNF formulas. In *Fifth International Symposium on the Theory and Applications of Satisfiability Testing, SAT-2002*, Cincinnati, USA (pp. 152-163).

- Åqvist, L. (1962). Reflections on the logic of nonsense. *Theoria, 28*, 138-158.

- Aristotle (DI/1941). *De Interpretatione*. Trans. E. M. Edgehill, in R. McKeon (ed.), *The Basic Works of Aristotle*. New York: Random House.

- Augusto, L. M. (2017). *Logical consequences. Theory and applications: An introduction*. London: College Publications.

- Augusto, L. M. (2019). *Formal logic: Classical problems and proofs*. London: College Publications.

- Augusto, L. M. (2020a). *Computational logic. Vol 1.: Classical deductive computing with classical logic*. 2nd ed. London: College Publications.

- Augusto, L. M. (2020b). *Languages, machines, and classical computation*. 2nd ed. London: College Publications.

- Baader, F. & Snyder, W. (2001). Unification theory. In A. Robinson & A. Voronkov (eds.), *Handbook of automated reasoning*, vol. 1 (pp. 441-526). Amsterdam: Elsevier / Cambridge, MA: MIT Press.

- Baaz, M., Fermüller, C. G., & Salzer, G. (2001). Automated deduction for many-valued logics. In A. Robinson & A. Voronkov (eds.), *Handbook of automated reasoning*, vol. 2 (pp. 1357-1400). Amsterdam: Elsevier / Cambridge, MA: MIT Press.

- Baaz, M., Fermüller, C. G., Salzer, G., & Zach, R. (1996). MULtlog 1.0: Towards an expert system for many-valued logics. *Lecture Notes in Artificial Intelligence, 1104*, 226-230.

- Beckert, B., Hähnle, R., & Manyà, F. (1999). Transformations between signed and classical clause logic. In *Proceedings of the 29th International Symposium on Multiple-Valued Logic* (pp. 248-255). Los Alamitos: IEEE CS Press.

- Beckert, B., Hähnle, R., & Manyà, F. (2000). The SAT problem of signed CNF formulas. In D. Basin et al. (eds.), *Labelled deduction* (pp. 59-80). Dordrecht: Kluwer.

- Beckert, B., Hähnle, R., & Schmitt, P. H. (1993). The *even more liberalized δ-rule in free variable semantic tableaux*. In G. Gottlob, A. Leitsch, & D. Mundici (eds.), *Proceedings of the third Kurt Gödel Colloquium KGC'93, Brno* (pp. 108-119). Springer.

- Bellman, R. E. & Zadeh, L. A. (1977). Local and fuzzy logics. In J. M. Dunn and G. Epstein (eds.), *Modern uses of multiple-valued logic* (pp. 105-165). Dordrecht: Reidel.

- Belnap, N. D. (1977). A useful four-valued logic. In J. M. Dunn & G. Epstein (eds.), *Modern uses of multiple-valued logic* (pp. 8-37). Dordrecht: Reidel.

- Bergmann, M. (2008). *An introduction to many-valued and fuzzy logic. Semantics, algebras, and derivation systems*. Cambridge: Cambridge University Press.

- Beth, E. W. (1955). Semantic entailment and formal derivability. *Mededlingen der Koninklijke Nederlandse Akademie van Wetenschappen, 18*, 309-342.

- Blake, A. (1937). *Canonical expressions in Boolean algebra*. Ph.D. thesis. University of Chicago, Illinois.

- Bochvar, D. A. (1938). On a three-valued calculus and its application to analysis of paradoxes of classical extended functional calculus. (Trans. title) *Matématičéskij Sbornik, 4*, 287-308. (Engl. trans. by M. Bergmann, *History and Philosophy of Logic, 2* (1981), 87-112.)

- Bolc, L. & Borowik, P. (1992). *Many-valued logics 1: Theoretical foundations*. Berlin, etc.: Springer.

- Bolc, L. & Borowik, P. (2003). *Many-valued logics 2: Automated reasoning and practical applications*. Berlin, etc.: Springer.

- Boolos, G. S., Burgess, J. P., & Jeffrey, R. C. (2007). *Computability and logic*. 5th ed. Cambridge, etc.: Cambridge University Press.

- Caleiro, C. & Marcos, J. (2009). Classic-like analytic tableaux for finite-valued logics. In H. Ono, M. Kanazawa, & R. de Queiroz (eds.), *Logic, Language, Information, and Computation. Proceedings of the 16th International Workshop, WoLLIC, Tokyo, Japan* (pp. 268-280). Springer.

- Carnielli, W. A. (1985). An algorithm for axiomatizing and theorem proving in finite many-valued propositional logics. *Logique et Analyse, 112*, 363-368.

- Carnielli, W. A. (1987). Systematization of finite many-valued logics through the method of tableaux. *Journal of Symbolic Logic, 52*, 473-493.

- Carnielli, W. A. (1991). On sequents and tableaux for many-valued logics. *Journal of Non-Classical Logic, 8*, 59-76.

- Chang, C.-L. & Lee, R. C.-T. (1973). *Symbolic logic and mechanical theorem proving.* New York & London: Academic Press.

- Chiswell, I. & Hodges, W. (2007). *Mathematical logic.* Oxford: Oxford University Press.

- Church, A. (1936a). An unsolvable problem of elementary number theory. *American Journal of Mathematics, 58*, 345-363.

- Church, A. (1936b). A note on the Entscheidungsproblem. *Journal of Symbolic Logic, 1*, 40-41.

- Cook, S. A. (1971). The complexity of theorem proving procedures. *Proceedings of the 3rd Annual ACM Symposium of Theory of Computing,* 151-158.

- D'Agostino, M. (1999). Tableau methods for classical propositional logic. In M. D'Agostino et al. (eds.), *Handbook of tableau methods* (pp. 45-123), Dordrecht: Kluwer.

- D'Avila Garcez, A. S., Lamb, L. C., & Gabbay, D. M. (2009). *Neural-symbolic cognitive reasoning.* Berlin, Heidelberg: Springer.

- Davis, M. (1957). A computer program for Presburger's algorithm. In *Summaries of Talks Presented at the Summer Institute for Symbolic Logic* (pp. 215-233). Princeton, NJ: Institute for Defense Analysis.

- Davis, M. & Putnam, H. (1958). Computational methods in the propositional calculus. Rensselaer Polytechnic Institute. (Unpublished report.)

- Davis, M. & Putnam, H. (1960). A computing procedure for quantification theory. *Journal of the ACM, 7*, 201-215.

- Davis, M., Logemann, G., & Loveland, D. (1962). A machine program for theorem-proving. *Communications of the ACM, 5*, 394-397.

- Di Lascio, L. (2001). Analytic fuzzy tableaux. *Soft Computing, 5*, 434-439.

- Di Zenzo, S. (1988). A many-valued logic for approximate reasoning. *IBM Journal of Research and Development, 32*, 552-565.

- Dunn, J. M. & Epstein, G. (1977). *Modern uses of multiple-valued logic.* Dordrecht: D. Reidel.

- Epstein, G. (1993). *Multiple-valued logic design: An introduction.* Bristol & Philadelphia: Institute of Physics Publishing.

- Esteva, F., Godo, L., Hájek, P., & Montagna, F. (2003). Hoops and fuzzy logic. *Journal of Logic and Computation, 13*, 532-555.

- Finn, V. K. (1972). An axiomatization of some propositional calculi and their algebras. (trans. title; Russian text) Vsiechsoiuznyi Institut Naučeskoi Informacii Academii Nauk SSSR, Moscow.

- Fitting, M. (1996). *First order logic and automated theorem proving.* 2nd ed. New York, etc.: Springer.

- Fitting, M. (1999). Introduction. In M. D'Agostino et al. (eds.), *Handbook of tableau methods* (pp. 1-44). Dordrecht: Kluwer.

- Fitting, M. & Orłowska, E. (eds.) (2003). *Beyond two: Theory and applications of multiple-valued logic.* Heidelberg: Physica-Verlag.

- Frege, G. (1892). Über Sinn und Bedeutung. *Zeitschrift für Philosophie und philosophische Kritik C*, 25-50.

- Gabbay, D. M. (1994). What is a logical system? In D. M. Gabbay (ed.), *What is a logical system?* (pp. 179-216). Oxford: OUP.

- Gabbay, D. M. (2014). What is a logical system? An evolutionary view: 1964-2014. In D. M. Gabbay, J. H. Siekmann, & J. Woods (eds.), *Handbook of the history of logic. Vol. 9: Computational logic* (pp. 41-132). Amsterdam, etc.: Elsevier.

- Gabbay, D. M. & Woods, J. (2001). The new logic. *Logic Journal of the IGPL, 9*, 141-174.

- Gabbay, D. M. & Woods, J. (2003). *A practical logic of cognitive systems. Vol. 1: Agenda relevance. A study in formal pragmatics.* Amsterdam, etc.: Elsevier.

- Gabbay, D. M. & Woods, J. (2005). *A practical logic of cognitive systems. Vol. 2: The reach of abduction: Insight and trial.* Oxford, etc.: Elsevier.

- Gentzen, G. (1934-5). Untersuchungen über das logische Schliessen. *Mathematische Zeitschrift, 39*, 176-210, 405-431. (Engl. trans.: Investigations into logical deduction. In M. E. Szabo (ed.), *The Collected Papers of Gerhard Gentzen* (pp. 68-131). Amsterdam: North-Holland.)

- Gerla, G. (2001). *Fuzzy logic: Mathematical tools for approximate reasoning.* Dordrecht: Springer.

- Gilmore, P. (1960). A proof method for quantification theory: Its justification and realization. *IBM Journal of Research and Development, 4*, 28-35.

- Gödel, K. (1930). Die Vollständigkeit der Axiome des logischen Funktionkalküls. *Monatshefte für Mathematik, 37*, 349-360. (Engl. trans.: The completeness of the axioms of the functional calculus of logic. In S. Feferman et al. (eds.), *Collected works. Vol. 1: Publications 1929-1936* (pp. 103-123). New York: OUP & Oxford: Clarendon Press, 1986.)

- Gottwald, S. (2001). *A treatise on many-valued logics.* Philadelphia: Research Studies Press.

- Grigolia, G. (1977). Algebraic analysis of Łukasiewicz-Tarski's $n$-valued logical systems. In R. Wójcicki & G. Malinowski (eds.), *Selected papers on Łukasiewicz sentential calculi* (p. 81-92). Wrocław: Ossolineum.

- Haack, S. (1974). *Deviant logic: Some philosophical issues.* Cambridge: Cambridge University Press.

- Hähnle, R. (1991). Towards an efficient tableau proof procedure for multiple-valued logics. In E. Börger et al. (eds.), *Selected papers from computer science logic, CLS'90, Heidelberg, Germany* (pp.248-260), Springer.

- Hähnle, R. (1994). *Automated deduction in multiple-valued logics.* Oxford: Oxford University Press.

- Hähnle, R. (1998). Commodious axiomatization of quantifiers in multiple-valued logic. *Studia Logica, 61*, 101-121.

- Hähnle, R. (1999). Tableaux for many-valued logics. In M. D'Agostino et al. (eds.), *Handbook of tableau methods* (pp. 529-580), Dordrecht: Kluwer.

- Hähnle, R. (2001). Advanced many-valued logics. In D. M. Gabbay & F. Guenthner (eds.), *Handbook of philosophical logic*. Vol. 2. 2nd ed. (pp. 297-395). Dordrecht: Reidel.

- Hähnle, R. & Schmitt, P. H. (1994). The liberalized $\delta$-rule in free-variable semantic tableaux. *Journal of Automated Reasoning, 13*, 211-221.

- Hähnle, R., Beckert, B., & Gerberding, S. (1994). $_3T^AP$: The many-valued theorem prover. 3rd ed. University of Karlsruhe.

- Hájek, P. (1997). Fuzzy logic and arithmetical hierarchy II. *Studia Logica, 58*, 129-141.

- Hájek, P. (1998). *Metamathematics of fuzzy logic*. Dordrecht: Kluwer.

- Halldén, S. (1949). The logic of nonsense. *Uppsala Universitets Årsskrift, 9*, 132 pp.

- Harrison, J. (2009). *Handbook of practical logic and automated reasoning*. Cambridge, etc.: CUP.

- Herbrand, J. (1930). *Recherches sur la théorie de la démonstration*. Thèses présentées à la Faculté des Sciences de Paris. 128 p.

- Heyting, A. (1930). Die formalen Regeln der intuitionistischen Logik. *Sitzungsberichte der Preussischen Akademie der Wissenschaften zu Berlin*, 42-56. (Engl. trans.: The formal rules of intuitionistic logic. In P. Mancosu (ed.), *From Brouwer to Hilbert: The debate on the foundations of mathematics in the 1920s* (pp. 311-327). Oxford & New York: Oxford University Press. 1998)

- Heyting, A. (1956). *Intuitionism: An introduction*. Amsterdam: North-Holland.

- Hilbert, D. (1900). Mathematische Probleme. Vortrag, gehalten auf dem internationalen Mathematiker-Kongress zu Paris 1900.

*Nachrichten von der Königl. Gesellschaft der Wissenschaften zu Göttingen. Mathematisch-Physikalische Klasse*, 253-297. (Engl. transl.: Mathematical problems. *Bulletin of the American Mathematical Society, 8* (1902), 437-479.)

- Hilbert, D. & Ackermann, W. (1928). *Grundzüge der theoretischen Logik.* Berlin: Springer.

- Hilbert, D. & Bernays, P. (1934). *Grundlagen der Mathematik. Vol. I.* Berlin.

- Hilbert, D. & Bernays, P. (1939). *Grundlagen der Mathematik. Vol. II.* Berlin.

- Hintikka, J. (1955). Form and content in quantification theory. *Acta Philosophica Fennica, 8*, 7-55.

- Hoder, K. & Voronkov, A. (2009). Comparing unification algorithms in first-order theorem proving. In *KI'09 Proceedings of the 22nd Annual German Conference on Advances in Artificial Intelligence* (pp. 435-443). Berlin, Heidelberg: Springer.

- Jaśkowski, S. (1934). On the rules of suppositions in formal logic. *Studia Logica, 1*, 5-32.

- Kalman, J. A. (2001). *Automated reasoning with OTTER.* Princeton, NJ: Rinton Press.

- Kleene, S. C. (1938). On a notation for ordinal numbers. *Journal of Symbolic Logic, 3*, 150-155.

- Kleene, S. C. (1952). *Introduction to metamathematics.* Amsterdam: North-Holland.

- Konikowska, B., Tarlecki, A., & Blikle, A. (1991). A three-valued logic for software specifications and validation. *Fundamenta Informaticae, XIV*, 411-453.

- Lee, R. C. T. (1972). Fuzzy logic and the resolution principle. *Journal of the ACM, 19*, 109-119.

- Leitsch, A. (1997). *The resolution calculus.* Berlin, etc.: Springer.

- Letz, R. (1999). First-order tableau methods. In M. D'Agostino et al. (eds.), *Handbook of tableau methods* (pp. 125-196), Dordrecht: Kluwer.

- Lis, Z. (1960). Logical consequence, semantic and formal. (Trans. title) *Studia Logica, 10*, 39-60.

- Lu, J. J., Murray, N. V., & Rosenthal, E. (1994). Signed formulas and fuzzy operator logics. In Z. W. Raś & M. Zemankova (eds.), *Methodologies for intelligent systems* (pp. 75-84). Berlin, etc.: Springer.

- Lu, J. J., Murray, N. V., & Rosenthal, E. (1998). A framework for automated reasoning in multiple-valued logics. *Journal of Automated Reasoning, 21*, 39-67.

- Łukasiewicz, J. (1920). O logice trójwartościowej. *Ruch Filozoficzny, 5*, 170-171. (Engl. trans.: On three-valued logic. In L. Borkowski (ed.), *Selected Works* (p. 87-88). Amsterdam: North-Holland.)

- Łukasiewicz, J. (1930). Philosophische Bemerkungen zu mehrwertigen Systemen des Aussagenkalküls. *Comptes Rendus des Scéances de la Société des Sciences et des Lettres de Varsovie CL. III, 23*, 51-77. (Engl. trans.: Philosophical remarks on many-valued systems of propositional logic. In S. McCall (ed.), *Polish Logic 1920-1939* (p. 40-65). Oxford: Clarendon Press. 1967.)

- Łukasiewicz, J., & Tarski, A. (1930). Untersuchungen über den Aussagenkalkül. *Comptes Rendus des Scéances de la Société des Sciences et des Lettres de Varsovie CL. III, 23*, 30-50.

- Malinowski, G. (1989). *Equivalence in intensional logics.* Warsaw: Polish Academy of Sciences, Institute of Philosophy and Sociology.

- Malinowski, G. (1993). *Many-valued logics.* Oxford: Clarendon Press.

- Malinowski, G. (2012). Multiplying logical values. *Logical Investigations, 18*, 292-308.

- McCarthy, J. (1963). A basis for a mathematical theory of computation. In P. Braffort & D. Hirshberg (eds.), *Computer programming and formal systems* (pp. 33-70). Amsterdam: North-Holland.

- Mendelson, E. (2009). *Introduction to mathematical logic.* 5th ed. Chapman and Hall / CRC.

- Metcalfe, G., Olivetti, N., & Gabbay, D. (2009). *Proof theory for fuzzy logics.* Springer.

- Morgan, C. G. (1976). A resolution principle for a class of many-valued logics. *Logique et Analyse, 19*, 311-339.

- Mostowski, A. (1957). On a generalization of quantifiers. *Fundamenta Mathematicae, 44*, 12-36.

- Mundici, D. & Olivetti, N. (1998). Resolution and model building in the infinite-valued calculus of Łukasiewicz. *Theoretical Computer Science, 200*, 335-366.

- Murray, N. V. & Rosenthal, E. (1993). Signed formulas: A liftable meta-logic for multiple-valued logics. In *Proceedings of the International Symposium on Methodologies for Intelligent Systems (ISMIS)*, Trondheim, Norway (pp. 275-284). Springer.

- Newell, A., Shaw, J., & Simon, H. (1957). Empirical explorations with the Logic Theory Machine. *Proceedings of the West Joint Computer Conference, 15*, 218-239.

- Novák, V. (1990). On the syntactic-semantical completeness of first-order fuzzy logic (Parts I and II). *Kybernetika, 26*, 47-66, 134-154.

- Novák, V., Perfilieva, I., & Močkoř, J. (1999). *Mathematical principles of fuzzy logic.* Boston: Kluwer.

- Olivetti, N. (2003). Tableaux for Łukasiewicz infinite-valued logic. *Studia Logica, 73*, 81-111.

- Orłowska, E. (1978). The resolution principle for ω+-valued logic. *Fundamenta Informaticae, 2*, 1-15.

- Pavelka, J. M. (1979). On fuzzy logic (Parts I-III). *Zeitschrift für mathematische Logik und Grundlagen der Mathematik, 25*, 45-52, 119-134, 447-464.

- Peirce, C. S. (1909/1966). Logic notebook (MS 339). In *The Charles S. Peirce papers*. Microfilm edition. Cambridge, MA: Harvard University Library.

- Piróg-Rzepecka, K. (1977). *Systems of nonsense logics.* (trans. title) Warsaw: PWN.

- Post, E. L. (1921). Introduction to a general theory of elementary propositions. *American Journal of Mathematics*, *43*, 163-185.

- Prawitz, D. (1960). An improved proof procedure. *Theoria*, *26*, 102-139.

- Prawitz, D. (1965). *Natural deduction. A proof-theoretical study*. Stockholm: Almqvist & Wiksell.

- Priest, G. (2008). *An introduction to non-classical logic*. Cambridge: Cambridge University Press.

- Quine, W. V. O. (1955). A proof procedure for quantification theory. *Journal of Symbolic Logic*, *20*, 141-149.

- Rasiowa, H. (1974). *An algebraic approach to non-classical logics*. Amsterdam: North-Holland.

- Reichenbach, H. (1946). *Philosophical foundations of quantum mechanics*. Berkeley, CA: University of California Press.

- Rescher, N. (1969). *Many-valued logic*. McGraw-Hill.

- Robinson, A. J. (1965). A machine-oriented logic based on the resolution principle. *Journal of ACM*, *12*, 23-41.

- Rosser, J. B. & Turquette, A. R. (1952). *Many-valued logics*. Amsterdam: North-Holland.

- Rybakov, V. V. (1997). *Admissibility of logical inference rules*. Studies in Logic and the Foundations of Mathematics, vol. 136, Elsevier.

- Salzer, G. (2000). Optimal axiomatizations of finitely valued logics. *Information and Computation*, *162*, 185-205.

- Scarpelini, B. (1962). Die Nichtaxiomatisierbarkeit des unendlichwertigen Prädikatenkalküls von Łukasiewicz. *Journal of Symbolic Logic*, *17*, 159-170.

- Schmitt, P. H. (1986). Computational aspects of three-valued logic. In J. H. Siekmann (ed.), *Proceedings of the 8th International Conference on Automated Deduction*, Oxford, England (pp. 190-198). Berlin, Heidelberg: Springer.

- Schütte, K. (1956). Ein System des verknüpfenden Schließens. *Archiv für mathematische Logik und Grundlagen der Wissenschaft*, *2*, 55-67.

- Segerberg, K. (1965). A contribution to nonsense logic. *Theoria*, *31*, 199-217.

- Skolem, T. (1920). Logisch-kombinatorische Untersuchungen über die Erfüllbarkeit oder Beweisbarkeit mathematischer Sätze. *Skrifter utgit av Videnskapsselskapet i Kristiania*, *4*, 4-36.

- Słupecki, J. (1936). Der volle dreiwertige Aussagenkalkül. *Comptes Rendus des Scéances de la Société des Sciences et des Lettres de Varsovie Cl. III*, *29*, 9-11. (English trans.: The full three-valued propositional calculus. In S. McCall (ed.), *Polish Logic 1920-1939* (p. 335-337). Oxford: Clarendon Press. 1967.)

- Słupecki, J. (1939). A criterion of completeness of many-valued systems of propositional logic. (Trans. title) *Comptes Rendus des Scéances de la Société des Sciences et des Lettres de Varsovie Cl. III*, *32*, 102-109.

- Smullyan, R. M. (1968). *First-order logic*. Mineola, NY: Dover.

- Stachniak, Z. (1996). *Resolution proof systems: An algebraic approach*. Kluwer.

- Straccia, U. (2001). Reasoning within fuzzy description logics. *Journal of Artificial Intelligence Research*, *14*, 137-166.

- Surma, S. J. (1977). An algorithm for axiomatizing every finite logic. In D. C. Rine (ed.), *Computer science and multiple-valued logics. Theory and applications.* (pp. 137-143), Amsterdam, etc.: North-Holland.

- Suszko, R. (1957). A formal theory of logical values. (trans. title) *Studia Logica*, *6*, 145-320.

- Suszko, R. (1977). The Fregean axiom and Polish mathematical logic in the 1920's. *Studia Logica*, *36*, 373-380.

- Tarski, A. (1930). Fundamentale Begriffe der Methodologie der deduktiven Wissenschaften. I. *Monatshefte für Mathematik und Physik*, *37*, 361-404. (Engl. trans.: Fundamental concepts of the methodology of the deductive sciences. In A. Tarski, *Logic, semantics, metamathematics: Papers from 1923 to 1938* (pp. 60-109). Oxford: Clarendon Press, 1956.)

- Tarski, A. (1935). Der Wahrheitsbegriff in formalisierten Sprachen. *Studia Philosophica*, *1*, 261-405 (Engl. trans.: The concept of truth in formalized languages. In A. Tarski, *Logic, semantics, metamathematics: Papers from 1923 to 1938* (pp. 152-278). Trans. by J. H. Woodger. Oxford: Clarendon Press, 1956) (Originally published in Polish in 1933.)

- Troelstra, A. S. & Schwichtenberg, H. (2000). *Basic proof theory.* 2nd ed. Cambridge: CUP.

- Tseitin, G. S. (1968). On the complexity of derivations in the propositional calculus. In A. O. Slisenko (ed.), *Studies in constructive mathematics and mathematical logic. Part 2. Seminar in mathematics* (pp. 115-125). Steklov Mathematical Institute.

- Turing, A. (1936-7). On computable numbers, with an application to the Entscheidungsproblem. *Proceedings of the London Mathematical Society, Series 2, 41,* 230-265.

- Wajsberg, M. (1931). Aksjomatyzacja trójwartościowego rachunku zdań. *Comptes Rendus des Scéances de la Société des Sciences et des Lettres de Varsovie Cl. III, 24,* 126-148. (Engl. trans.: Axiomatization of the three-valued propositional calculus. In S. McCall (ed.), *Polish Logic 1920-1939* (p. 264-284). Oxford: Clarendon Press. 1967.)

- Walther, C. (1985). A mechanical solution of Schubert's Steamroller by many-sorted resolution. *Artificial Intelligence, 26,* 217-224.

- Whitehead, A. N. & Russell, B. (1910). *Principia mathematica. Vol. I.* Cambridge: Cambridge University Press.

- Wójcicki, R. (1969). Logical matrices strongly adequate for structural sentential calculi. *Bulletin de l'Académie Polonaise des Sciences, Série des sci. math. astr. et physiques, 6,* 333-335.

- Wójcicki, R. (1988). *Theory of logical calculi: Basic theory of consequence operations.* Dordrecht: Kluwer.

- Zach, R. (1993). *Proof theory of finite-valued logics.* Diplomarbeit / Technical report TUW-E185.2-Z.1-93, Institut für Algebra und diskrete Mathematik, TU Wien.

- Zadeh, L. A. (1965). Fuzzy sets. *Information and Control, 8,* 338-353.

- Zadeh, L. A. (1975). Fuzzy logic and approximate reasoning. *Synthese, 30*, 407-428.

# Index

This index is a hybrid of an *Index nominum* and an *Index rerum*. With respect to the latter, only the most general, positive, concept is given; for example, for *(un)satisfiability* and *(un)satisfiable*, look under *satisfiability*. (An exception is *indeterminacy*.) For complex expressions (e.g., *many-valued satisfiability problem*), look under the final noun (e.g., *Problem, Many-valued satisfiability*). Exceptions are complex expressions beginning with one or more proper nouns, which should be looked under the (first) name (e.g., *Belnap's 4-valued system*; *Hintikka set*; *Church-Turing theorem*) and expressions beginning with central concepts (e.g., *Truth-value gap*). For most concepts, only the page where they are defined or first occur significantly is given. Abbreviations are provided in brackets, and recurrent abbreviations are also listed as entries (e.g., CNF). Logical systems are listed both by the first name/concept and by the first letter denoting them (e.g, both *Post's n-valued system* and "$P_n$" are distinct entries in the Index). As for the *Index nominum*, only proper nouns associated with fundamental aspects (e.g., theorems, problems, structures) and/or logical systems are listed; for these, all pages of occurrence are given. The letters Å and Ł are alphabetically processed as A and a variant of L, respectively, and the Greek letter Π is so as P.

# Index

2-SAT, 84
3-SAT, 84
$_3T^AP$ (Tableaux-based prover), 240

**A**
Adequateness, 73
Adjunction, 78
Algebra of formulae, 63
Algebra of signs, 253
Algebra, Absolutely free, 322
Algebra, Abstract, 321
Algebra, Boolean, 58, 323
Algebra, G- (Gödel algebra), 325
Algebra, Many-valued (MV-algebra), 324
Algebra, MV- (Many-valued algebra), 324
Algebra, Π- (Product algebra), 325
Algebra, Product (Π-algebra), 325
Algebras, BL-, 324
Algorithm, 6
Algorithm, DPLL, 107
An$_3$ (Åqvist's 3-valued system), 178
Analytic (proof procedure), 8
Åqvist, L., 178
Åqvist's 3-valued system (An$_3$), 178
Argument, Logical, 51
Aristotle, 151

Assistant, Proof, 7
Assumption, 97
Atom, 14
Axiom, 67
Axiom, Derived, 163
Axiomatization, 73

**B**
$B_3^E$ (Bochvar's external 3-valued system), 173
$B_3^I$ (Bochvar's internal 3-valued system), 171
$B_4$ (Belnap's 4-valued system), 183
Belnap, N. D., 183–185
Belnap's 4-valued system ($B_4$), 183
Bernays, P. I., 197
Beth, E. W., 86, 108
Bivalence, 2
BL (Basic logic), 191
BL* (FO basic logic), 202
Blake, A., 87
Bochvar, D. A., 171, 172, 174, 176, 177, 196
Bochvar's 3-valued system (see also: $B_3^E$, $B_3^I$), 171
Boole, G., 86, 151
Brouwer, L. E. J., 3

**C**
Calculus, 67

# Index

Calculus, Logical, 15
Calculus, Sequent, 101
CFOL (Classical first-order logic), 23
Church, A., 82, 86
Church-Turing theorem, 82
Church-Turing thesis, 86
CL (Classical logic), 22
Clause, 25
Clause, Definite, 25
Clause, Dual-Horn, 25
Clause, Empty, 25
Clause, Empty signed, 39
Clause, Ground, 25
Clause, Horn, 25
Clause, Parent, 125
Clause, Regular Horn, 39
Clause, Signed, 39
Clause, Unit, 25
Closure (under a rule $r$), 67
Closure, Existential, 20
Closure, Quantifier, 19
Closure, Universal, 20
CNF (Conjunctive normal form), 32
cnf (minimal CNF for signed formulae), 228
CNF-SAT, 84
Combination, Boolean positive, 272
Compactness, 71
Complementarity, 25
Complementarity, $T$-, 302
Complementarity, $\varepsilon$-, 296
Completeness, 72
Completeness, Functional, 23
Completeness, Refutation, 292
Complexity, Computational, 82
Conclusion (of an argument), 51
Condensation, 132
Confidence, 300
Conjunction (Connective), 22

Conjunction, Bold/Strong, 207
Connective, Logical, 18
Consequence, Disjunctive (Rule; DCn), 165
Consequence, Logical, 52
Consistency, 69
Constant, Logical, 13
Constant, Non-logical, 13
Contingency, 60
Contradiction, 60
Contradiction, Quasi-, 157
Contraposition (Rule; ConP), 165
Contraposition, Generalized (Rule; GConP), 165
Cook, S. A., 84
Correlate, Semantical, 55
Countermodel, 59
Counter-proof, 69
CPL (Classical propositional logic), 23

**D**

Davis, M., 86, 87, 104, 123, 137
De Morgan's laws (DM), 27
Decidability, 81
Decisiveness, 161
Decomposition, Unicity of, 14
Deduction, $Res <_A$, 133
Deduction, Resolution, 125
Denotation, 55
Derivability, 69
Disjunction (Connective), 22
Disjunction, Bold/Strong, 207
Distribution function, 15
Distributive laws, 34
DM (De Morgan's laws), 27
DNF (Disjunctive normal form), 32
dnf (minimal DNF for signed formulae), 228
Domain (of discourse), 20, 314

## E

Entailment, 60
Entailment, Degree-, 159
Entailment, Fuzzy, 190
Entailment, Quasi-, 159
Entscheidungsproblem, 81
Equisatisfiability, 30
Equivalence, Logical, 17
Ex falso quodlibet, 78
Expression, 14
Expression, Signed formula (SFE), 39

## F

$F_3$ (Finn's 3-valued system), 176
Factor, 125
Factoring, Signed, 284
Filter, 319, 330
Finn, V. K., 176, 177
Finn's 3-valued system ($F_3$), 176
FO (First-order), 18
Form, Conjunctive normal (CNF), 32
Form, Disjunctive normal (DNF), 32
Form, Negation normal (NNF), 26
Form, Partial normal (PaNF), 198
Form, Prenex normal (PNF), 28
Form, Skolem normal (SNF), 30
Formula, 14
Formula, Closed, 19
Formula, Complementary, 37
Formula, Open, 19
Formula, Quantified, 19
Formula, Regular Horn, 39
Formula, Signed, 37
Formula, Signed 2-CNF, 39
Formula, Signed CNF, 39
Formula, Well-formed, 18
Frame, 20
Frame, Constraint, 270

Free for, 24
Frege system, 96
Frege, F. L. G., 3, 55, 86, 95, 151
Fregean axiom, 55
Frege-Łukasiewicz axiom system, 95
Function, 316
Function, Boolean, 56
Function, Characteristic, 320
Function, Membership, 320

## G

$G_3$ (Gödel's 3-valued system), 189
$G_3^*$ (Gödel's FO 3-valued system), 233
$G_\aleph$ (Gödel's fuzzy system), 192
$G_\aleph^*$ (Gödel's FO fuzzy system), 205
$G_n$ (Gödel's $n$-valued system), 188
$G_n^*$ (Gödel's FO $n$-valued system), 233
GEN-SAT$_\infty$ (Generalized satisfiability problem), 296
Gentzen, G., 97, 100, 240
Gilmore, P. C., 87, 94
Gödel algebra (G-algebra), 325
Gödel t-norm, 191
Gödel, K., 71, 73, 76, 181, 188, 223
Gödel's 3-valued system ($G_3$), 189
Gödel's completeness theorem, 76
Gödel's FO 3-valued system ($G_3^*$), 233
Gödel's FO fuzzy system ($G_\aleph^*$), 205
Gödel's FO $n$-valued system ($G_n^*$), 233
Gödel's fuzzy system ($G_\aleph$), 192
Gödel's $n$-valued system ($G_n$), 188
Graph, 330
Grelling, K., 171

# Index

## H
Halldén, S., 177–180
Halldén's 3-valued system ($Hn_3$), 177
Herbrand base, 90
Herbrand instance, 90
Herbrand interpretation (H-interpretation), 90
Herbrand model (H-model), 91
Herbrand semantics, 87
Herbrand universe, 88
Herbrand, J., 85, 87, 88, 93
Herbrand's theorem, 88
Heyting, A., 181, 182
Heyting's 3-valued system (see $G_3$), 182
Hilbert system, 96
Hilbert, D., 3, 86, 96, 197
H-interpretation (Herbrand interpretation), 90
Hintikka set, 115
Hintikka sign set, 260
Hintikka, J., 86, 108
Hintikka's lemma, 115
H-model (Herbrand model), 91
$Hn_3$ (Halldén's 3-valued system), 177
HORN-SAT, 84
HORN-SAT$_\infty$, 296
Hypothetical syllogism, 76

## I
Ideal, 319, 329
Indeterminacy, 151
Inference, 54
Inference, Deductive, 59
Instance, 41
Instance, Ground, 41
Interpretation, 20
Interpretation, Signature, 20
Involution law (LDN), 27

## J
Jaśkowski, S., 97

## K
$K_3^S$ (Kleene's strong 3-valued system), 174
$K_3^W$ (Kleene's weak 3-valued system), 175
Kleene, S. C., 96, 152, 174, 175, 261, 269
Kleene's strong 3-valued system ($K_3^S$), 174
Kleene's weak 3-valued system ($K_3^W$), 175
K-regularity, 161
KROM-SAT$_\infty$, 296
$k$-SAT, 84

## L
$\mathcal{L}$ (a propositional axiom system), 79
$\mathcal{L}^*$ (FO extension of $\mathcal{L}$), 96
$L_3$ (3-valued system [R. Hähnle's]), 256
$L_\star$ (Continuum-valued logic), 191
$L_\star^*$ (FO continuum-valued logic), 202
Language, First-order, 18
Language, Formal, 13
Language, Logical, 13
Language, Object, 13
Language, Propositional, 18
Lattice, 325
Law of contraposition, 76
Law of identity, 76
LDN (Double negation law), 27
Leibniz, G. W. von, 86
Lifting lemma, 128
Lindenbaum bundle, 66
Lindenbaum condition, 168
Lindenbaum matrix, 66
Lindenbaum, A., 66, 153

Lis, Z., 243
Literal, 25
Literal, Signed, 38
Logemann, G., 87, 104
Logic, A, 61, 69
Logic, Basic (BL), 191
Logic, Classical first-order (CFOL), 23
Logic, Classical propositional, 23
Logic, Computational, 6
Logic, Continuum-valued ($L_\star$), 191
Logic, Deductive, 1
Logic, FO basic (BL*), 202
Logic, FO continuum-valued ($L_\star^*$), 202
Logic, FO product ($\Pi_\aleph^*$), 205
Logic, Fuzzy (FL), 189
Logic, Fuzzy operator, 298
Logic, Mathematical, 3
Logic, Product (fuzzy) ($\Pi_\aleph$), 193
Logic, Signed, 37
Loveland, D., 87, 104
Łoś, J., 70
$Ł_3$ (Łukasiewicz's 3-valued system), 161
$Ł_n$ (Łukasiewicz's $n$-valued system), 169
$Ł_n^*$ (Łukasiewicz's FO $n$-valued system), 196
$Ł_\aleph$ (Łukasiewicz's fuzzy system), 169
$Ł_\aleph^*$ (Łukasiewicz's FO fuzzy system), 196
Łukasiewicz, J., 95, 152, 161, 162, 166–171, 174, 188, 202, 223, 275, 281, 295, 307
Łukasiewicz's 3-valued system ($Ł_3$), 161
Łukasiewicz's $n$-valued system ($Ł_n$), 169
Łukasiewicz's FO $n$-valued system ($Ł_n^*$), 196
Łukasiewicz's fuzzy system ($Ł_\aleph$), 169
Łukasiewicz's FO fuzzy system ($Ł_\aleph^*$), 196
Łukasiewicz t-norm, 190

**M**

Malinowski, G., xiv, 65, 66, 155, 162, 169, 186, 197
Material equivalence (Connective), 22
Material implication (Connective), 22
Matrix (of a formula in PNF), 28
Matrix representation, 65
Matrix, Logical, 63, 64
Matrix, Minimal, 179
McCarthy, J., 261
McCarthy-Kleene's many-valued system (MK), 261
Meaninglessness, 177
Metalanguage, 13
Metalogic, 5
MGU (Most general unifier), 41
MK (McCarthy-Kleene's many-valued system), 261
Model, 59
Model (algebraic structure), 321
Model theory, 55
Modus ponens (MP), 68
Modus ponens, Generalized (Rule; GMP), 166
Modus tollens, 76
Monotonicity, 54
Monotonicity, Weak, 78
Morphisms, 322
Mostowski, A., 196, 198
MP (Modus ponens), 68
MULtlog, 235
MV-SAT (Many-valued satisfiability problem), 217

## Index

MV-SAT$_\infty$ (Problem for infinitely many-valued logics, Satisfiability), 294
MV-VAL (Many-valued validity problem), 7

### N
Negation (Connective), 22
Negation law, Double (LDN), 27
Nelson, L., 171
Newell, A., 86
NNF (Negation normal form), 26
Nonsense (Logical), 177
Nonsense, Logics of, 177
Normality, 160

### O
Operation, 316
Operation, Consequence, 52
Operation, Finitary consequence, 53
Operation, Idle consequence, 53
Operation, Inconsistent/Trivial consequence, 53
Operation, Inference, 54
Operation, Matrix consequence, 64
Operation, Standard consequence, 53
Operation, Structural consequence, 53
Operator, Fuzzy, 298
Order ($n$-th), 17
Order, Partial, 317
Ordering, Atom (or A-), 130

### P
P$_3^*$ (Post's FO 3-valued system), 233
P$_n$ (Post's $n$-valued system), 186
P$_n^*$ (Post's FO $n$-valued system), 233
Paradox, Semantical, 171
Paradox, Sorites, 177
Parameter, 97
Pavelka-style system, 202
Peirce, C. S., 151, 152
PEM (Principle of excluded middle), 2
$\Pi_\aleph$ (Product (fuzzy) logic), 193
PI-clash, 135
Piróg-Rzepecka, K., 180
Piróg-Rzepecka's 3-valued system (Rn$_3$), 180
PNC (Principle of non-contradiction), 76
PNF (Prenex normal form), 28
Poset, 317
Post algebra, 58
Post, E. L., 57, 58, 152, 157, 186, 187
Post's FO 3-valued system (P$_3^*$), 233
Post's FO $n$-valued system (P$_n^*$), 233
Post's $n$-valued system (P$_n$), 186
$\mathbf{P} \stackrel{?}{=} \mathbf{NP}$, 82
Prawitz, D., 87, 97, 99
Premise, 51
Presburger, M., 86
Principle of bivalence, 76
Principle of excluded middle (PEM), 2
Principle of explosion, 78
Principle of extensionality, 55
Principle of non-contradiction (PNC), 76
Principle, Interpretation, 155
Principle, Resolution, 123
Problem for infinitely many-valued logics (MV-SAT$_\infty$), Satisfiability, 294
Problem, Boolean satisfiability (SAT; $\overline{\text{VAL}}$), 83

Problem, Decision, 7
Problem, Deduction, 6
Problem, Generalized satisfiability (GEN-SAT$_\infty$), 296
Problem, Many-valued satisfiability (MV-SAT), 217
Problem, Many-valued validity (MV-VAL), 7
Problem, Signed satisfiability, 220
Problem, Unification, 42
Problem, Validity (VAL; $\overline{\text{SAT}}$), 6
Procedure, Decision, 7
Procedure, Refutation, 8
Product t-norm, 191
Product, Fuzzy, 300
Proof, 68
Proof theory, 66
Proof, Tableau, 108
Provability, 70
Prover, Theorem, 7
Prover9-Mace 4 (Resolution-based prover), 137
Putnam, H., 87, 104, 123, 137

**Q**

$Q_3$ (Reichenbach's 3-valued system), 182
QN (Quantifier duality), 27
Quantification, Existential, 21
Quantification, Trivial, 19
Quantification, Universal, 21
Quantifier, 19
Quantifier duality (QN), 27
Quantifier, Distribution, 201
Quantifier, Existential, 22
Quantifier, Generalized, 194
Quantifier, Universal, 22
Quantifiers, Mutual definability of the (QD), 197
Quine, W. O., 86

**R**

Reasoning, Automated, 7
Reductio ad absurdum, 76
Reductio ad absurdum, Strong, 78
Redundancy, 78
Reflexivity, 54
Refutation, 69
Reichenbach, H., 182
Reichenbach's 3-valued system ($Q_3$), 182
Relation, 316
Relation, Binary, 316
Relation, Consequence, 53
Relation, Fuzzy binary, 205
Relation, Inference, 54
Relation, Matrix consequence, 64
Relation, Semantical consequence, 60
Relation, Syntactical consequence, 69
Renaming, Variable, 41
Representation, Signed CNF/DNF, 222
Rescher, N., xiv, 151, 152, 154, 157–160, 169, 186
Residuation, Law of, 329
Resolution refinement, 129
Resolution, Binary, 123
Resolution, Hyper-, 134
Resolution, Lattice-regular positive unit, 308
Resolution, Macro-, 134
Resolution, Mono-signed/regular binary, 284
Resolution, Negative hyper-, 134
Resolution, PI-, 136
Resolution, Positive hyper-, 134
Resolution, Regular, 308
Resolution, Regular negative hyper-, 308

Resolution, Regular positive unitary, 298
Resolution, Semantic, 134
Resolution, Signed $<_A$-ordering macro-, 287
Resolution, Signed binary, 283
Resolution, Signed hyper-, 287
Resolution, Signed macro-, 287
Resolution, Signed many-valued, 288
Resolution, Unit-resulting, 141
Resolvent, 125
Resolvent, Hyper-, 134
Resolvent, Macro-, 134
Resolvent, $\mathcal{I}$-semantic, 135
Resolvent, PI-, 135
Resolvent, Signed $<_A$-ordering macro-, 287
Resolvent, Signed hyper-, 287
Resolvent, Signed macro-, 287
Resolvent, $T$-, 302
Rn$_3$ (Piróg-Rzepecka's 3-valued system), 180
Robinson algorithm, 42
Robinson, A. J., 42, 87
Robinson, G., 87
Rosser, J. B., xiv, 156, 157, 198, 200
Rosser-Turquette (RT) generalized quantifier, 198
Rule of Inference, 67
Rule of universal generalization (GEN∀), 202
Rule, Analytic, 112
Rule, Condensation, 132
Rule, Cut, 104
Rule, Derived, 165
Rule, Generalization (GEN), 96
Rule, Generalization (GEN'), 197
Rule, Lattice-regular reduction, 308
Rule, One-literal, 123

Rule, Propositional CNF/DNF transformation, 224
Rule, Quantifier CNF/DNF transformation, 225
Rule, Reduction, 224
Rule, Regular reduction, 298
Rule, Rewriting, 26
Rule, Sequent, 101
Rule, Signed merging, 285
Rule, Structural inference, 67
Rule, Substitution (SBT), 166
Rule, Substitution (SUB), 68
Rule, Transformation, 224
Rule, Transposition (TR), 165
Rule, Unification, 43
Russell, B., 86, 171, 186

**S**

SAT (Boolean satisfiability problem), 83
SAT, Signed (Signed satisfiability problem), 220
Satisfiability, 59
Satisfiability, $D$-, 218
Satisfiability, ∀-, 122
Satisfiability, Fuzzy, 296
Satisfiability, $S$-, 219
Satisfiability, $T$-, 301
Satisfiability, $\varepsilon$-, 296
Satisfiability-equivalence, 30
Schema, Axiom, 67
Schubert's steamroller, 139
SCNFF (Signed CNF formula), 39
Scope, Quantifier, 19
Segerberg, K., 179
Segerberg's 3-valued system (Sn$_3^i$; $i = 1, 2, 3$), 179
Semantic clash, 135
Semantics, 15, 59
Semantics, Signed formulae, 219

Semantics,Tableaux-manageable, 270
Semi-decidability, 82
Sentence, 19
Sequent, 100
Set, 313
Set complementarity, 314
Set of signs, Complete, 254
Set, Contradiction, 258
Set, Crisp, 319
Set, Disagreement, 42
Set, Downward saturated, 115
Set, Fuzzy, 320
SFE (Signed formula expression), 39
Sheffer stroke, 78
Sign (of a formula), 37
Sign, Empty, 39
Sign, Positive/Negative regular, 302
Sign, Regular, 38
Signature, 19
Signed predicate, 304
Skolem constant, 30
Skolem function, 30
Skolem, T., 30, 87
Słupecki, J., 57, 166
Smullyan, R. M., 97, 108, 109, 111, 116, 117, 240, 242
$Sn_3^i$ ($i = 1, 2, 3$; Segerberg's 3-valued system), 179
SNF (Skolem normal form), 30
Soundness, 72
SUB (Substitution rule), 68
Subformula property, 101
Subformula, Immediate, 14
Substitutable for, 24
Substitution, 15
Substitution lemma, 74
Substitution, Empty, 41
Substitution, Ground, 41
Substitutions, Composition of, 41

Subsumption, 39
Surma, S. J., 240, 241, 244, 245, 247, 249, 251, 252
Surma's algorithm, 244
Suszko, R., 64, 70, 153–155, 160
Suszko's Thesis, 153
Syllogism, Generalized hypothetical (Rule; GHS), 165
Syllogism, Hypothetical (Rule; HS), 165
Syntax, 15
System, Axiom, 96
System, Deductive, 71
System, Inference, 54
System, Logical, 52
System, Proof, 67
System, Tableaux proof, 108

**T**

Tableau, 108
Tarski, A., 15, 71, 168, 169
Tarskian logic, 153
Tarskian semantics, 87
Tarski-style conditions, 78
Tautology, 60
Tautology, Quasi-, 157
Term, 14
Theorem, 68
Theorem, Compactness, 71
Theorem, Deduction, 71
Theorem, Deduction-detachment, 80
Theorem, Modified deduction-detachment, 208
Theorem, Quasi-deduction, 208
Theory, 81
t-norm, 190
Tractability, 82
Transitivity, 54
Tree, 331
Tree, Deduction, 125
Tree, Refutation, 125

*Index*

Tree, Semantic, 92
Truth function, 15
Truth table, 15
Truth value, 15
Truth value, Anti-designated, 158
Truth value, Designated, 63
Truth value, Undesignated, 157
Truth-functionality, 55
Truth-functionality, Quasi-, 158
Truth-preservation, 59
Truth-value gap, 161
Truth-value glut, 171
Tseitin algorithm, 35
Turing machine, 82
Turing, A., 82, 86
Turquette, A. R., xiv, 156, 157, 198, 200

**U**
Unification, 41
Unifier, 41
Unifier, Most general (MGU), 41
Uniformity, 160
Unit deletion, 137
Universe, 314

**V**
Vagueness, 189
VAL (Validity problem), 6
Validity, 60
Validity, $D$-, 217
Validity, Degree-, 159
Validity, Fuzzy, 189
Validity, Quasi-, 160
Validity, $S$-, 219
Valuation, 15
Valuation, Boolean, 2
Variable, Bound, 19
Variable, Free, 19
Variant, $x$-, 74

**W**
Wajsberg, M., 163, 169

Whitehead, A. N., 86, 186
Wójcicki, R., 55, 66
Wos, L., 87

**Z**
$Z_\aleph$ (Zadeh's fuzzy system), 272
Zadeh, L. A., 189, 190, 295
Zadeh's fuzzy system ($Z_\aleph$), 272